Stress analysis of polymers

Stress analysis of polymers

J. G. Williams

Reader in Polymer Engineering,
Mechanical Engineering Department,
Imperial College, London

A Halsted Press Book

John Wiley & Sons
New York

Published in the U.S.A., Philippine Islands only by Halsted Press, a Division of John Wiley & Sons, Inc., New York

First published 1973

ISBN 0 471 94860 8

Library of Congress Catalog Card no:
LC 73–39326

Printed in Great Britain

Contents

Introduction

Polymers are being used increasingly as engineering materials and this has resulted in people without the necessary basic training being required to use them in a wide range of applications. These include engineers from many specialisations, whose training has been concerned almost entirely with metals and, while their knowledge of linear elastic stress analysis is usually good, they have little experience of large strain behaviour and time dependence which are essential for design with polymeric materials. In addition, many chemists with no training in engineering need to understand stress analysis in order to foresee applications and deal with technical service enquiries. This book aims to meet both needs. It is a self-contained course in stress analysis which starts from first principles and includes standard work familiar in any strength of materials course. As such it will provide a basic text for those with no previous knowledge in the field. The examples and emphasis differ from the usual texts in that they are directed towards topics of particular relevance to polymers. The book also covers, in detail, those topics which engineers will require when using these materials.

The level of mathematics used has been limited so that those with an engineering or science degree should find few unfamiliar topics. Where methods of analysis which do not always appear in undergraduate courses are essential to the development of the subject, e.g. Laplace transforms, they are described in the text. Avoiding more advanced mathematical concepts has led to a loss of conciseness in the early chapters of the book. No mention is made of tensor or matrix representation, for example, which, if used, greatly reduces the amount of algebra and can elucidate the relationships involved. However, it was considered that the effort involved in understanding unfamiliar conventions would offset any advantage gained and that a somewhat longer text in more familiar mathematics is justified in an introductory work.

The general approach is that of a macro theory in that no consideration is given to micro-structure or the molecular nature of the materials. The analysis will be developed as a branch of continuum mechanics so that polymers are assumed to be infinitely divisible and elements of material can be considered to be infinitely small. The results of the analysis are therefore at their most reliable when describing the behaviour of bodies whose dimensions are large compared with their structure. This includes most cases encountered in practice but care must be exercised when very localised effects, such as deformation around crack tips, are considered since the material may often not be regarded as behaving as a continuum locally. The study of continuum mechanics involves solids, liquids and gases and any particular branch of the subject is defined by the properties of the material considered. Plastics may be regarded as that group of materials which have a long chain molecular structure plus additives. Polymers are the basic material and are used in quite pure forms, such as rubbers, while plastics are polymers plus additives which give a wide range of properties over and above those of the basic polymer. In terms of continuum theory a practically useful description of the properties of both may be obtained with elastic behaviour (both small and large strain), plastic flow and visco-elastic behaviour. The mathematical descriptions used are idealisations but they are of considerable utility in solving practical problems. Continuum mechanics provides a powerful method of analysis which is used as the scheme for this text. Problems are solved in continuum mechanics using four steps; the analysis of forces, the analysis of the displacements they produce, the constitutive relationships (material properties) and the boundary conditions of the particular problem. Each step is entirely separate and each is the basis of distinct sections of the book. Thus the first two steps form chapter one, the third makes up chapters two and three and the fourth covers chapters four, five and six. A list of further reading is given at the end of each chapter.

Chapter one analyses the forces acting on an element of material and defines the stress components and the relationships between them. This is a conventional development of the subject with particular emphasis on concepts such as the transformation of coordinates and invariance used in later sections. In addition the deformation of an element is considered resulting in the definition of strain components. Special emphasis is laid on large strain behaviour since it is of particular importance in polymer stress analysis. Chapter two deals with time independent material properties as described in the theories of elastic deformation and plastic flow. The familiar linear elasticity relationships are developed including anisotropy and the concept of strain energy. The latter is then employed to develop the finite strain elasticity equations used to describe rubbery behaviour. Yield criteria are also considered in this chapter together with the plastic flow equations. Chapter three describes time dependent behaviour and develops the general theory of linear visco-elasticity. In addition, the properties of practically useful model materials are examined in some detail.

Chapters four, five and six are collections of problems of particular interest in connection with polymeric materials which are grouped according to the simplifications used in their solution. Problems involving bending as the main deformation mode make up chapter four and include simple beam theory with its extension to plates and curved rods. Topics such as large deflections, sandwich beams and vibrations are also given because of their relevance to polymeric materials. Chapter five deals with the solution of problems involving bodies with an axis of symmetry and is concerned with both the small and large strain deformation of cylinders and circular discs. Axisymmetric membranes are analysed with particular reference to large displacements. The subject of stress concentrations forms the basis of chapter six which leads to the development of fracture mechanics. This is a topic of increasing interest for polymeric materials as they are used in more critical applications.

The author wishes to express his thanks to many colleagues for their help and advice when preparing this book. In particular, special thanks are due to Richard Ogorkiewicz for his continued encouragement and constructive criticism and to Robert Ferguson for his help in proof reading and compiling the index. Paul Ewing's work in preparing the drawings is also appreciated.

1 The analysis of stress and strain

Introduction

This chapter is intended to lay the foundations on which the subsequent ones are built. The general approach of a continuum mechanics solution will be followed throughout the book in that problems will be solved in four stages, i.e. by considering equilibrium, compatibility, constitutive relationships and boundary conditions. Here we will consider the first two, which are independent of material properties or the type of loading system or geometry. The consideration of the equilibrium condition throughout a material leads to the analysis of stress and this will be dealt with at some length since it is a convenient way of introducing concepts such as the transformation of coordinates, principal values and invariants. The independent requirement that the material remains a continuous whole (i.e. a continuum) results in relationships between the displacements which are the compatibility conditions and form the basis of strain analysis. The results of both analyses are of fundamental importance in stress analysis and will be used in all subsequent chapters.

1.1. The analysis of stress at a point

1.1.1. *Definitions*

Stress at a point is defined as the force per unit area acting on a given plane passing through that point. It is not possible to define the stress without specifying its magnitude, direction and the plane on which it acts. There is thus a whole range of stresses at any point depending on the plane considered.

Consider an arbitrary plane ABCD drawn through a point O at which the stress state is to be examined as shown in Figure 1.1. The stress acting on this plane is shown as S and is in some arbitrary direction relative to the plane. The plane is defined by the direction of its normal, shown as direction a. Clearly S can be split into components normal and parallel to the plane. Thus there is a normal component in direction a and the component parallel to the plane may be expressed as two components in arbitrary (but mutually perpendicular) directions b and c. The total stress S on any plane can therefore be expressed in terms of two sorts of components:

(i) Normal stress—the component normal to the plane.

(ii) Shear stresses—the components parallel to the plane.

Since stress components act in a particular direction and on a particular plane it is necessary to specify both the direction of the components and the normal of the plane when defining a stress component. The convention is to use a double suffix notation of the form:

$p_{\alpha\beta}$ (the symbol p will be used for stress components throughout the text)

where α is the direction in which the stress component acts and β is the direction of the normal of the plane on which the stress component acts.

In Figure 1.1 the normal to the plane acts in direction a and hence the normal stress component is given by:

$$p_{aa}$$

since it acts in direction a also. Clearly, all normal stresses have the two suffixes the same as they are, by definition, normal to the planes on which they act.

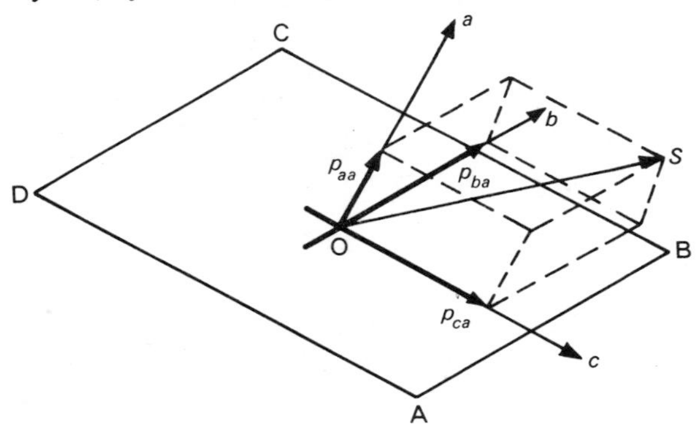

Figure 1.1 Components of stress on a plane.

Shear stresses on the other hand always have suffixes which differ from each other as they are parallel to the plane on which they act and therefore in some direction other than the normal, e.g. p_{ba} is in direction b on the plane whose normal is in the direction a.

1.1.2. *Components*

We now consider the components necessary to define a state of stress at a point. A small element of material at the point must be in translational equilibrium and hence to specify this condition in three dimensions we may consider the equilibrium of forces in three arbitrary but mutually perpendicular directions, say x, y and z as shown in Figure 1.2. Since we are concerned with stress at a point we may consider a very small rectangular element at O with sides normal to the coordinate directions as shown in Figure 1.2. The stress on each face can be expressed as a normal component and shear components in the other two coordinate directions. Since the element is very small the stresses may be assumed not to vary from one side of the element to the other and it is clear that the translational equilibrium condition is achieved by putting the normal stress components in opposite senses on opposing faces. The shear components are drawn in the same way in Figure 1.2 and they also comply with the condition

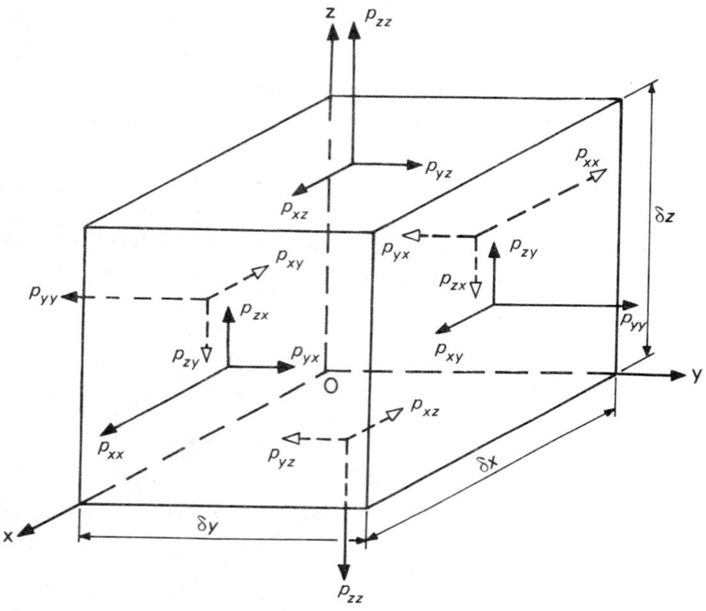

Figure 1.2 Components at a point O in the orthogonal directions x, y and z.

of equilibrium of forces, e.g. the force from p_{zy} on the exposed face is $p_{zy}\delta z\ \delta x$ and it is balanced by an equal and opposite one on the hidden face. However it is clear that a further condition must be considered in that the element must also be in equilibrium with regard to the moments which act upon it and while each pair of shear stresses result in no force they do give a moment, e.g. for p_{zy} it is $p_{zy}\ \delta z\ \delta x\ \delta y$.

If we consider rotational equilibrium about the z axis the components producing moments are shown in Figure 1.3 and for equilibrium of the moments we have:

$$-p_{xx}\,(\delta y\ \delta z)\,\frac{\delta y}{2}+p_{xx}\,(\delta y\ \delta z)\,\frac{\delta y}{2}-p_{yy}\,(\delta x\ \delta z)\,\frac{\delta x}{2}$$

$$+p_{yy}\,(\delta x\ \delta z)\,\frac{\delta x}{2}+p_{yx}\,(\delta y\ \delta z)\,\delta x-p_{xy}(\delta x\ \delta z)\,\delta y$$

$$+p_{yz}(\delta y\ \delta x)\frac{\delta x}{2} - p_{yz}\ (\delta y\ \delta x)\frac{\delta y}{2} + p_{xz}(\delta y\ \delta x)\frac{\delta x}{2}$$

$$-p_{xz}(\delta y\ \delta x)\frac{\delta y}{2} = 0$$

where anticlockwise moments are positive. The components p_{yz} and p_{xz} are not shown in Figure 1.3 but their moments may be deduced from Figure 1.2. The

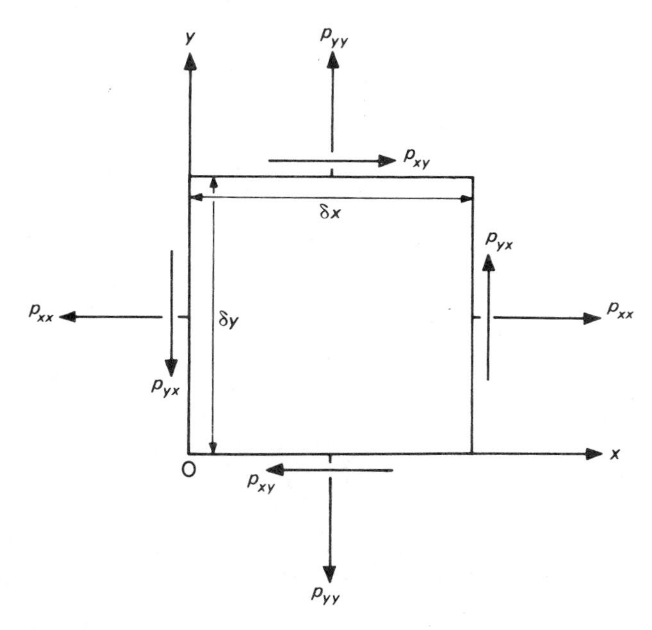

Figure 1.3 The element at a point O viewed from the z direction.

terms in brackets are the areas on which stresses act, the other element length is the moment arm. As most of the terms cancel out the equation reduces to:

$$p_{yx} = p_{xy}$$

and similarly by considering equilibrium about the x and y axes we have:

$$p_{yz} = p_{zy} \text{ and } p_{xz} = p_{zx}.$$

This equality of the shear stresses acting on orthogonal planes is referred to as the complementary shear stress condition.

Thus the state of stress at a point can be defined in terms of six stress components

three normal stresses, p_{xx}, p_{yy}, p_{zz}
and three shear stresses, p_{xy}, p_{yz}, p_{zx}

1.1.3. *Stress on a general plane*

The state of stress at a point is defined in terms of components referred to arbitrary coordinates x, y and z and we will now consider how the components on any other plane may be deduced. A general plane with its normal in direction N is shown in Figure 1.4 and the direction of this normal may be defined in terms

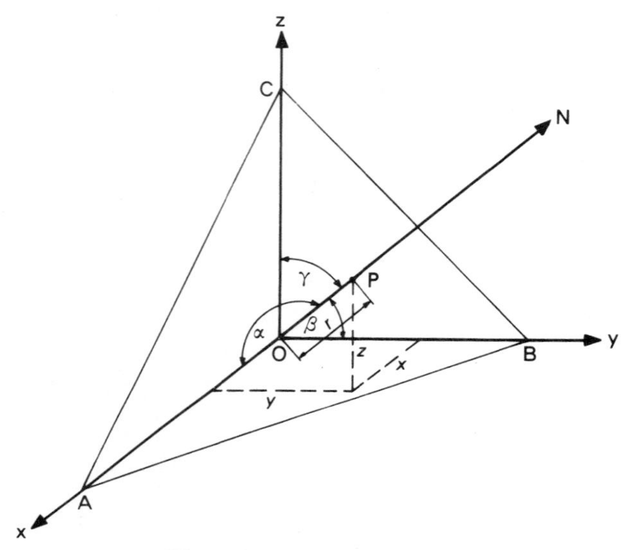

Figure 1.4 A general plane.

of the angles of inclination to the x, y and z directions, i.e. α, β and γ respectively. If the distance OP from the origin O to the point P, where the normal through the origin cuts the plane, is r, then:

$$r^2 = x^2+y^2+z^2$$

$$\therefore \quad 1 = \left(\frac{x}{r}\right)^2+\left(\frac{y}{r}\right)^2+\left(\frac{z}{r}\right)^2$$

Now if $\cos\alpha = \frac{x}{r} = l$ say, $\cos\beta = \frac{y}{r} = m$ and $\cos\gamma = \frac{z}{r} = n$ then we have:

$$1 = l^2+m^2+n^2 \tag{1.1}$$

where l, m and n are called the direction cosines of the line OP. Any two are sufficient to define the position of the line as is exemplified in equation (1.1). A full account of the properties of direction cosines can be found in many mathematical texts and only two results will be given here for use in later sections.

If two lines have direction cosines l, m, n and l', m', n' respectively then the angle between them, ψ, is given by:

$$\cos\psi = ll'+mm'+nn'$$

For perpendicular lines $\psi = \frac{\pi}{2}$ and hence:

$$0 = ll' + mm' + nn' \tag{1.2}$$

If we consider the areas of the faces of the element shown in Figure 1.4 then it can be shown that:

$$\frac{\text{Area OBC}}{\text{Area ABC}} = l, \quad \frac{\text{OAC}}{\text{ABC}} = m \quad \text{and} \quad \frac{\text{OAB}}{\text{ABC}} = n \tag{1.3}$$

Figure 1.5 shows the total stress S on the general plane and illustrates two possible sets of components. Firstly S may be resolved in the x, y and z directions

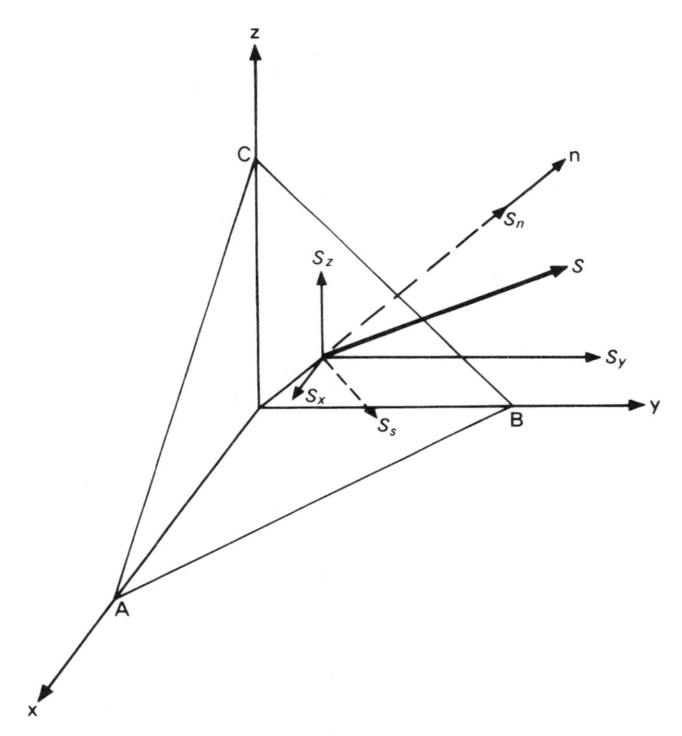

Figure 1.5 Stress components on a general plane.

and secondly it may be resolved in directions normal and parallel to the plane to give:

$$\left.\begin{aligned} S^2 &= S_x{}^2 + S_y{}^2 + S_z{}^2 \\ \text{and} \qquad S^2 &= S_n{}^2 + S_s{}^2 \end{aligned}\right\} \tag{1.4}$$

If we now consider the equilibrium of the element shown in Figure 1.5 in the

x direction and let the area ABC $= \delta A$ then noting equation (1.3) we have:

$$S_x \, \delta A = p_{xx} \, \delta A \, l + p_{xy} \, \delta A \, m + p_{xz} \, \delta A \, n$$

$$\left.\begin{aligned} &\text{and hence} && S_x = p_{xx} \, l + p_{xy} \, m + p_{xz} \, n \\ &\text{Similarly} && S_y = p_{yx} \, l + p_{yy} \, m + p_{yz} \, n \\ &\text{and} && S_z = p_{zx} \, l + p_{zy} \, m + p_{zz} \, n \end{aligned}\right\} \quad (1.5)$$

by considering equilibrium in the y and z directions respectively.

Resolving S_x, S_y and S_z in the normal direction we have:

$$S_n = S_x l + S_y \, m + S_z \, n \quad (1.6)$$

and by substituting from equations (1.5) we have:

$$S_n = p_{xx} l^2 + p_{yy} \, m^2 + p_{zz} \, n^2 + 2p_{xy} lm + 2p_{yz} mn + 2p_{zx} nl \quad (1.7)$$

Similarly the shear stress on the general plane may be found from equations (1.4), (1.5) and (1.6).

If a state of stress is defined with reference to one set of coordinates with the appropriate six components it is therefore possible to deduce the appropriate components for the state of stress in terms of any other set of coordinates. For example if we have coordinates a, b and c with direction cosines $l \, m \, n$, $l'm'n'$ and $l''m''n''$ with respect to x, y and z then the new normal stress components may be deduced from equation (1.7) where $S_n = p_{aa}$ with similar expressions for p_{bb} and p_{cc}. The appropriate shear stress expressions may be deduced from the expression for S_s using the result in equation (1.2) since a, b and c are orthogonal. These are the transformation relationships which describe the transformation from one coordinate system to another.

1.1.4. *Principal stresses*

Since the values of the stress components depend on the planes used to describe the state of stress at a point then there is a range of possible values. The determination of the principal stresses defines limiting values of the normal stress components within this range. The nature of the variation in the normal stress is shown schematically in Figure 1.6 where S_n, as described in equation (1.7), is plotted radially from the origin over the whole range of values of l, m and n. Thus in the coordinate directions the values are p_{xx}, p_{yy} and p_{zz} as shown. The surface generated is an ellipsoid and there are three extreme values on the major axes which define the limits of the normal stresses. These extreme conditions are called stationary values and are defined by the condition:

$$\frac{\partial S_n}{\partial l} = \frac{\partial S_n}{\partial m} = \frac{\partial S_n}{\partial n} = 0$$

where $\dfrac{\partial}{\partial l}$, etc., the partial derivatives, are the rates of change with respect to one direction cosine only and the condition is equivalent to the usual definition of maxima used in calculus. As l, m and n are not independent only two of the

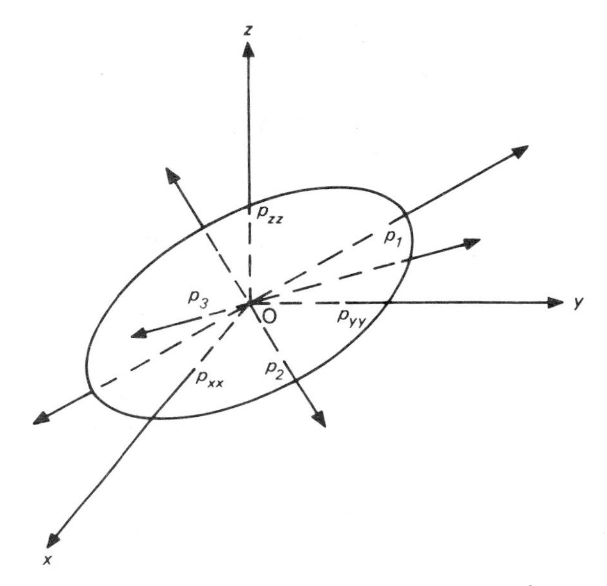

Figure 1.6 The stress ellipsoid showing the principal values.

conditions need be specified and by combining equations (1.1) and (1.6) we have:

$$S_n = S_x\, l + S_y\, m + S_z\, (1 - l^2 - m^2)^{\frac{1}{2}}$$

Taking the partial derivative with respect to l gives:

$$\frac{\partial S_n}{\partial l} = S_x + S_z \frac{1}{2}(1 - l^2 - m^2)^{-\frac{1}{2}}(-2l) + l\frac{\partial S_x}{\partial l} + m\frac{\partial S_y}{\partial l} + n\frac{\partial S_z}{\partial l}$$

and hence:

$$\frac{S_x}{l} = \frac{S_z}{n}$$

A similar condition may be obtained using $\dfrac{\partial}{\partial m}$ and the result may be written as:

$$\frac{S_x}{l} = \frac{S_y}{m} = \frac{S_z}{n} = p \text{ say.} \tag{1.8}$$

Substituting in equations (1.5) we have:

$$\left.\begin{aligned} 0 &= (p_{xx} - p)\, l + p_{xy}\, m + p_{xz}\, n \\ 0 &= p_{yx}\, l + (p_{yy} - p)\, m + p_{yz}\, n \\ 0 &= p_{zx}\, l + p_{zy}\, m + (p_{zz} - p)\, n \end{aligned}\right\} \tag{1.9}$$

The direction cosines l, m and n may be eliminated between these three equations to give an expression for p of the form:

$$p^3-(p_{xx}+p_{yy}+p_{zz})\,p^2+(p_{xx}p_{yy}+p_{yy}p_{zz}+p_{zz}p_{xx}$$
$$-p_{xy}{}^2-p_{yz}{}^2-p_{zx}{}^2)\,p-(p_{xx}p_{yy}p_{zz}+2p_{xy}p_{yz}p_{zx}$$
$$-p_{xx}p_{yz}{}^2-p_{yy}p_{zx}{}^2-p_{zz}p_{xy}{}^2) = 0 \qquad (1.10)$$

which is called the stress cubic. Returning to equation (1.8) and substituting from equation (1.6) gives the values of S_n at the stationary condition:

$$S_n = p\,l^2+p\,m^2+p\,n^2 = p$$

i.e. p is the value of S_n at the extreme condition. It can be shown that the stress cubic always has three real roots, which are conventionally written as:

$$p_1 > p_2 > p_3$$

and these are the three principal values. Clearly p_1 is the largest value of normal stress which can exist at a point and p_3 is the smallest. It can also be shown that the three directions of the principal values are mutually orthogonal as shown in Figure 1.6.

The shear stresses on the principal planes (i.e. the planes on which the principal stresses act) are of interest and may be determined by considering the total stress S on the planes which can be deduced from equations (1.4) and (1.8):

$$S^2 = p^2l^2+p^2m^2+p^2n^2$$

i.e.

$$S = p$$

Again from equations (1.4) we have:

$$S_s{}^2 = S^2-S_n{}^2 = p^2-p^2 = 0$$

and hence the shear stress on the principal planes is zero which is sometimes used as a definition.

Since p_1, p_2 and p_3 are mutually perpendicular it is possible to use their directions as the coordinates of the stress system. The transformation equation is then derived from equation (1.7) by replacing p_{xx}, p_{yy} and p_{zz} by p_1, p_2 and p_3 respectively with $p_{xy} = p_{yz} = p_{zx} = 0$ giving:

$$S_n = p_1l^2+p_2m^2+p_3n^2 \qquad (1.11)$$

where l, m and n are now measured from the three principal stress directions. The transformation equation for the shear stress is useful in terms of principal values and is:

$$S_s{}^2 = (p_1-p_2)^2\,l^2m^2+(p_2-p_3)^2\,m^2n^2+(p_3-p_1)^2\,n^2l^2 \qquad (1.12)$$

1.1.5. *Limiting shear stresses*

As with normal stresses the shear stresses also have limiting values and they may be determined, in a manner similar to that used for principal stresses, from equation (1.12). The stationary condition is $\frac{\partial S_s}{\partial l} = 0$ and $\frac{\partial S_s}{\partial m} = 0$ and substituting from equation (1.1) into equation (1.12) and differentiating gives:

$$\left.\begin{aligned} 0 &= 2l(p_3 - p_1)\,[2m^2(p_1 - p_3) - (1 - 2l^2)(p_1 - p_3)] \\ 0 &= 2m(p_3 - p_2)\,[2l^2(p_2 - p_3) - (1 - 2m^2)(p_2 - p_3)] \end{aligned}\right\} \quad (1.13)$$

Since both equations must be satisfied at the stationary points the solution for one must satisfy the other. A solution for the first equation is $l = 0$ and, since $p_2 \neq p_3$, from the second equation we have:

$$(1 - 2m^2) = 0$$

i.e.

$$m^2 = \tfrac{1}{2}$$

and equation (1.1) gives $n^2 = \frac{1}{2}$. Similarly from the second equation $m = 0$ is a solution and the first equation gives $l^2 = \frac{1}{2}$, and hence $n^2 = \frac{1}{2}$. The direction cosines, measured from the principal directions, of all the planes at the stationary condition are given by:

$$\left.\begin{aligned} l^2 &= 0,\ m^2 = \tfrac{1}{2},\ n^2 = \tfrac{1}{2} \\ l^2 &= \tfrac{1}{2},\ m^2 = 0,\ n^2 = \tfrac{1}{2} \\ l^2 &= \tfrac{1}{2},\ m^2 = \tfrac{1}{2},\ n^2 = 0 \end{aligned}\right\} \quad (1.14)$$

These conditions define twelve separate planes which are parallel to one principal axis and inclined at 45° to the other two. The values of S_s are obtained by substituting in equation (1.12) and they are:

$$\pm\frac{p_1 - p_2}{2},\ \pm\frac{p_2 - p_3}{2},\ \pm\frac{p_3 - p_1}{2}$$

for the $n = 0$, $l = 0$ and $m = 0$ planes respectively. The direction cosines of the shear stresses may be obtained by considering a further static equilibrium condition in the general plane shown in Figure 1.5. Resolving S_n and S_s in the x coordinate direction gives:

$$S_x = S_s\,l_s + S_n\,l$$

i.e.

$$l_s = \frac{S_x - S_n\,l}{S_s}$$

and similarly:

$$m_s = \frac{S_y - S_n\,m}{S_s}$$

and

$$n_s = \frac{S_z - S_n\,n}{S_s} \quad (1.15)$$

where l_s, m_s and n_s are the direction cosines of S_s on the plane whose normal has direction cosines l, m and n.

The normal stresses on the planes of limiting shear stresses are given by equation (1.11); i.e. for $l^2 = 0$, $m^2 = n^2 = \frac{1}{2}$ we have:

$$S_n = \frac{p_2 + p_3}{2}$$

and similarly for the other two conditions:

$$\frac{p_1 + p_2}{2} \text{ and } \frac{p_1 + p_3}{2}$$

Unlike the principal planes, for which the shear stress was zero for limiting values of normal stress, the normal stress has a value for planes of limiting shear stresses.

The directions of the various components with respect to the principal directions is illustrated in Figure 1.7, where four of the planes (two each from

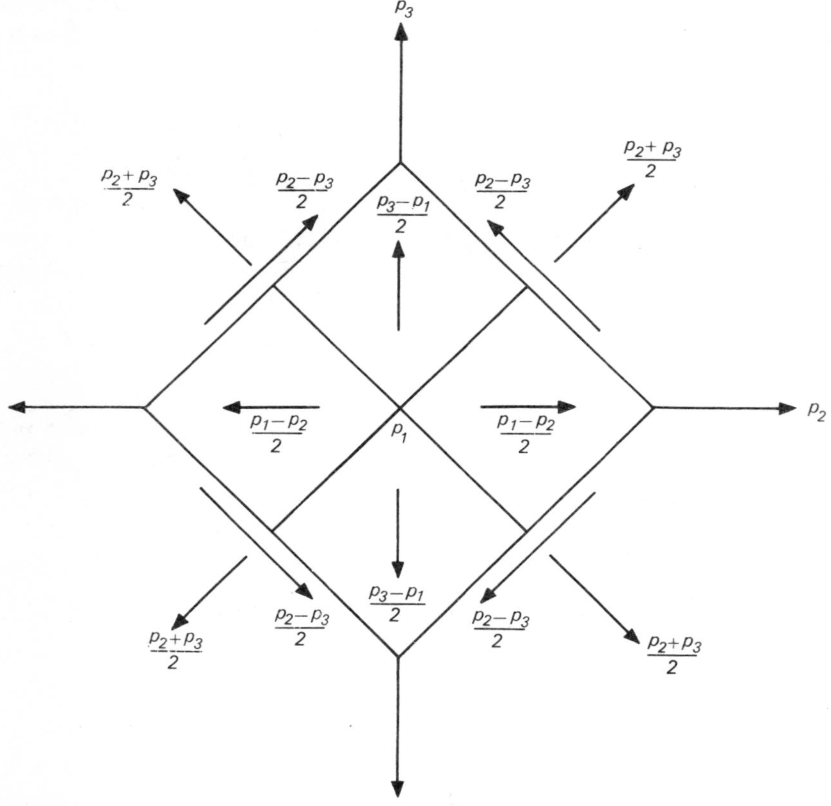

Figure 1.7 The directions of the maximum shear stresses with respect to the principal directions.

the $m = 0$ and $n = 0$ condition) are shown viewed from the direction of p_1. The directions of the shear stresses shown are determined from equations (1.15).

1.1.6. *Invariants*

Many physical phenomena which are governed by the state of stress are not functions of the components defined with respect to some arbitrary coordinate system. Such phenomena, which include yielding for example, must be described in terms of functions of the stress components which are independent of the coordinate system used. These functions must therefore be invariant with respect to the coordinate system and there are three such functions which may be regarded as the basic forms. They are given by the coefficients of the stress cubic, equation (1.10), which may be written as:

$$p^3 - J_1 p^2 + J_2 p - J_3 = 0$$

where:

$$\begin{aligned} J_1 &= p_{xx} + p_{yy} + p_{zz} \\ J_2 &= p_{xx}p_{yy} + p_{yy}p_{zz} + p_{zz}p_{xx} - p_{xy}^2 - p_{yz}^2 - p_{zx}^2 \\ J_3 &= p_{xx}p_{yy}p_{zz} + 2p_{xy}p_{yz}p_{zx} - p_{xx}p_{yz}^2 - p_{yy}p_{zx}^2 - p_{zz}p_{xy}^2 \end{aligned}$$

and J_1, J_2 and J_3 are the invariants of the stress system at the point.

The invariant nature of the functions may be illustrated with J_1 which may be derived for a new set of coordinates a, b and c with direction cosines $l\,m\,n$, $l'm'n'$ and $l''m''n''$ measured from x, y and z respectively. From equation (1.7) the three new normal stress components become:

$$\begin{aligned} p_{aa} &= p_{xx}l^2 + p_{yy}m^2 + p_{zz}n^2 + 2p_{xy}lm + 2p_{yz}mn + 2p_{zx}nl \\ p_{bb} &= p_{xx}l'^2 + p_{yy}m'^2 + p_{zz}n'^2 + 2p_{xy}l'\,m' + 2p_{yz}m'n' + 2p_{zx}n'l' \\ p_{cc} &= p_{xx}\,l''^2 + p_{yy}m''^2 + p_{zz}n''^2 + 2p_{xy}l''\,m'' + 2p_{yz}m''n'' + 2p_{zx}n''l'' \end{aligned}$$

The first invariant is obtained by summing the three equations to give:

$$\begin{aligned} J_1 = p_{aa} + p_{bb} + p_{cc} &= p_{xx}(l^2 + l'^2 + l''^2) + p_{yy}(m^2 + m'^2 + m''^2) \\ &+ p_{zz}(n^2 + n'^2 + n''^2) + 2p_{xy}(lm + l'm' + l''m'') + 2p_{yz}(mn + m'n' + m''n'') \\ &+ 2p_{zx}(nl + n'l' + n''l'') \end{aligned}$$

Now l, l' and l'' are the direction cosines of a, b and c with respect to the x direction and could therefore be regarded in the reverse sense as describing the position of x with respect to a, b and c so that we may write:

$$l^2 + l'^2 + l''^2 = 1$$

and similarly for the expressions in m and n. If we examine the expression:

$$lm + l'm' + l''m''$$

then by the same argument this is the cosine of the angle between the directions x and y and is zero since they are perpendicular (see equation (1.2)). Similarly the expressions in m, n and n, l are also zero and the expression reduces to:

$$J_1 = p_{aa} + p_{bb} + p_{cc} = p_{xx} + p_{yy} + p_{zz}$$

and J_1 is independent of the coordinate system used. A similar, though more cumbersome, argument may be applied to J_2 and J_3.

For phenomena which take place at constant volume, such as plastic flow and yielding, the stresses are frequently analysed in terms of deviatoric stresses which are the difference between the actual stress system and the mean or hydrostatic stress given by:

$$p_m = \frac{p_{xx}+p_{yy}+p_{zz}}{3} = \frac{J_1}{3} \tag{1.16}$$

Thus the deviatoric stress becomes, for example:

$$p'_{xx} = p_{xx}-p_m = \frac{2}{3}\left[p_{xx}-\frac{1}{2}(p_{yy}+p_{zz})\right] \tag{1.17}$$

and similarly for p_{yy} and p_{zz}. The shear stresses are not affected by introducing p_m, i.e. $p'_{xy} = p_{xy}$, and a stress cubic for the deviatoric stress system may be derived giving invariants of the form:

$$\begin{aligned} J_1' &= p'_{xx}+p'_{yy}+p'_{zz} = 0 \\ J_2' &= p'_{xx}p'_{yy}+p'_{yy}p'_{xx}+p'_{xx}p'_{zz}-p_{xy}{}^2-p_{yz}{}^2-p_{zx}{}^2 \\ &= -\tfrac{1}{6}\left[(p_{xx}-p_{yy})^2+(p_{yy}-p_{zz})^2+(p_{zz}-p_{xx})^2+6p_{xy}{}^2+6p_{yz}{}^2+6p_{zx}{}^2\right] \\ J_3' &= p'_{xx}p'_{yy}p'_{zz}+2p_{xy}p_{yz}p_{zx}-p'_{xx}p_{yz}{}^2-p'_{yy}p_{zx}{}^2-p'_{zz}p_{xy}{}^2 \end{aligned} \tag{1.18}$$

1.1.7. *Graphical representation*

It is possible to represent the state of stress at a point in a simple graphical form which can help to clarify the properties of a state of stress. The representation is called Mohr's stress circle and it may be derived by considering the equations for the normal and shear stresses on a general plane in terms of the principal stresses:

$$\begin{aligned} S_n &= p_1l^2+p_2m^2+p_3n^2 \\ \text{and} \quad S_s{}^2 &= p_1{}^2l^2+p_2{}^2m^2+p_3{}^2n^2-S_n{}^2 \end{aligned} \tag{1.19}$$

in conjunction with equation (1.1). These may be manipulated to eliminate m and n to give:

$$\left\{S_n-\left(\frac{p_3+p_2}{2}\right)\right\}^2+S_s{}^2 = l^2(p_2-p_1)(p_3-p_1)+\left(\frac{p_2-p_3}{2}\right)^2 \tag{1.20}$$

If a graph is plotted of S_n versus S_s for any particular value of l then equation (1.20) is that of a circle of the form:

$$(S_n-A)^2+S_s{}^2 = R_1{}^2$$

where A is the distance of the centre from the origin along the S_n axis and R_1 is the radius given by:

$$R_1 = \left[l^2(p_2-p_1)(p_3-p_1)+\left(\frac{p_2-p_3}{2}\right)^2\right]^{\frac{1}{2}}$$

Since $p_1 > p_2 > p_3$, both terms are positive and the smallest value R_1 can have is when $l = 0$. Therefore all values of R_1 must be greater than:

$$\hat{R}_1 = \frac{p_2-p_3}{2}$$

This circle is drawn in Figure 1.8 together with some general value for a particular value of l $(0 < l < 1)$ and the stress state on the plane whose direction cosine is l lies on the circle of radius R_1.

Similarly by eliminating l and m from the three equations we have;

$$\left\{S_n-\left(\frac{p_1+p_2}{2}\right)\right\}^2+S_s^2 = n^2(p_1-p_3)(p_2-p_3)+\left(\frac{p_1-p_2}{2}\right)^2$$

In this case the centre is at $S_n = (p_1+p_2)/2$ and the minimum radius $\hat{R}_2$ is $(p_1-p_2)/2$. The intercepts of the minimum radii on the S_n axis are p_3 and p_2 for the R_1 circle and p_2 and p_1 for the $\hat{R}_2$ circle. The state of stress lies on the general

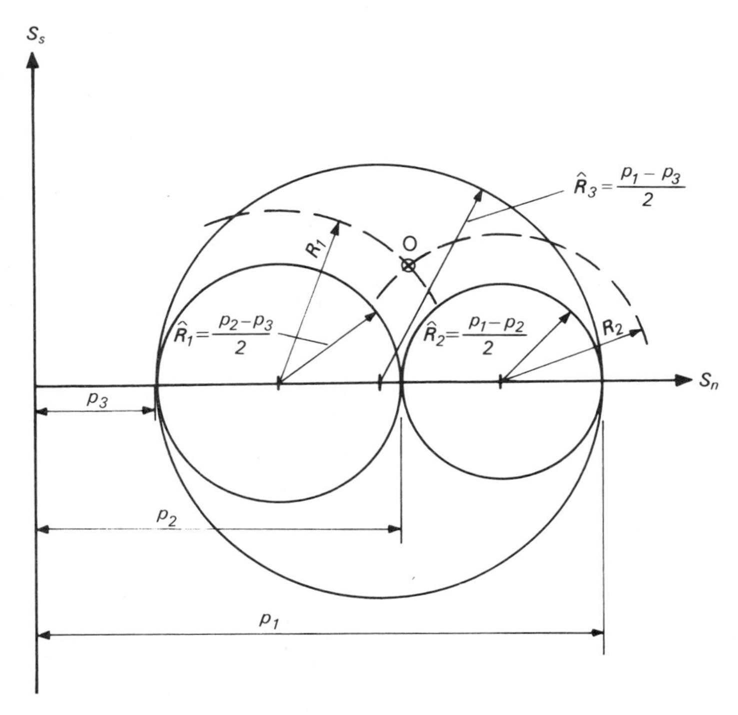

Figure 1.8 The graphical construction of a state of stress at a point.

radius R_2 and where R_1 and R_2 intercept, O in Figure 1.8, is the stress state on the plane whose normal has the direction cosines l and n.

A third equation is derived by eliminating l and n and we have:

$$\left\{S_n-\left(\frac{p_3+p_1}{2}\right)\right\}^2+S_s^{\,2} = m^2(p_3-p_2)(p_1-p_2)+\left(\frac{p_3-p_1}{2}\right)^2$$

The centre distance is now $(p_3+p_1)/2$ but the radius is different from the preceding cases. The first term is negative and the second positive so that $m = O$ is the maximum radius possible and all other value must lie within this circle. The radius is $\hat{R}_3 = (p_1-p_3)/2$ and is shown on Figure 1.8. The radius for m corresponding to l and n (i.e. $m = (1-l^2-n^2)^{\frac{1}{2}}$) must pass through O as shown.

This general form is of limited value for problem solving but does illustrate the main properties of a state of stress at a point. For a stress state with three principal values p_1, p_2 and p_3 the three circles are constructed as shown in Figure 1.9. All other planes at that point must have values of S_n and S_s which

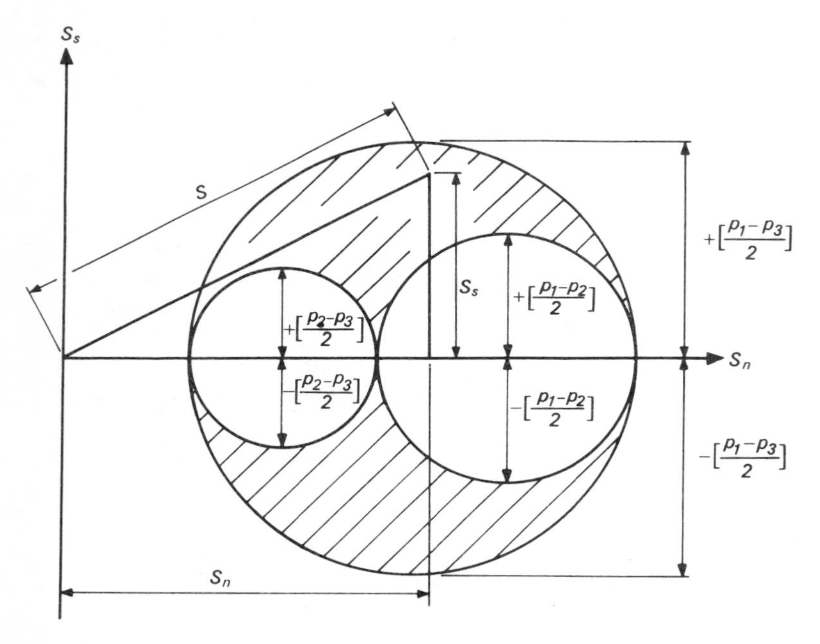

Figure 1.9 The graphical representation of a state of stress at a point showing the maximumum shear stresses.

lie within the shaded area. The principal values lie on the S_s axis illustrating the zero shear stress condition. The limiting values of shear stress are also apparent as the radii of the circles and the total stress S on any plane may be constructed as shown.

1.1.8. *Special cases*

Some stress systems are frequently encountered in practice and are employed in analysis and are worthy of further explanation.

(i) *Plane stress*

Many practical problems deal with bodies in which there is a stress free surface. If the dimension normal to this surface is very much less than those in the surface we obtain the equivalent of a thin sheet. It is assumed that the normal stress on planes parallel to the surface is zero on the argument that since it is zero on both faces it will not vary appreciably over the small thickness and in any case will be very much less than the normal stresses on planes normal to the surface of the sheet, i.e. stresses acting in the plane of the sheet. For the sheet shown in Figure 1.10, z the through thickness direction is a principal direction

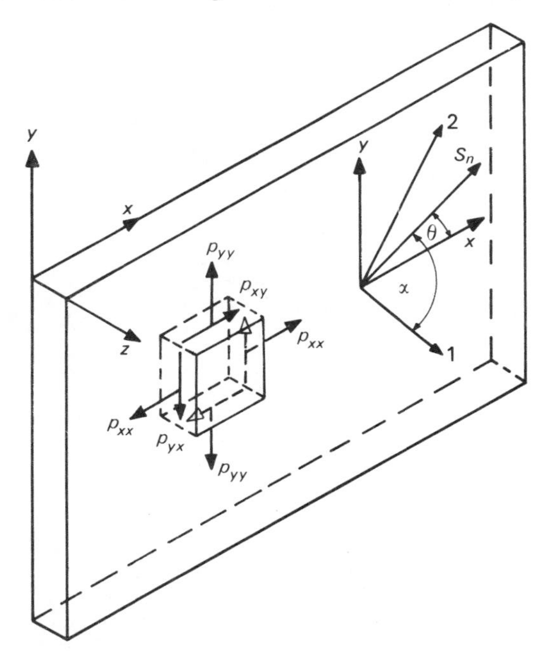

Figure 1.10 A state of plane stress in a thin sheet.

because the shear stresses are zero; i.e. $p_{yz} = p_{xz} = 0$, with a principal value $p_{zz} = 0$. If p_1 and p_2 are the principal stresses acting on planes normal to the surface then the Mohr's circles are as shown in Figure 1.11 where $p_{zz} = p_3 = 0$. The shaded area represents stresses acting on planes through the thickness but inclined at angles to the surface. Interest is usually confined to planes normal to the surface, i.e. the stresses acting in the plane of the sheet, which is the $n = 0$ circle as shown in Figure 1.11. The equation for the normal and shear stresses acting on all these planes may be derived from the $n = 0$ condition and for S_n, from equation (1.7), we have:

$$S_n = p_{xx}l^2 + p_{yy}m^2 + 2p_{xy}lm$$

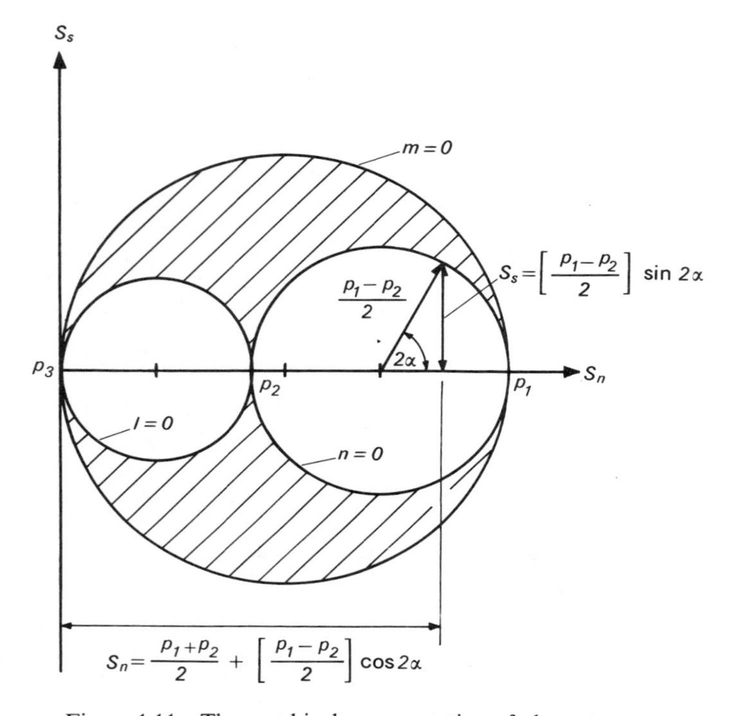

Figure 1.11 The graphical representation of plane stress.

If the angle of S_n to the x direction is taken as θ as shown in Figure 1.10 then $l = \cos\theta$ and since $l^2 + m^2 = 1$ for $n = 0$ then:

$$S_n = p_{xx}\cos^2\theta + p_{yy}\sin^2\theta + 2p_{xy}\sin\theta\cos\theta$$

This equation may be rearranged in a more convenient form giving:

$$S_n = \left(\frac{p_{xx} + p_{yy}}{2}\right) + \left(\frac{p_{xx} - p_{yy}}{2}\right)\cos 2\theta + p_{xy}\sin 2\theta \tag{1.21}$$

and similarly we have:

$$S_s = \left(\frac{p_{xx} - p_{yy}}{2}\right)\sin 2\theta - p_{xy}\cos 2\theta \tag{1.22}$$

If the coordinate directions are taken as the principal stress directions as shown in Figure 1.10 and α is the angle between S_n and p_1 then the relationships become:

$$S_n = \left(\frac{p_1+p_2}{2}\right)+\left(\frac{p_1-p_2}{2}\right)\cos 2\alpha \tag{1.23}$$

and

$$S_s = \left(\frac{p_1-p_2}{2}\right)\sin 2\alpha \tag{1.24}$$

The angle α, together with S_n and S_s, is shown in Figure 1.11 and the graphical form of equations (1.23) and (1.24) is apparent.

The principal stresses, p_1 and p_2, may be deduced from the stress cubic, equation (1.10), by noting that:

$$p_{zz} = p_{yz} = p_{zx} = 0$$

giving, a quadratic of the form:

$$p^2-(p_{xx}+p_{yy})p+(p_{xx}p_{yy}-p_{xy}{}^2) = 0$$

The roots of this are the principal values:

$$p_{1,2} = \left(\frac{p_{xx}+p_{yy}}{2}\right)\pm\sqrt{\left(\frac{p_{xx}-p_{yy}}{2}\right)^2+p_{xy}{}^2} \tag{1.25}$$

The same result may be obtained by differentiating equation (1.21) with respect to θ and using the condition $dS_n/d\theta = 0$.

Because the stresses acting in the plane of the sheet are represented by a circle the graph may be used to determine the principal values. For example if S_n is known on two orthogonal planes, say S_{n1} ($\equiv p_{xx}$) and S_{n2} ($\equiv p_{yy}$), and if S_s ($\equiv p_{xy}$) on one of them is also known then from the complimentary shear stress condition it must be the same on the other. The circle may be constructed as shown in Figure 1.12. The two values of S_n are set off on the S_n axis and vertical lines drawn. The value of S_s is set off above and below the S_n axis and the intersection points marked A, B, C and D. Now as S_{n1} and S_{n2} are orthogonal they must be at 180° on the circle and therefore AC and DB are diameters. The circle is drawn through the centre E and the principal values read from the S_n axis.

The other two circles are not usually drawn but it should always be remembered that the other stresses do exist, particularly when calculating maximum shear stresses. If p_1 and p_2 have different signs then $\pm(p_1-p_2/2)$ are the maximum values but if they are the same sign then $p_3 = 0$ must be considered giving either $\pm\frac{1}{2}p_1$ or $\pm\frac{1}{2}p_2$ as maxima.

(ii) *Simple tension*

Simple tension is one of the simplest stress systems to produce experimentally since it may be achieved by pulling a rod or bar. There are no applied shear stresses and only one normal stress giving:

$$p_{xx} = p_T, \ p_{yy} = p_{zz} = p_{xy} = p_{yz} = p_{zx} = 0$$

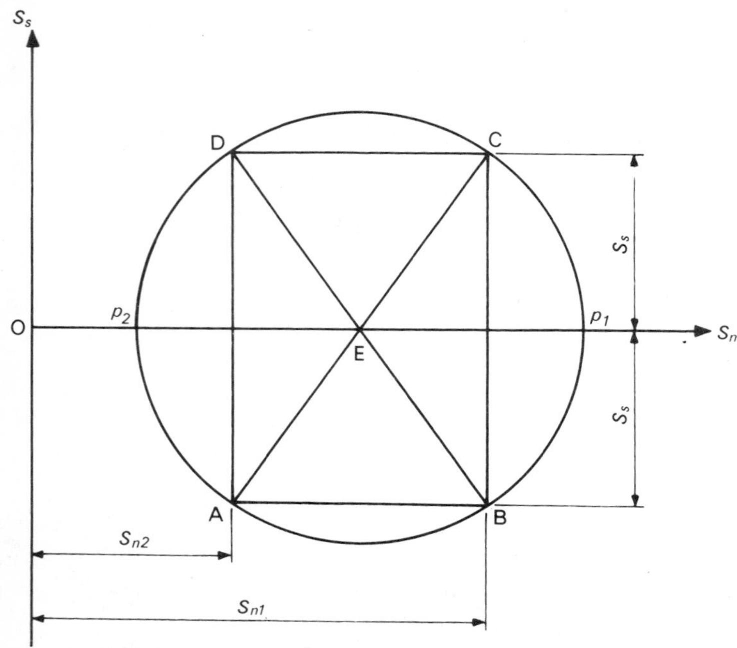

22 Figure 1.12 The construction of the plane stress Mohr's circle.

and hence it is a special case of plane stress. The Mohr's circle is shown in Figure 1.13(a) and is a single circle since $p_2 = p_3 = 0$. p_T is a principal value and the maximum shear stress is $\frac{1}{2}p_T$ and acts at 45° to p_T. The positions of four possible planes of maximum shear are shown in Figure 1.13(b) and will be recognised as the planes of shear failure frequently seen in practice.

(iii) *Simple shear*

Simple shear is produced by applying only a shear stress to a body as in twisting a rod about its axis. Again we have a special case of plane stress since all the normal stresses and the remaining two shear stresses are zero, i.e.

$$p_{xy} = p_s, \ p_{xx} = p_{yy} = p_{zz} = p_{yz} = p_{zx} = 0$$

The Mohr's circle is shown in Figure 1.14 with principal stresses of $\pm p_s$ at 45° to the shear stresses.

(iv) *Hydrostatic stress*

Let us consider a stress system in which the three principal stresses are equal:

i.e.
$$p_1 = p_2 = p_3 = p$$

The Mohr's circle is thus a single point on the S_n axis and on all planes at that point the normal stress is p. It can be seen clearly from the expression for S_n, equation (1.11), and from equation (1.12) for S_s that the shear stresses are zero on all planes. Thus for a hydrostatic stress system defined as one with three orthogonal normal stresses equal, there are no shear components and hence hydrostatic stress would not be expected to produce any shearing action and would result only in volume changes.

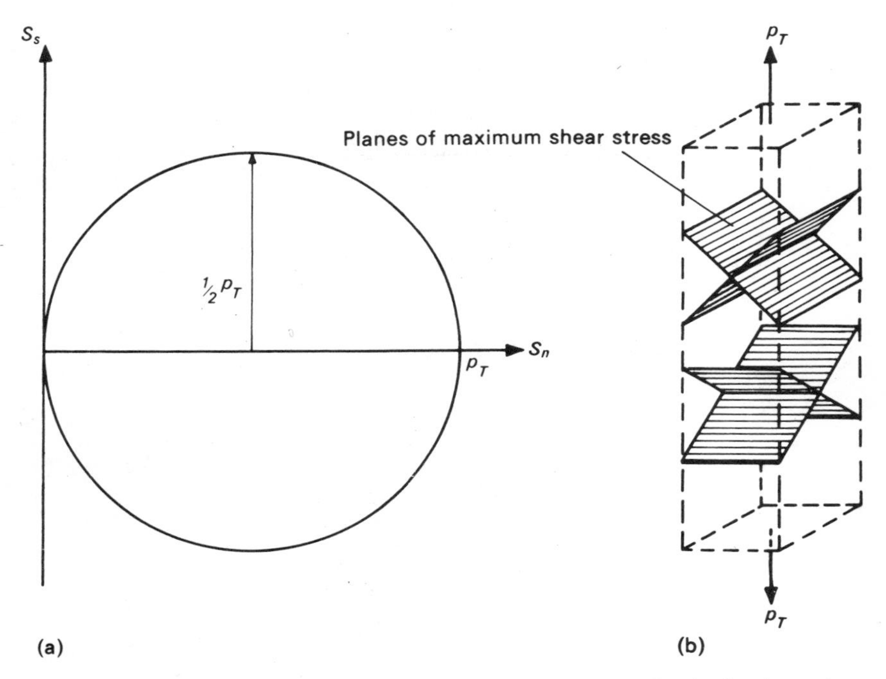

Figure 1.13 Mohr's circle and the planes of maximum shear stress in simple tension.

An example of the effect of hydrostatic stress is provided by a stress system in which the principal stresses are:

$$p_1,\ p_2 = \frac{p_1 + p_3}{2},\ p_3$$

The hydrostatic stress for this system is:

$$p_m = \frac{1}{3}\left(p_1 + \frac{p_1 + p_3}{2} + p_3\right) = \frac{p_1 + p_3}{2} = p_2$$

and the various components are shown drawn in Figure 1.15. Comparison of the diagram with that for simple shear shows that this is the same stress state with a superimposed hydrostatic stress. The deviatoric stress system is therefore identical with simple shear. Such a system can be employed for investigating the effects of hydrostatic stress on shear dependent phenomena and is also used

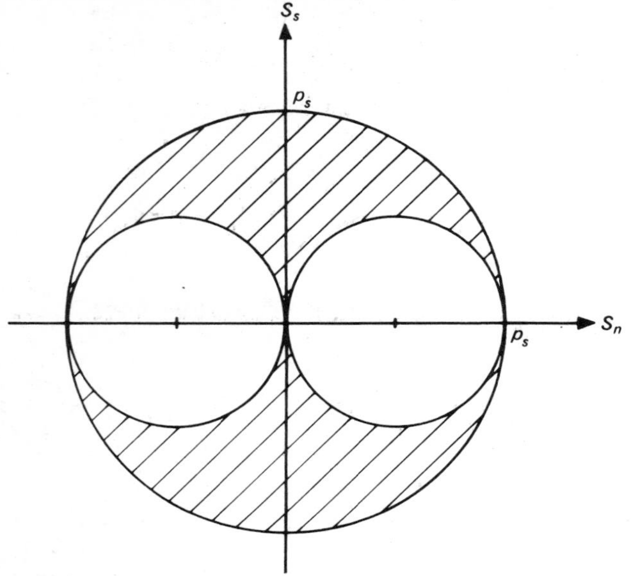

Figure 1.14 Mohr's circles for a simple shear stress state.

in testing materials which deform at constant volume since the condition $p_2 = (p_1 + p_3)/2$ is that for plane strain for these cases.

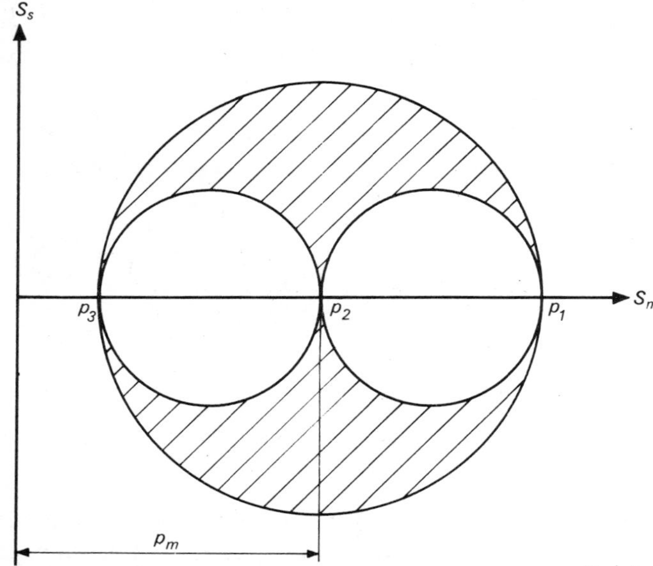

Figure 1.15 Mohr's circles for a stress system in which $p_2 = \dfrac{p_1 + p_3}{2}$

1.2. Stress at adjacent points and the equilibrium equations

The foregoing discussion has been concerned with the equilibrium of the stress components on various planes at a point. The equilibrium of a stress system implies more than this, however, since the way in which stresses vary from one point to another is considered. Clearly, for the whole system to be in equilibrium, the stresses must be in equilibrium with the external forces. In addition to this, however, the variations of stresses from any point to an adjacent point within the body are also governed by the requirement of local equilibrium.

1.2.1. *Cartesian coordinates*

We now consider a rectangular element as shown in Figure 1.16 which does not represent a point, but a small element. If, for example, the stress on the left hand

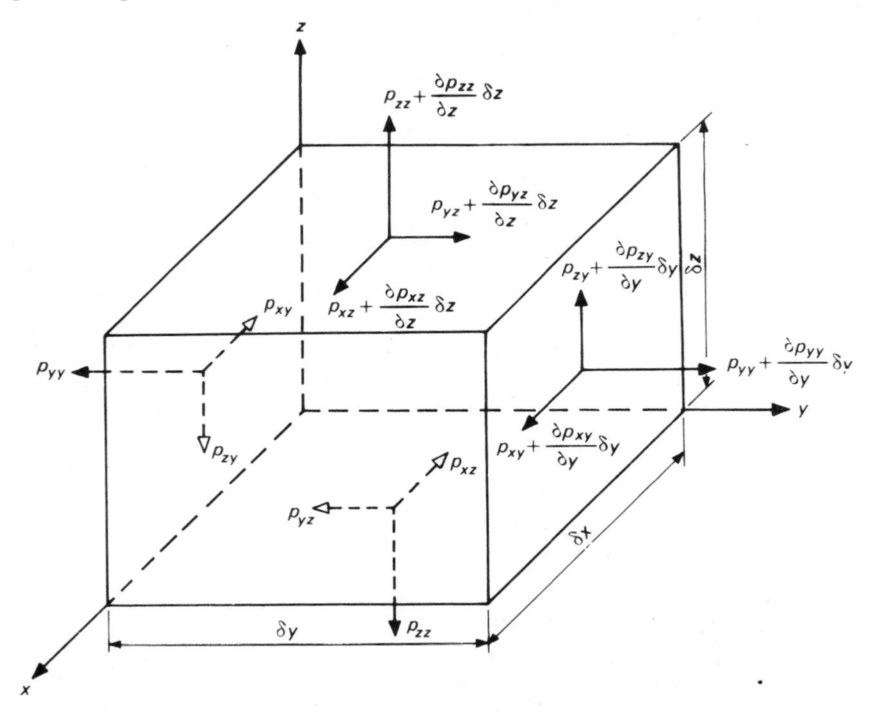

Figure 1.16 Variations of the stress components across a small element in the y and z directions.

face is p_{yy} then on the parallel right-hand face a distance δy away the value will have changed to:

$$p_{yy} + \frac{\partial p_{yy}}{\partial y}\,\delta y$$

In Figure 1.2 there was considered to be no change in p_{yy} because the element represented a point. Here we have a small element as before but this time it is not reduced to a point and the variations of stress along the three axes are included in the analysis. As the element is small it is assumed that the rate of change may be taken as constant over the small distance δy and the change in p_{yy} may be written as the product of the rate of change and the distance over which it occurs. This change is then added to the original p_{yy} value.

The variations of the stress components in the y and z directions are also shown in Figure 1.16 (those in the x direction are omitted for clarity). This system must satisfy the equilibrium conditions in the three coordinate directions. Figure 1.17 shows all the components which act in the y direction,

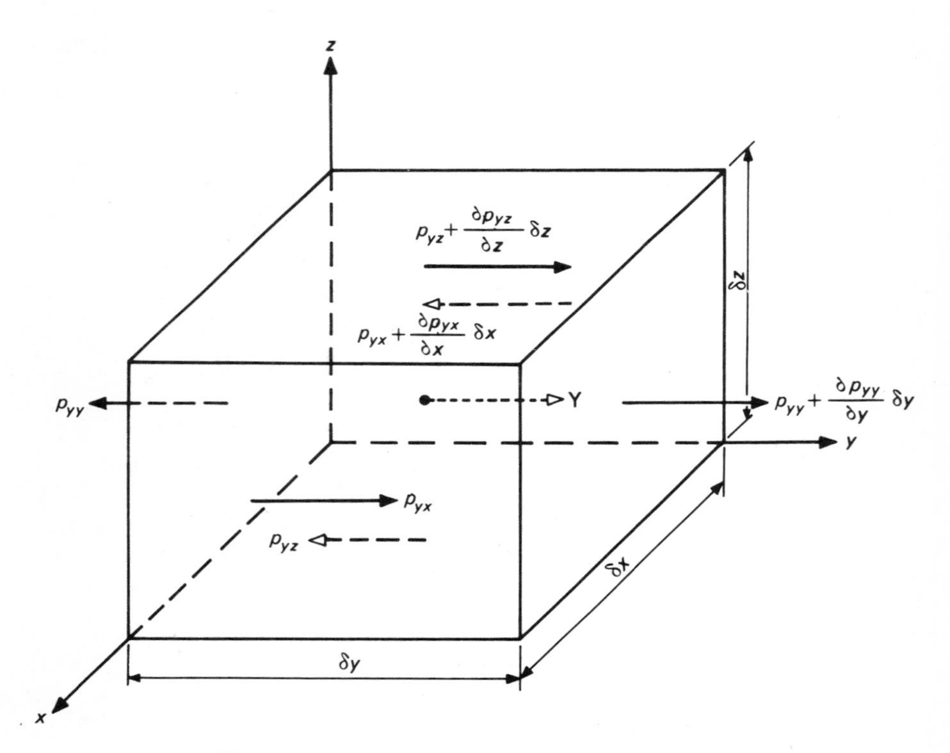

Figure 1.17 Components of stress in the y direction.

together with a body force per unit volume Y. This latter represents the forces which may be generated within the body of the material such as gravity, centrifugal forces, etc. The total force on the element will result in an acceleration in the y direction, f_y.

The total force is given by:

$$\left(p_{yy}+\frac{\partial p_{yy}}{\partial y}\,\delta y\right)\delta z\,\delta x-(p_{yy})\,\delta z\,\delta x$$
$$+\left(p_{yz}+\frac{\partial p_{yz}}{\partial z}\,\delta z\right)\delta y\,\delta x-(p_{yz})\,\delta y\,\delta x$$
$$+\left(p_{yx}+\frac{\partial p_{yx}}{\partial x}\,\delta x\right)\delta y\,\delta z-(p_{yx})\,\delta y\,\delta z$$
$$+\delta x\,\delta y\,\delta z\,Y$$

By Newton's laws of motion this may be equated to the product of mass times acceleration for the element; i.e. $\delta x\,\delta y\,\delta z\,\rho f_y$ where ρ is the density.

Simplifying this equation and dividing through by $\delta x\,\delta y\,\delta z$ gives:

$$\frac{\partial p_{yy}}{\partial y}+\frac{\partial p_{yx}}{\partial x}+\frac{\partial p_{yz}}{\partial z}+Y = \rho f_y \qquad (1.26(a))$$

Similarly by considering equilibrium in the other two directions we have:

$$\frac{\partial p_{xx}}{\partial x}+\frac{\partial p_{xy}}{\partial y}+\frac{\partial p_{xz}}{\partial z}+X = \rho f_x \qquad (1.26(b))$$

$$\frac{\partial p_{zz}}{\partial z}+\frac{\partial p_{zx}}{\partial x}+\frac{\partial p_{zy}}{\partial y}+Z = \rho f_z \qquad (1.26(c))$$

In most of the applications to be considered in later sections only static equilibrium will be involved and hence:

$$f_x = f_y = f_z = 0$$

These equations are frequently used, particularly for determining stress distributions and are of major importance. The special case of plane stress ($p_{zz} = p_{zx} = p_{zy} = 0$) is often used:

$$\left.\begin{aligned}\frac{\partial p_{xx}}{\partial x}+\frac{\partial p_{xy}}{\partial y}+X &= 0\\ \frac{\partial p_{yy}}{\partial y}+\frac{\partial p_{xy}}{\partial x}+Y &= 0\end{aligned}\right\} \qquad (1.27)$$

1.2.2. *Cylindrical polar coordinates*

Many engineering components such as round rods and tubes are solids or surfaces of revolution about an axis of geometric symmetry. A convenient coordinate system for describing the position of an element in such a body is r, θ and z as shown in Figure 1.18(a) where Oz is the axis of symmetry and the angle θ is measured from some line. The three general equations of equilibrium for the element shown may be determined by considering equilibrium in the r, θ and z

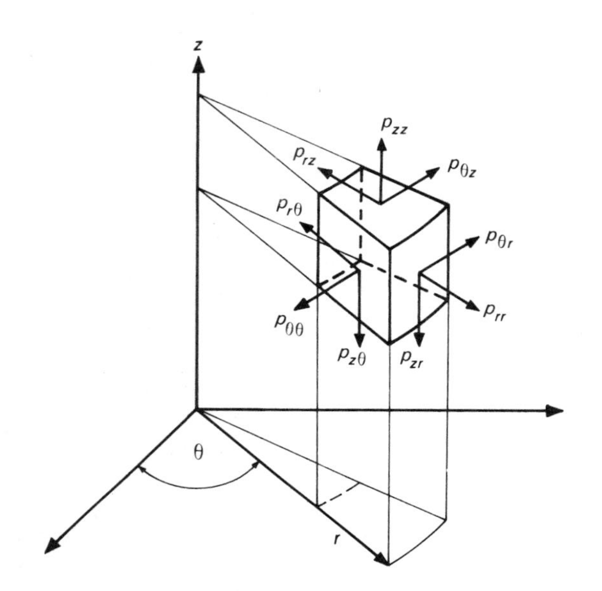

(a) Cylindrical polar coordinates

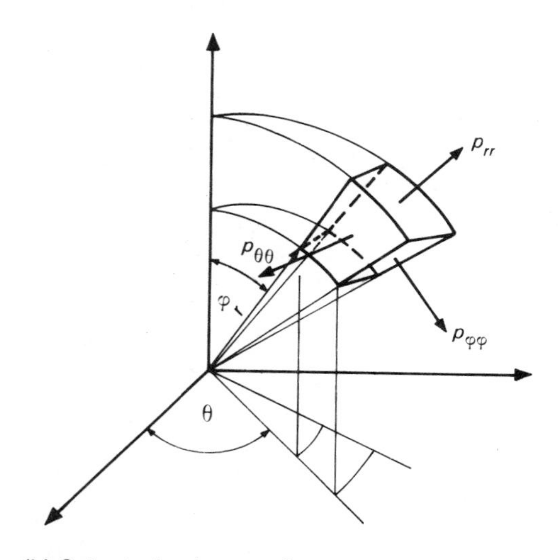

(b) Spherical polar coordinates

Figure 1.18 Coordinate systems.

directions to give the results:

$$\left.\begin{aligned}
\frac{\partial p_{rr}}{\partial r}+\frac{\partial p_{rz}}{\partial z}+\frac{1}{r}\frac{\partial p_{r\theta}}{\partial \theta}+\frac{p_{rr}-p_{\theta\theta}}{r}+F_r &= 0\\
\frac{\partial p_{r\theta}}{\partial r}+\frac{1}{r}\frac{\partial p_{\theta\theta}}{\partial \theta}+\frac{\partial p_{\theta z}}{\partial z}+\frac{2p_{r\theta}}{r}+F_\theta &= 0\\
\frac{\partial p_{rz}}{\partial r}+\frac{1}{r}\frac{\partial p_{\theta z}}{\partial \theta}+\frac{p_{rz}}{r}+\frac{\partial p_{zz}}{\partial z}+F_z &= 0
\end{aligned}\right\} \qquad (1.28)$$

A useful special case occurs when the stress system is symmetrical about the z axis so that all $\partial/\partial\theta$ terms become zero and the shear stress $p_{r\theta} = 0$ also. In addition variations of the stress system along the axis of symmetry are frequently zero giving the condition $p_{rz} = p_{\theta z} = 0$ so that equations (1.28) reduce to:

$$\frac{dp_r}{dr}+\frac{p_r-p_\theta}{r}+F_r = 0 \qquad (1.29)$$

Since no shear stresses appear in the equation the notation is abbreviated for the axisymmetric case so that $p_{rr} = p_r$ and $p_{\theta\theta} = p_\theta$.

1.2.3. *Spherical polar coordinates*

For problems concerning parts of spherical sections having symmetry about a single point a spherical polar coordinate system as shown in Figure 1.18(b) is the most convenient. For the case of spherical symmetry of the stresses $p_{\varphi\varphi} = p_{\theta\theta}$, and the equivalent equation to (1.29) is:

$$\frac{dp_r}{dr}+\frac{2}{r}(p_r-p_\theta)+F_r = 0 \qquad (1.30)$$

1.3. Deformation and strain

1.3.1. *Definitions*

The reaction of any solid continuum to an imposed stress system is to develop a state of strain. This is brought about by the displacement of points in the continuum relative to adjacent points. Clearly displacement itself is not sufficient to produce strain since strain implies the movement of a point relative to some other point. If we consider a small element of length l_0 in a continuum which, when a stress system is applied, attains a length of l then the displacement of one end relative to the other is given by:

$$u = l-l_0 \qquad (1.31)$$

Since strain is a definition of relative movement it is convenient to define it as a dimensionless function which may be written as:

$$\frac{l}{l_0} = f(\xi) \tag{1.32}$$

Thus any function f of the strain ξ will fulfil the condition of defining the change of length per unit length and may be chosen for convenience. Several are in use in different specialisations and will be used in later sections. The most common are:

(a) *Engineer's strain e*
This is used mainly for describing small strains (to be discussed later) and particularly in linear elastic analysis. It is defined as:

$$\frac{l}{l_0} = 1+e \tag{1.33}$$

for the simple element discussed previously.

(b) *Extension ratio λ*
This definition is used in large strain analysis and particularly in problems involving rubber elasticity. It is:

$$\frac{l}{l_0} = \lambda \tag{1.34}$$

(c) *Green's strain ε′*
Another large strain definition which may be written as:

$$\frac{l}{l_0} = [1+2\varepsilon']^{\frac{1}{2}} \tag{1.35}$$

(d) *Natural strain ε*
A further large strain definition developed for use in connection with plasticity theory but of more general use is couched in terms of its change as:

$$\mathrm{d}\varepsilon = \frac{\mathrm{d}l}{l}$$

This may be integrated to give:

$$\frac{l}{l_0} = \exp \varepsilon \tag{1.36}$$

Clearly all four may be converted from one to the other:

$$\lambda = 1+e = \frac{\mathrm{d}\lambda}{\mathrm{d}\varepsilon} = \frac{\mathrm{d}\varepsilon'}{\mathrm{d}\lambda}$$

1.3.2. *The general equations in two dimensions*

The definitions given so far are obviously inadequate for general use in that they take no account of directional changes. The forms of the general strain equations are best illustrated by considering a two-dimensional case as shown in Figure 1.19.

Consider an element with sides δx and δy originally orthogonal and intersecting at a point O. Suppose now that O is displaced to a point O′ by displacements of u and v in the x and y directions respectively. If we now assume that

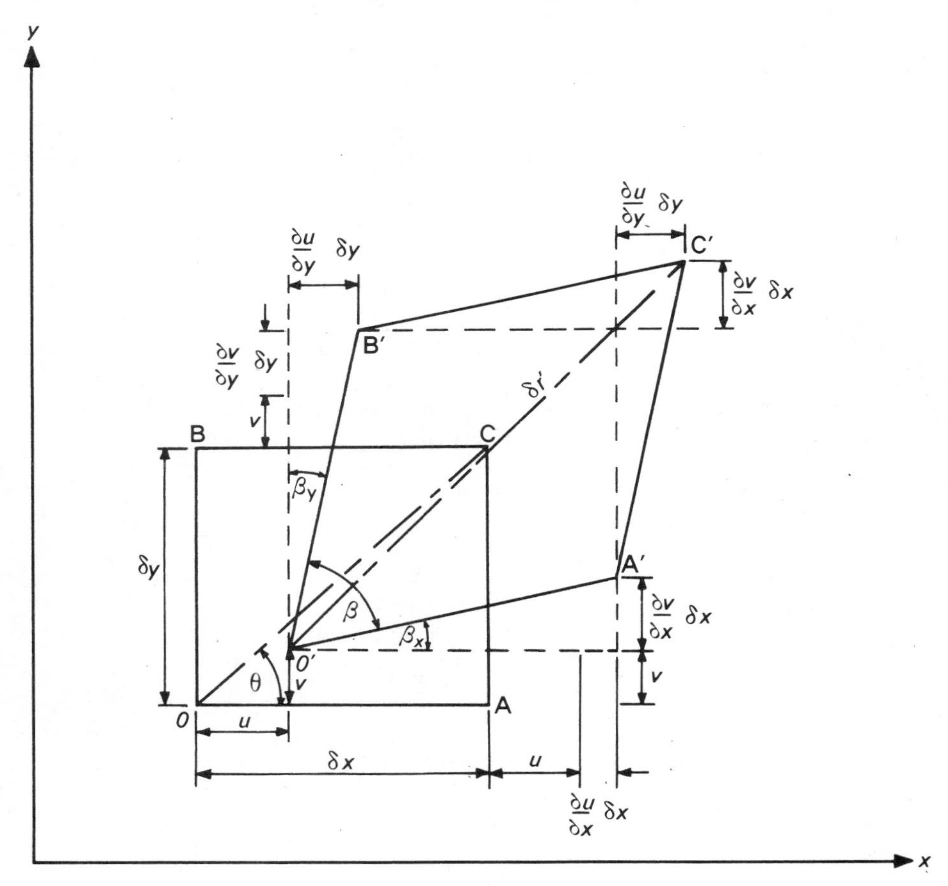

Figure 1.19 Strains in two dimensions.

the strain is homogeneous, and hence does not vary over the lengths δx and δy, then we may write down expressions for the movements of the opposite ends of the sides, A and B. A will move in the x direction by an amount u as did O but A is a distance δx from O and u will vary with x. Thus the rate of change of u with respect to x may be taken as $\partial u/\partial x$ and over the small distance δx the change in u will be $(\partial u/\partial x)\,\delta x$. The total movement of A in the x direction will be:

$$u+\frac{\partial u}{\partial x}\,\delta x$$

Similarly v will vary and A will move in the y direction by an amount:

$$v+\frac{\partial v}{\partial x}\,\delta x$$

The new length of the element OA is therefore given by:

$$(\delta x')^2 = \left(\delta x+\frac{\partial u}{\partial x}\,\delta x\right)^2+\left(\frac{\partial v}{\partial x}\,\delta x\right)^2$$

i.e.

$$\left(\frac{\delta x'}{\delta x}\right)^2 = \left(1+\frac{\partial u}{\partial x}\right)^2+\left(\frac{\partial v}{\partial x}\right)^2 \qquad (1.37)$$

Similarly we have:

$$\left(\frac{\mathrm{O'B'}}{\mathrm{OB}}\right)^2 = \left(\frac{\delta y'}{\delta y}\right)^2 = \left(1+\frac{\partial v}{\partial y}\right)^2+\left(\frac{\partial u}{\partial y}\right)^2 \qquad (1.38)$$

The distortion of the element can be defined by the change of the right angle AOB. This now becomes a new angle A′O′B′, say β, and the distortion may be described in terms of the cosine of this angle. If the angles made with the x and y axes are taken as β_x and β_y respectively then:

$$\beta = \frac{\pi}{2}-(\beta_x+\beta_y)$$

$$\therefore \quad \cos\beta = \sin\beta_x\cos\beta_y+\sin\beta_y\cos\beta_x$$

Now from Figure 1.19 we have:

$$\sin\beta_x = \frac{\partial v}{\partial x}\,\frac{\delta x}{\delta x'}, \qquad \cos\beta_x = \left(1+\frac{\partial u}{\partial x}\right)\frac{\delta x}{\delta x'}$$

$$\sin\beta_y = \frac{\partial u}{\partial y}\,\frac{\delta y}{\delta y'}, \qquad \cos\beta_y = \left(1+\frac{\partial v}{\partial y}\right)\frac{\delta y}{\delta y'}$$

and hence:

$$\cos\beta = \frac{\delta x}{\delta x'}\,\frac{\delta y}{\delta y'}\left(\frac{\partial u}{\partial y}+\frac{\partial v}{\partial x}+\frac{\partial v}{\partial x}\,\frac{\partial v}{\partial y}+\frac{\partial u}{\partial y}\,\frac{\partial u}{\partial x}\right) \qquad (1.39)$$

A further quantity of interest is the length of a line in any direction of length δr where δr is taken as OC in Figure 1.19 and its position is fixed by the choice of δx and δy. From Figure 1.19 it can be seen that B′C′ is parallel to O′A′ and that A′C′ is parallel to O′B′. Thus we may write:

$$\delta r'^2 = \delta x'^2+\delta y'^2+2\delta x'\,\delta y'\cos\beta$$

and this may be rewritten as:

$$\left(\frac{\delta r'}{\delta r}\right)^2 = \left(\frac{\delta x'}{\delta x}\right)^2\left(\frac{\delta x}{\delta r}\right)^2 + \left(\frac{\delta y'}{\delta y}\right)^2\left(\frac{\delta y}{\delta r}\right)^2 + 2\left(\frac{\delta x'}{\delta x}\right)\left(\frac{\delta y'}{\delta y}\right)\left(\frac{\delta x}{\delta r}\right)\left(\frac{\delta y}{\delta r}\right)\cos\beta$$

If θ is taken as the angle between the original direction of the line and the x axis then:

$$\cos\theta = \frac{\delta x}{\delta r} \quad \text{and} \quad \sin\theta = \frac{\delta y}{\delta r}$$

and,

$$\left(\frac{\delta r'}{\delta r}\right)^2 = \left(\frac{\delta x'}{\delta x}\right)^2\cos^2\theta + \left(\frac{\delta y'}{\delta y}\right)^2\sin^2\theta + 2\left(\frac{\delta x'}{\delta x}\right)\left(\frac{\delta y'}{\delta y}\right)\sin\theta\cos\theta\cos\beta$$

In terms of extension ratios in the coordinate directions and in the general direction this becomes:

$$\lambda_r^{\,2} = \lambda_{xx}^{\,2}\cos^2\theta + \lambda_{yy}^{\,2}\sin^2\theta + \lambda_{xx}\lambda_{yy}\cos\beta . 2\sin\theta\cos\theta \tag{1.40}$$

where $\lambda_r = \dfrac{\delta r'}{\delta r}$, $\lambda_{xx} = \dfrac{\delta x'}{\delta x}$ and $\lambda_{yy} = \dfrac{\delta y'}{\delta y}$.

It is now appropriate to compare this equation with that for the normal stress in a general direction for the two-dimensional case (plane stress):

$$S_n = p_{xx}\cos^2\theta + p_{yy}\sin^2\theta + p_{xy} . 2\sin\theta\cos\theta \tag{1.41}$$

The similarity in form is apparent and equation (1.40) is the transformation equation for λ^2 from one coordinate system to another just as equation (1.41) is that for stress. Exact parity in form is achieved by defining a shear extension ratio of the form:

$$\lambda_{xy}^{\,2} = \lambda_{xx}\lambda_{yy}\cos\beta$$

and the derivations for the stress system may be used for the extension ratios by replacing the stress by the appropriate extension ratio squared. Thus we may derive principal values $\lambda_1^{\,2}$, $\lambda_2^{\,2}$ and $\lambda_3^{\,2}$, invariants, graphical constructions, etc., by direct substitution.

For a general definition of strain ξ in terms of λ we may write:

$$\lambda = f(\xi)$$

where f is a general function in which $f(0) = 1$. It is clear from equation (1.40) that any function transforms in the appropriate manner for the first two terms. Any other function could be used to define the shear strain but in order to effect a transformation it would have to be substituted into equation (1.40). It is more convenient to define the shear strain in such a way that it does transform directly and the rest of the analysis may be derived by a direct substitution in the stress equations. To achieve this we may define shear strain as:

$$f^2(\xi_{xy}) - 1 = f(\xi_{xx})f(\xi_{yy})\cos\beta \tag{1.42}$$

Thus for any definition of strain appropriate shear strains may be defined and since $f(\xi)$ transforms with λ then principal strains may be defined from:

$$\lambda_1 = f(\xi_1), \quad \lambda_2 = f(\xi_2), \quad \lambda_3 = f(\xi_3)$$

These principal strains, ξ_1, ξ_2 and ξ_3 are themselves invariant in the sense that they are a particular set of coordinates independent of any arbitrarily chosen set but no single value is invariant since a particular direction would have to be chosen. However all symmetrical functions of the principal values are invariant and the three simplest forms are:

$$\begin{aligned} I_1 &= \xi_1 + \xi_2 + \xi_3 \\ I_2 &= \xi_1\xi_2 + \xi_2\xi_3 + \xi_3\xi_1 \\ I_3 &= \xi_1\xi_2\xi_3 \end{aligned} \tag{1.43}$$

1.3.3. *Extension ratios*

The extension ratios may be written in terms of displacements from, for example, equations (1.37) and (1.39), extended by analogy to three dimensions giving:

$$\lambda_{xx}^2 = \left(1 + \frac{\partial u}{\partial x}\right)^2 + \left(\frac{\partial v}{\partial x}\right)^2 + \left(\frac{\partial w}{\partial x}\right)^2$$

and

$$\lambda_{xy}^2 = \frac{\partial u}{\partial y} + \frac{\partial v}{\partial x} + \frac{\partial v}{\partial x}\frac{\partial v}{\partial y} + \frac{\partial u}{\partial x}\frac{\partial u}{\partial y} + \frac{\partial w}{\partial x}\frac{\partial w}{\partial y}$$

where w is the displacement in the z direction.

The invariants may be obtained by substitution in the appropriate stress equations. To further illustrate the analogy with the stress equations it is of interest to draw the two dimensional Mohr's circle for extension ratios as shown in Figure 1.20. By analogy with equations (1.23) and (1.24) extension ratios on any plane with a normal at an angle α to the λ_1 direction are given by:

$$\lambda_r^2 = \left(\frac{\lambda_1^2 + \lambda_2^2}{2}\right) + \left(\frac{\lambda_1^2 - \lambda_2^2}{2}\right)\cos 2\alpha$$

and

$$\lambda_{xy}^2 = \lambda_{xx}\lambda_{yy}\cos\beta = \left(\frac{\lambda_1^2 - \lambda_2^2}{2}\right)\sin 2\alpha$$

The angle between any two originally orthogonal lines is given by:

$$\cos\beta = \frac{(\lambda_1^2 - \lambda_2^2)\sin 2\alpha}{[(\lambda_1^2 + \lambda_2^2)^2 - (\lambda_1^2 - \lambda_2^2)\cos^2 2\alpha]^{\frac{1}{2}}}$$

It should be noted that the principal directions ($\alpha = 0$) remain orthogonal as expected.

The principal values are given by:

$$\lambda_{1,2}^2 = \frac{\lambda_{xx}^2 + \lambda_{yy}^2}{2} \pm \sqrt{\left[\frac{\lambda_{xx}^2 - \lambda_{yy}^2}{2}\right]^2 + \lambda_{xy}^4}$$

1.3.4. *Natural strain*

The normal strains are defined as:

$$\varepsilon_{xx} = \ln \lambda_{xx},\ \varepsilon_{yy} = \ln \lambda_{yy},\ \varepsilon_{zz} = \ln \lambda_{zz} \tag{1.44}$$

so that the function f becomes:

$$f(\varepsilon_{xx}) = \lambda_{xx} = \exp(\varepsilon_{xx})$$

The shear strain may then be deduced from equation (1.42) giving:

$$\exp(2\varepsilon_{xy}) - 1 = \exp(\varepsilon_{xx}) \exp(\varepsilon_{yy}) \cos \beta$$

i.e.

$$\varepsilon_{xy} = \tfrac{1}{2}\ln[1 + \lambda_{xx}\lambda_{yy} \cos \beta] \tag{1.45}$$

The invariants are obtained directly from equations (1.43):

$$I_1 = \varepsilon_1 + \varepsilon_2 + \varepsilon_3,\ I_2 = \varepsilon_1\varepsilon_2 + \varepsilon_2\varepsilon_3 + \varepsilon_3\varepsilon_1,\ I_3 = \varepsilon_1\varepsilon_2\varepsilon_3 \tag{1.46}$$

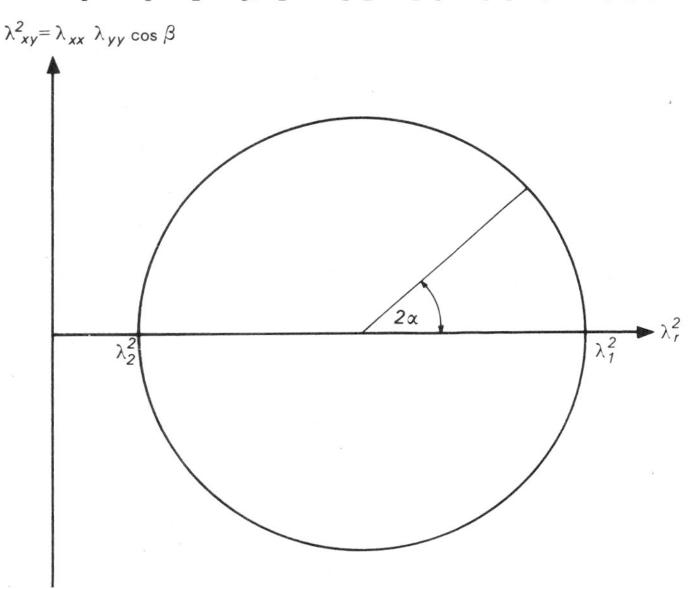

Figure 1.20 Mohr's circle for extension ratios in two dimensions

In terms of displacements the strains have the forms:

$$\left.\begin{aligned} \varepsilon_{xx} &= \tfrac{1}{2}\ln\left[\left(1 + \frac{\partial u}{\partial x}\right)^2 + \left(\frac{\partial v}{\partial x}\right)^2 + \left(\frac{\partial w}{\partial x}\right)^2\right] \\ &\text{and} \\ \varepsilon_{xy} &= \tfrac{1}{2}\ln\left[1 + \frac{\partial u}{\partial y} + \frac{\partial v}{\partial x} + \frac{\partial v}{\partial x}\frac{\partial v}{\partial y} + \frac{\partial u}{\partial x}\frac{\partial u}{\partial y} + \frac{\partial w}{\partial x}\frac{\partial w}{\partial y}\right] \end{aligned}\right\} \tag{1.47}$$

1.3.5. *Green's strain*

By definition we have:

$$\varepsilon'_{xx} = \tfrac{1}{2}[\lambda_{xx}^2 - 1] \tag{1.48}$$

so that

$$f(\varepsilon'_{xx}) = \lambda_{xx} = (1 + 2\varepsilon'_{xx})^{\frac{1}{2}}$$

and

$$f^2(\varepsilon'_{xy}) - 1 = 2\varepsilon'_{xy} = \lambda_{xx}\lambda_{yy}\cos\beta \tag{1.49}$$

Direct substitution into equation (1.40) gives:

$$\varepsilon'_r = \varepsilon'_{xx}\cos^2\theta + \varepsilon'_{yy}\sin^2\theta + \varepsilon'_{xy}\, 2\sin\theta\cos\theta$$

which is exactly the same form as the stress equation and is the reason the definition is used. In terms of displacements we have:

$$\left.\begin{aligned} \varepsilon'_{xx} &= \frac{1}{2}\left[\left(1 + \frac{\partial u}{\partial x}\right)^2 + \left(\frac{\partial v}{\partial x}\right)^2 + \left(\frac{\partial w}{\partial x}\right)^2 - 1\right] \\ &\text{and} \\ \varepsilon'_{xy} &= \frac{1}{2}\left[\frac{\partial u}{\partial y} + \frac{\partial v}{\partial x} + \frac{\partial v}{\partial x}\frac{\partial v}{\partial y} + \frac{\partial u}{\partial x}\frac{\partial u}{\partial y} + \frac{\partial w}{\partial x}\frac{\partial w}{\partial y}\right] \end{aligned}\right\} \tag{1.50}$$

1.3.6. *Engineer's and infinitesimal strain*

Engineer's strain may be defined as:

$$e_{xx} = \lambda_{xx} - 1$$

so that:

$$f(e_{xx}) = \lambda_{xx} = 1 + e_{xx}$$

and

$$f^2(e_{xy}) - 1 = (1 + e_{xy})^2 - 1 = (1 + e_{xx})(1 + e_{yy})\cos\beta$$

In terms of displacements these definitions become:

$$\left.\begin{aligned} e_{xx} &= \left[\left(1 + \frac{\partial u}{\partial x}\right)^2 + \left(\frac{\partial v}{\partial x}\right)^2 + \left(\frac{\partial w}{\partial x}\right)^2\right]^{\frac{1}{2}} - 1 \\ &\text{and} \\ e_{xy} &= \left[1 + \frac{\partial u}{\partial y} + \frac{\partial v}{\partial x} + \frac{\partial v}{\partial x}\frac{\partial v}{\partial y} + \frac{\partial u}{\partial x}\frac{\partial u}{\partial y} + \frac{\partial w}{\partial x}\frac{\partial w}{\partial y}\right]^{\frac{1}{2}} - 1 \end{aligned}\right\} \tag{1.51}$$

In this form it is not a particularly useful definition which is why the previously mentioned definitions have been used. However if we now limit our attention to small strains and assume that all the derivatives of displacement are very much less than unity then we may ignore the second-order terms and equations (1.51) reduce to:

$$e_{xx} = \frac{\partial u}{\partial x} \quad \text{and} \quad e_{xy} = \frac{1}{2}\left(\frac{\partial u}{\partial y} + \frac{\partial v}{\partial x}\right)$$

Examination of equations (1.47) and (1.50) for natural and Green's strain respectively show that both definitions reduce to the same form for infinitesimal strains. Both are better finite strain definitions than engineer's strain but the latter has arisen historically because it is simple to apply directly to the infinitesimal case. To derive the infinitesimal strain definitions we may return to Figure 1.19 and note that if the change of length of $\delta x'$ due to its change of direction is small then the contribution of the term $\partial v/\partial x \, \delta x$ in equation (1.37) may be ignored and we have:

$$\frac{\delta x'}{\delta x} = 1 + \frac{\partial u}{\partial x}$$

so that:

$$e_{xx} = \frac{\delta x'}{\delta x} - 1 = \frac{\partial u}{\partial x}$$

Similarly since the angular changes are small:

$$\tan \beta_x \simeq \beta_x \simeq \frac{\partial v}{\partial x} \quad \text{and} \quad \beta_y \simeq \frac{\partial u}{\partial y}$$

The mean change in the right angle is therefore given by:

$$\frac{\beta_x + \beta_y}{2} = \frac{1}{2}\left(\frac{\partial v}{\partial x} + \frac{\partial u}{\partial y}\right) = e_{xy}$$

In three dimensions the equations become:

$$\begin{aligned} e_{xx} &= \frac{\partial u}{\partial x}, \; e_{yy} = \frac{\partial v}{\partial y}, \; e_{zz} = \frac{\partial w}{\partial z} \\ e_{xy} &= \frac{1}{2}\left(\frac{\partial u}{\partial y} + \frac{\partial v}{\partial x}\right), \; e_{yz} = \frac{1}{2}\left(\frac{\partial v}{\partial z} + \frac{\partial w}{\partial y}\right), \; e_{zx} = \frac{1}{2}\left(\frac{\partial w}{\partial x} + \frac{\partial u}{\partial z}\right) \end{aligned} \tag{1.52}$$

In this form the transformation equations, invariants, etc., are obtained by direct substitution into the stress equations. Unfortunately, for historical reasons the $\frac{1}{2}$ factor is omitted in the shear strains when engineer's strains are used and so in order to conform with other engineering texts shear strain will be written as:

$$e_{xy} = \frac{\partial u}{\partial y} + \frac{\partial v}{\partial x}, \quad e_{yz} = \frac{\partial v}{\partial z} + \frac{\partial w}{\partial y}, \quad e_{zx} = \frac{\partial w}{\partial x} + \frac{\partial u}{\partial z} \tag{1.53}$$

By substitution in the stress equations we may deduce a strain cubic of the form:

$$e^3 - I_1 e^2 + I_2 e - I_3 = 0$$

where:

$$\begin{aligned} I_1 &= e_{xx} + e_{yy} + e_{zz} \\ I_2 &= e_{xx}e_{yy} + e_{yy}e_{zz} + e_{zz}e_{xx} - \frac{e^2{}_{xy}}{4} - \frac{e^2{}_{yz}}{4} - \frac{e^2{}_{xz}}{4} \\ I_3 &= e_{xx}e_{yy}e_{zz} + \frac{1}{4}e_{xy}e_{yz}e_{zx} - \frac{1}{4}e_{xx}e^2_{yz} - \frac{1}{4}e_{yy}e^2_{zx} - \frac{1}{4}e_{zz}e^2_{xy} \end{aligned} \tag{1.54}$$

i.e. all terms involving shear strains must be modified by a factor of $\frac{1}{2}$ for each shear strain. For the two-dimensional case, the cubic reduces to:

$$e^2 - (e_{xx} + e_{yy})\, e + \left(e_{xx}e_{yy} - \frac{e^2_{xy}}{4}\right) = 0$$

so that the principal values become:

$$e_{1,2} = \frac{e_{xx} + e_{yy}}{2} \pm \sqrt{\left[\frac{e_{xx} - e_{yy}}{2}\right]^2 + \frac{e^2_{xy}}{4}} \tag{1.55}$$

This is the same form as equation (1.25) except for the factor of $\frac{1}{2}$ on each e_{xy} term. The Mohr's circle is drawn in the same way as for stress except that the shear axis is halved.

1.3.7. *Compatibility equations*

Since there are six components of strain defined in terms of three displacements it is possible to derive equations which relate the strain components by eliminating the displacements. Such relationships, called the compatibility equations, must be satisfied at all points in a continuum. For example, from equations (1.52) and (1.53) for infinitesimal strains we have:

$$e_{xx} = \frac{\partial u}{\partial x},\ e_{yy} = \frac{\partial v}{\partial y} \text{ and } e_{xy} = \frac{\partial u}{\partial y} + \frac{\partial v}{\partial x}$$

so that:

$$\frac{\partial^2 e_{xy}}{\partial x \partial y} = \frac{\partial^3 u}{\partial x \partial y^2} + \frac{\partial^3 v}{\partial x^2 \partial y} = \frac{\partial^2 e_{xx}}{\partial y^2} + \frac{\partial^2 e_{yy}}{\partial x^2} \tag{1.56}$$

and two similar forms for e_{yz} and e_{zx}. There are three equations of the form:

$$\frac{\partial^2 e_{xx}}{\partial y \partial z} = \frac{\partial^3 u}{\partial x \partial y \partial z} = \frac{1}{2}\frac{\partial}{\partial x}\left(\frac{\partial e_{xy}}{\partial z} + \frac{\partial e_{xz}}{\partial y} - \frac{\partial e_{yz}}{\partial x}\right) \tag{1.57}$$

1.3.8. *Cylindrical and spherical polar coordinates*

The cylindrical polar coordinate system discussed in section 1.2.2. and shown in Figure 1.18(a) may also be used to describe the strain system. We may define the displacements in the coordinate directions as u_r, u_θ and u_z and for infinitesimal strains we have:

$$\begin{aligned}
e_{rr} &= \frac{\partial u_r}{\partial r}, \; e_{\theta\theta} = \frac{u_r}{r}+\frac{\partial u_\theta}{r\partial\theta}, \; e_{zz} = \frac{\partial u_z}{\partial z} \\
e_{r\theta} &= \frac{\partial u_\theta}{\partial r}-\frac{u_\theta}{r}+\frac{1}{r}\frac{\partial u_r}{\partial\theta}, \; e_{rz} = \frac{\partial u_r}{\partial z}+\frac{\partial u_z}{\partial r} \\
e_{\theta z} &= \frac{\partial u_\theta}{\partial z}+\frac{1}{r}\frac{\partial u_z}{\partial\theta}
\end{aligned} \tag{1.58}$$

A common use of these is for axial symmetry of the strain system so that $\partial/\partial\theta = 0$ and $u_\theta = 0$ resulting in:

$$\begin{aligned}
e_r &= \frac{\partial u_r}{\partial r}, \; e_\theta = \frac{u_r}{r}, \; e_z = \frac{\partial u_z}{\partial z} \\
e_{rz} &= \frac{\partial u_z}{\partial r}+\frac{\partial u_r}{\partial z}
\end{aligned} \tag{1.59}$$

Note that a single suffix is used for the normal strain in axial symmetry as in the stress equation (1.29).

Finite strains may also be expressed in cylindrical polar coordinates and are particularly useful when there is geometric and strain symmetry about the z axis. The extension ratios may then be written as λ_r, λ_θ and λ_z and the usual invariant conditions, etc., may be employed.

For problems involving spherical symmetry it is useful to use spherical polar coordinates giving the infinitesimal strains.

$$e_{rr} = \frac{\partial u_r}{\partial r}, \; e_{\theta\theta} = e_{\phi\phi} = \frac{u_r}{r} \tag{1.60}$$

1.3.9. *Volume changes*

The three invariants of the finite strain state in terms of extension ratios are:

$$\begin{aligned}
I_1 &= \lambda_{xx}^2+\lambda_{yy}^2+\lambda_{zz}^2 \\
I_2 &= \lambda_{xx}^2\lambda_{yy}^2+\lambda_{yy}^2\lambda_{zz}^2+\lambda_{zz}^2\lambda_{xx}^2-\lambda_{xy}^4-\lambda_{yz}^2-\lambda_{zx}^4 \\
I_3 &= \lambda_{xx}^2\lambda_{yy}^2\lambda_{zz}^2+2\lambda_{xy}^2\lambda_{yz}^2\lambda_{zx}^2-\lambda_{xx}^2\lambda_{yz}^4-\lambda_{yy}^2\lambda_{zx}^4-\lambda_{zz}^2\lambda_{xy}^4
\end{aligned} \tag{1.61}$$

I_1 may be regarded as a measure of the overall length change in the element. It is not, however, the length of the diagonal of the distorted element but the sum of the squares of the side lengths. The area of the distorted element face in two dimensions, a parallelogram, shown in Figure 1.19 is given by:

$$A_{xy} = \lambda_{xx}\lambda_{yy}\sin\beta$$

If we consider the whole element there are three distortion angles which we will give the suffixes of the two directions which define them, i.e. β_{xy} is between λ_{xx} and λ_{yy}. Thus we may write down the sum of the squares of the areas of the parallelepiped formed by deformation giving:

$$A_{xy}^2 + A_{yz}^2 + A_{zx}^2 = \lambda_{xx}^2\lambda_{yy}^2\sin^2\beta_{xy} + \lambda_{yy}^2\lambda_{zz}^2\sin^2\beta_{yz} + \lambda_{zz}^2\lambda_{xx}^2\sin^2\beta_{zx}$$

Substituting for the shear extension ratios we have:

$$A_{xy}^2 + A_{yz}^2 + A_{zx}^2 = \lambda_{xx}^2\lambda_{yy}^2 + \lambda_{yy}^2\lambda_{zz}^2 + \lambda_{zz}^2\lambda_{xx}^2 - \lambda_{xy}^4 - \lambda_{yz}^4 - \lambda_{zx}^4 = I_2 \qquad (1.62)$$

Thus I_2 may be regarded as a measure of overall area change of the element. The square of the volume of the parallelepiped is given by:

$$V^2 = \lambda_{xx}^2\lambda_{yy}^2\lambda_{zz}^2\,[1 - \cos^2\beta_{xy} - \cos^2\beta_{yz} - \cos^2\beta_{zx} + 2\cos\beta_{xy}\cos\beta_{yz}\cos\beta_{zx}]$$

which on substitution for the shear extension ratios gives:

$$V^2 = \lambda_{xx}^2\lambda_{yy}^2\lambda_{zz}^2 - \lambda_{zz}^2\lambda_{xy}^4 - \lambda_{yy}^2\lambda_{xz}^4 - \lambda_{xx}^2\lambda_{yz}^4 + 2\lambda_{xy}^2\lambda_{yz}^2\lambda_{zx}^2 = I_3 \qquad (1.63)$$

Thus I_3 is the square of the volume of the element which is of importance, since, if the deformation takes place at constant volume then:

$$V^2 = I_3 = 1$$

under all circumstances. It is of interest to note that if constant volume is specified for the two-dimensional case then the area of the parallelogram A_{xy} must remain constant and hence:

$$A_{xy} = \lambda_{xx}\lambda_{yy}\sin\beta_{xy} = \text{constant}$$

Thus if the angle is changed then λ_{xx} or λ_{yy} or both must change, illustrating the interdependence of angular changes and stretching.

For infinitesimal strains we may neglect terms of higher order than the first so that:

$$\lambda_{xx}^2 \simeq 1 + 2e_{xx} \text{ and } \lambda_{xy}^2 \simeq e_{xy} \text{ for example.}$$

From equation (1.63) we have:

$$V^2 = 1 + 2(e_{xx} + e_{yy} + e_{zz})$$

If we define an engineer's volumetric strain such that:

$$V = 1 + \Delta$$

then for infinitesimal volume changes we have:

$$\Delta = e_{xx} + e_{yy} + e_{zz} \qquad (1.64)$$

This is the first invariant of the infinitesimal strains given in equations (1.54).

The first invariant for natural strains in terms of principal values (equation (1.46)) is also a measure of volume change since:

$$\begin{aligned} I_1 &= \varepsilon_1+\varepsilon_2+\varepsilon_3 = \ln\lambda_1+\ln\lambda_2+\ln\lambda_3 \\ &= \ln\lambda_1\lambda_2\lambda_3 = \ln V \end{aligned}$$

For constant volume $V = 1$ and hence:

$$\varepsilon_1+\varepsilon_3+\varepsilon_3 = 0 \tag{1.65}$$

Bibliography

1. Love, A.E.H. *Theory of Elasticity*, Cambridge University Press, London (1926).
2. Ford, H. *Advanced Mechanics of Materials*, Longmans, London (1963).
3. Nadai, A. *Theory of Flow and Fracture of Solids*, Vol. 1. McGraw-Hill, New York (1950).

2 Time independent behaviour

Introduction

Chapter one dealt with the first two stages of a continuum mechanics analysis by considering the consequences of equilibrium and compatibility. The equilibrium condition gave rise to six components of stress to describe the state of stress at a point and three relationships between them as determined by the equilibrium of forces in the three coordinate directions, i.e. six unknowns and three equations. Compatibility considerations resulted in three displacements in the coordinate directions and six components of strain with six equations relating the strains to the displacements, i.e. nine unknowns and six equations. Thus the analysis so far has resulted in the definition of fifteen unknowns and only nine relationships between them so that to define a solution completely six more equations must be obtained. The presence of a state of stress at a point results in a state of strain at that point and it is the relationships between the six components of one with the components of the other which provide the missing six equations. These equations, called the constitutive relationships, are determined by the nature of the material and are the means by which the properties of particular materials are introduced into the continuum solution.

Continua range in form from hard crystalline solids, such as rocks and metals, to liquids and gases and include a considerable variation of physical properties between the extremes. Any one material may exhibit a wide range of behaviour depending on its temperature and so a constitutive relationship capable of describing a material completely would have to include every possibility. In principle there is no reason why a constitutive equation capable of describing such a wide range of properties should not exist but it is clear that it would be extremely complicated and that in a particular set of circumstances only a small part of the equation would be used. In practice, therefore, constitutive equations are used to describe properties under a very limited set of conditions so that a simple representation may be used. In order to achieve mathematical simplicity, idealisations are employed and it is the relationships based on these which are used in the various branches of continuum theory. Thus classical small strain linear-elasticity is based on Hooke's law and is a reasonably realistic representation for crystalline solids at low stress levels. The assumption of Newtonian viscosity is the basis of fluid mechanics and is a good representation for low viscosity fluids. Because the constitutive equations are idealisations, no real material ever obeys the assumptions involved exactly. However many materials can be successfully described with quite a limited number of assumptions and a useful representation of their behaviour is provided.

The two extremes of material behaviour may be taken as:

(a) The rigid solid; i.e. a material which does not deform under any stress state so that all the strains are zero giving the six equations:

$$\xi_{xx} = \xi_{yy} = \xi_{zz} = \xi_{xy} = \xi_{yz} = \xi_{zx} = 0$$

The stress distribution in the material is thus solely a consequence of equilibrium.

(b) The perfect fluid; i.e. a material which may withstand a hydrostatic pressure but cannot sustain any shear stresses so that we have the six equations:

$$p'_{xx} = p'_{yy} = p'_{zz} = p_{xy} = p_{yz} = p_{zx} = 0$$

The displacement conditions in this case are solely a consequence of compatibility requirements. All real materials lie somewhere between these two extremes and it is useful at this point to consider the behaviour of polymers to determine where it lies in the spectrum between the two. In this book we shall be concerned only with the solid state and so will not consider problems of fluid flow. The effect of temperature changes is of major importance, however, in that there is a particular temperature, called the glass transition temperature, above which polymers behave as rubbers. In a simple tension test the behaviour on loading and unloading is as shown in Figure 2.1(a) where the strains reached may be several hundreds of percent. The curve is markedly non-linear but the strain is almost completely and immediately recoverable so that an assumption of perfect elasticity, i.e. a unique relationship between stress and strain which is independent of the time scale and history of loading, is a good representation. The shape of the curve changes with temperature and as the temperature increases the polymer will eventually become a fluid. No attempt will be made in this text to include temperature in the analysis although it is assumed that the parameters involved in each representation may vary with temperature.

Below the glass transition temperature (the glassy state) there is a marked change in the nature of polymer behaviour and the simple tension nominal stress-strain curve now has the form as shown in Figure 2.1(b). At low strains the curve rises more steeply when compared with the rubbery behaviour although there is always some degree of curvature. The strains are recoverable, but only over a period of time, and the curve is sensitive to strain rate and the loading history. This region is termed visco-elastic since it incorporates some of the characteristics of both viscous flow and perfect elasticity. For strains less than about 0·5%, many materials may be described as linearly visco-elastic (see chapter 3) but above this value there are marked non-linearities. At strains of about 5–10%, point A in Figure 2.1(b), there is a peak in the stress-strain curve

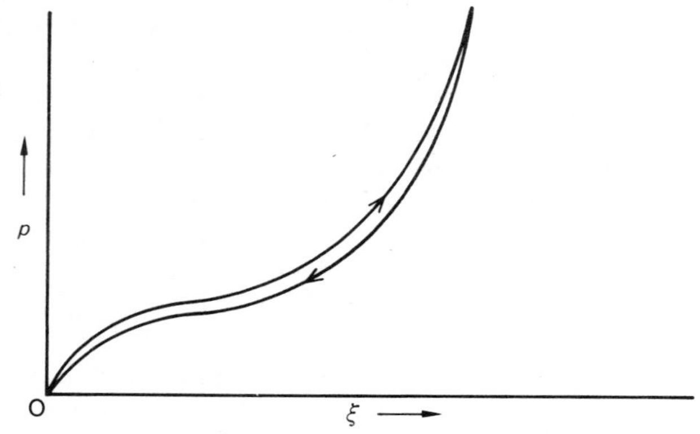

(a) Rubbery behaviour of polymers above the glass transition temperature

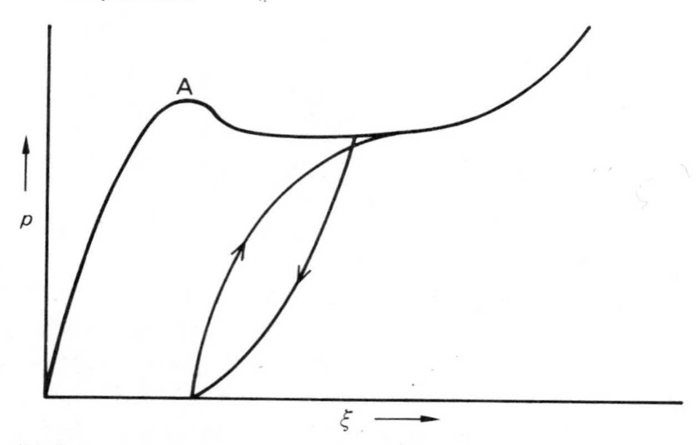

(b) Behaviour below the glass transition temperature

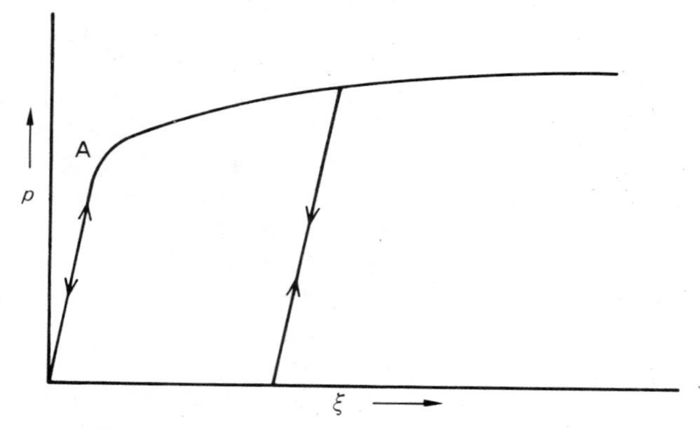

(c) A linear-elastic-plastic material

Figure 2.1 Simple tension stress-strain behaviour.

beyond which there is much less recovery, and unloading and reloading shows a marked hysteresis. Point A may be reasonably regarded as a yield point since the deformations thereafter are only recoverable over a very protracted period and are essentially permanent.

Figure 2.1(c) shows the stress-strain curve for a linear-elastic-plastic material in which there is perfectly linear-elastic behaviour up to the yield point A after which permanent deformation sets in. Unloading results in elastic recovery and reloading along a line parallel to the original elastic line. For most crystalline solids such as metals, this is close representation of what actually occurs. Also, since the strain at A is usually around 0·1 % then the assumptions of classical small strain elasticity: i.e. infinitesimal strains, a linear relationship between stress and strain and perfect elasticity combine to provide a useful theory. Linear-elasticity theory is well developed and is the basis of all stress analysis methods and although it is not an exact description for polymers, it does describe many aspects of their low strain behaviour. In this text the linear-elasticity theory will be considered first and visco-elasticity regarded as a development therefrom. Plasticity theory applies to the post yield regime and assumes completely irreversible deformation which is time independent. Again this is a reasonable representation for crystalline solids but is less precise for polymers. However it does provide a framework in which to examine polymer behaviour where there is a long delay in recovery and will be considered in some detail for that purpose.

The constitutive equations to be described here will therefore be idealisations which, while not entirely appropriate for polymers, will be useful for their description and for obtaining solutions to practical problems. Chapter two contains the three main time independent theories, namely, small strain linear elasticity, rubber elasticity and plasticity. Chapter three will consider time dependent behaviour and describe visco-elasticity. The application of the constitutive equations to particular boundary conditions will be considered in subsequent chapters.

2.1. Small strain linear-elasticity

2.1.1. *Definitions and assumptions*

An elastic material is one in which there is a unique relationship between stress and strain which is independent of the sequence of events and the time scale of the loading history. Thus all strain changes resulting from stress changes are instantaneous and depend only on the current stress and strain state. Since the system is completely reversible all the input energy is stored and is recovered on unloading. In the general case the stresses and strains may interact, such as for large deformations, where the changes of shape result in changes in the stresses arising from the applied loads. In small strain theory it is assumed that the shape changes are small and are therefore ignored so that the stresses are independent of the strains they produce. In consequence we may write an elastic, one-dimensional, relationship of the form:

$$e = \phi(p) \tag{2.1}$$

where ϕ may be any function of p. It is further assumed in the linear theory that this relationship is linear giving:

$$e = ap$$

where a is a constant, which is Hooke's law in one dimension. If we consider the general three-dimensional form, then each strain component could be influenced by all six stress components so that the six constitutive equations are:

$$\begin{aligned}
e_{xx} &= a_{11}p_{xx}+a_{12}p_{yy}+a_{13}p_{zz}+a_{14}p_{xy}+a_{15}p_{yz}+a_{16}p_{zx}\\
e_{yy} &= a_{21}p_{xx}+a_{22}p_{yy}+a_{23}p_{zz}+a_{24}p_{xy}+a_{25}p_{yz}+a_{26}p_{zx}\\
e_{zz} &= a_{31}p_{xx}+a_{32}p_{yy}+a_{33}p_{zz}+a_{34}p_{xy}+a_{35}p_{yz}+a_{36}p_{zx}\\
e_{xy} &= a_{41}p_{xx}+a_{42}p_{yy}+a_{43}p_{zz}+a_{44}p_{xy}+a_{45}p_{yz}+a_{46}p_{zx}\\
e_{yz} &= a_{51}p_{xx}+a_{52}p_{yy}+a_{53}p_{zz}+a_{54}p_{xy}+a_{55}p_{yz}+a_{56}p_{zx}\\
e_{zx} &= a_{61}p_{xx}+a_{62}p_{yy}+a_{63}p_{zz}+a_{64}p_{xy}+a_{65}p_{yz}+a_{66}p_{zx}
\end{aligned} \tag{2.2}$$

The 36 constants used here may be reduced in number by examining the consequences of the assumptions in more detail. Since the behaviour is assumed to be reversible and linear then clearly the order in which the stresses are applied does not affect the resulting strain, i.e. if we apply p_{xx} and p_{yy}, e_{xx} will be the same whether p_{xx} or p_{yy} is applied first. Thus the response to a particular stress is a simple addition, be it positive or negative, and the resulting strain may be defined by superimposing the strain on the existing strain. This is the principle of superposition which is a considerable aid in obtaining solutions to elasticity problems. Suppose that a normal stress p_{xx} is applied to an element of side lengths δx, δy and δz. The resulting deflection in the x direction is therefore $e_{xx}\,\delta x$ and since the applied force is $p_{xx}\,\delta y\,\delta z$ then the work done in this linear system is given by:

$$\tfrac{1}{2}\,p_{xx}\,e_{xx}\,\delta x\,\delta y\,\delta z$$

This work is stored as elastic energy. If a normal stress p_{yy} is now applied then the energy increases by:

$$\tfrac{1}{2}\,p_{yy}e_{yy}\,\delta x\,\delta y\,\delta z$$

and in addition, since p_{xx} is acting while p_{yy} is applied, there is an additional amount given by:

$$p_{xx}\,e'_{xx}\,\delta x\,\delta y\,\delta z$$

where e'_{xx} is the strain in the x direction produced by p_{yy}. Referring to equations (2.2) we see that:

$$e_{xx} = a_{11}\, p_{xx}, \qquad e_{yy} = a_{22} p_{yy} \qquad \text{and} \quad e'_{xx} = a_{12} p_{yy}$$

The principle of superposition is applied here since the constants a_{22} and a_{12} are used to describe the response to p_{yy} when p_{xx} is already acting. The total stored energy now becomes:

$$U = (\tfrac{1}{2}a_{11}p_{xx}{}^2 + \tfrac{1}{2}a_{22}p_{yy}{}^2 + a_{12}p_{xx}p_{yy})\, \delta x\, \delta y\, \delta z$$

If the analysis is repeated with p_{yy} applied first and then p_{xx} the total stored energy is given by:

$$U' = (\tfrac{1}{2}a_{22}p_{yy}{}^2 + \tfrac{1}{2}a_{11}p_{xx}{}^2 + a_{21}p_{xx}p_{yy})\, \delta x\, \delta y\, \delta z$$

The constant on the cross product term is now a_{21} since it arises from the strain in the y direction resulting from p_{xx}. Since superposition applies there can be no difference in the two end conditions so that $U = U'$ and hence:

$$a_{12} = a_{21}$$

A similar argument may be applied to all the constants with different suffixes so that there are fifteen equalities:

$$a_{12} = a_{21}, \qquad a_{13} = a_{31}, \qquad a_{14} = a_{41}, \qquad \text{etc.}$$

resulting in a reduction in the number of constants from 36 to 21. These 21 constants will describe any general material no matter how complex are the variations of elastic properties with direction. It is rare to require the use of all of them however and some special cases are of practical importance.

2.1.2. *Isotropic materials*

An isotropic material is one in which the properties are the same in all directions which imposes special limitations on the 21 constants. Firstly there can be no interaction between shear and normal stress effects; i.e. shear stresses do not produce direct strains and vice versa. Thus nine of the constants are zero:

$$a_{41} = a_{42} = a_{43} = a_{51} = a_{52} = a_{53} = a_{61} = a_{62} = a_{63} = 0$$

Because the responses are the same in all directions then the responses to direct stresses must be symmetrical giving:

$$a_{11} = a_{22} = a_{33} \qquad \text{and} \quad a_{12} = a_{13} = a_{32}$$

In addition shear stresses do not give shear strains in directions other than those in which they act in isotropic materials and the shear stress response must also be symmetrical giving:

$$a_{54} = a_{64} = a_{65} = 0 \qquad \text{and} \quad a_{44} = a_{55} = a_{66}$$

Thus there are three constants giving six equations of the form:

$$e_{xx} = Ap_{xx} + Bp_{yy} + Bp_{zz}$$
$$e_{yy} = Bp_{xx} + Ap_{yy} + Bp_{zz}$$
$$e_{zz} = Bp_{xx} + Bp_{yy} + Ap_{zz}$$
$$e_{xy} = Cp_{xy}, \qquad e_{yz} = Cp_{yz}, \qquad e_{zx} = Cp_{zx}$$

A further relationship may also be derived by considering a plane stress system in which only a shear stress p_{xy} is applied. The normal stresses in two orthogonal directions a and b inclined at angles of θ and $\theta + (\pi/2)$ to the x direction respectively are given by equation (1.21):

$$p_{aa} = p_{xy} \sin 2\theta \qquad \text{and} \quad p_{bb} = -p_{xy} \sin 2\theta$$

The strain in direction a is therefore:

$$e_{aa} = Ap_{aa} + Bp_{bb} = (A - B)p_{xy} \sin 2\theta$$

The strain may also be deduced from the strain transformation equation:

$$e_{aa} = \frac{e_{xy}}{2} \sin 2\theta$$

and hence:

$$e_{aa} = \frac{C}{2} p_{xy} \sin 2\theta$$

Comparing the two expressions for e_{aa} we have:

$$C = 2(A - B)$$

The two remaining constants are usually defined in terms of the simple tension stress state so that Young's modulus is given by:

$$E = \frac{p_{xx}}{e_{xx}} = \frac{1}{A}, \qquad \therefore \quad A = \frac{1}{E}$$

for $p_{yy} = p_{zz} = p_{xy} = p_{yz} = p_{zx} = 0$. The ratio of lateral to axial strains in simple tension is called Poisson's ratio and is given by:

$$\nu = \frac{-e_{yy}}{e_{xx}} = \frac{-Bp_{xx}}{Ap_{xx}}, \qquad \therefore \quad B = \frac{-\nu}{E}$$

The shear modulus is defined in a simple shear system:

$$G = \frac{p_{xy}}{e_{xy}} = \frac{1}{C} = \frac{E}{2(1+\nu)} \tag{2.3}$$

and hence:

$$e_{xx} = \frac{1}{E}[p_{xx} - \nu(p_{yy} + p_{zz})]$$

$$e_{yy} = \frac{1}{E}[p_{yy} - \nu(p_{xx} + p_{zz})] \tag{2.4}$$

$$e_{zz} = \frac{1}{E}[p_{zz} - \nu(p_{xx} + p_{yy})]$$

$$e_{xy} = \frac{2(1+\nu)}{E}p_{xy}, \qquad e_{yz} = \frac{2(1+\nu)}{E}p_{yz} \qquad \text{and} \qquad e_{zx} = \frac{2(1+\nu)}{E}p_{zx}$$

A bulk modulus, which describes volumetric strain, is sometimes used and it may be defined from equation (1.64):

$$\Delta = e_{xx} + e_{yy} + e_{zz} = \left(\frac{1-2\nu}{E}\right)(p_{xx} + p_{yy} + p_{zz})$$

Hence we have:

$$K = \frac{p_m}{\Delta} = \frac{p_{xx} + p_{yy} + p_{zz}}{3\Delta} = \frac{E}{3(1-2\nu)} \tag{2.5}$$

For simple tension:

$$e_{yy} = e_{zz} = -\nu e_{xx}$$

and hence:

$$\Delta = (1-2\nu)e_{xx}$$

For most polymers below the glass transition temperature ν is in the range 0·35–0·40 so that Δ is positive indicating an increase in volume during tensile straining. Constant volume is given by $\Delta = 0$; i.e. $\nu = \frac{1}{2}$ and for this case K is infinite.

Equations (2.4) may be inverted to give the stresses in terms of strains and we have, for example:

$$p_{xx} = \frac{\nu E}{(1+\nu)(1-2\nu)}\Delta + \frac{E}{(1+\nu)}e_{xx} \tag{2.6}$$

and similar expressions for p_{yy} and p_{zz}.

2.1.3. *Strain energy in isotropic materials*

The strain energy stored in an isotropic linear-elastic material may be deduced from the work done by applying a stress system. If we consider an element of side lengths δx, δy and δz and apply a stress p_{xx} then the displacement of the force $p_{xx}\,\delta y\,\delta z$ is $e_{xx}\,\delta x$. Thus the work done on this linear system is:

$$\tfrac{1}{2}p_{xx}e_{xx}\,\delta x\,\delta y\,\delta z$$

This may be considered as the stored strain energy since the strain is, by definition, completely reversible and hence the strain energy per unit volume is given by:

$$\tfrac{1}{2}p_{xx}e_{xx}$$

If a stress p_{yy} is now applied we have an energy of $\frac{1}{2}p_{yy}\,e_{yy}$ and, in addition, since p_{xx} is present when p_{yy} is applied, a term of the form:

$$-\nu e_{yy}\,p_{yy}$$

Similarly we may consider the action of p_{zz} and deduce total strain energy due to the direct stresses by addition. The contribution of the shear stresses is simply derived in the form $\frac{1}{2}p_{xy}e_{xy}$, etc. Substituting for the strains in terms of stresses from equations (2.4) we have the final result:

$$W = \frac{1}{2E}(p_{xx}{}^2+p_{yy}{}^2+p_{zz}{}^2)-\frac{\nu}{E}(p_{xx}p_{yy}+p_{yy}p_{zz}+p_{zz}p_{xx})$$
$$+\frac{1}{2G}(p_{xy}{}^2+p_{yz}{}^2+p_{zx}{}^2) \qquad (2.7)$$

By substituting from equations (2.6) the expression in terms of strains may be obtained:

$$W = \frac{(1-\nu)E}{2(1+\nu)(1-2\nu)}(e_{xx}{}^2+e_{yy}{}^2+e_{zz}{}^2)$$
$$+\frac{\nu E}{(1+\nu)(1-2\nu)}(e_{xx}e_{yy}+e_{yy}e_{zz}+e_{zz}e_{xx})$$
$$+\frac{G}{2}(e_{xy}{}^2+e_{yz}{}^2+e_{zx}{}^2) \qquad (2.8)$$

It is easily verified from equation (2.8) that the rate of change of the energy with respect to any strain is the corresponding stress. For example:

$$\frac{\partial W}{\partial e_{xx}} = \frac{(1-\nu)E}{2(1+\nu)(1-2\nu)}\,2\,e_{xx}+\frac{\nu E}{(1+\nu)(1-2\nu)}(e_{yy}+e_{zz})$$

which may be rearranged to give:

$$\frac{\partial W}{\partial e_{xx}} = \frac{E}{(1+\nu)}e_{xx}+\frac{\nu E}{(1+\nu)(1-2\nu)}(e_{xx}+e_{yy}+e_{zz}) = p_{xx} \qquad (2.9)$$

from equation (2.6). Conversely from equation (2.7) we have:

$$\frac{\partial W}{\partial p_{xx}} = \frac{p_{xx}}{E}-\frac{\nu}{E}(p_{yy}+p_{zz}) = e_{xx} \qquad (2.10)$$

The strain energy must be independent of the coordinate system used and hence expressible solely in terms of invariants. This is verified in equation (2.8), for example, which may be written as:

$$W = \frac{(1-\nu)E}{2(1+\nu)(1-2\nu)} I_1{}^2 - \frac{E}{(1+\nu)} I_2 \tag{2.11}$$

where I_1 and I_2 are the first and second invariants of the strains, equations (1.54). The elastic equations may be derived from an analysis of this type by assuming limitations on the general form of the strain energy function which may be written as:

$$W(I_1, I_2, I_3)$$

and from equation (2.9) we have:

$$p_{xx} = \frac{\partial W}{\partial e_{xx}} = \frac{\partial W}{\partial I_1}\frac{\partial I_1}{\partial e_{xx}} + \frac{\partial W}{\partial I_2}\frac{\partial I_2}{\partial e_{xx}} + \frac{\partial W}{\partial I_3}\frac{\partial I_3}{\partial e_{xx}}$$

i.e.

$$p_{xx} = \frac{\partial W}{\partial I_1} + \frac{\partial W}{\partial I_2}(e_{yy}+e_{zz}) + \frac{\partial W}{\partial I_3}\left(e_{yy}\,e_{zz} - \frac{e_{yz}{}^2}{4}\right) \tag{2.12}$$

If, for example, it is assumed that all terms of higher order than e^2 may be ignored then W may be written as:

$$W = b_1 I_1 + b_2 I_1{}^2 + b_3 I_2$$

where b_1, b_2 and b_3 are constants. Differentiating and substituting in equation (2.12) we have:

$$p_{xx} = b_1 + b_2 I_1 + b_3(e_{yy}+e_{zz}) = b_1 + b_2\Delta - b_3 e_{xx}$$

Since $p_{xx} = 0$ when e_{xx} and $\Delta = 0$ then $b_1 = 0$ and we have Hooke's law in the form of equation (2.6). Higher order elasticity effects may be introduced as required; e.g. by including e^3 we have:

$$W = b_1 I_1 + b_2 I_1{}^2 + b_3 I_1{}^3 + b_4 I_2 + b_5 I_1 I_2 + b_6 I_3$$

where $b_1, b_2, \ldots b_6$ are constants. This results in a constitutive equation of the form:

$$p_{xx} = A\Delta + B\Delta^2 + Ce_{xx} + D\left(e_{yy}e_{zz} - \frac{e_{yz}{}^2}{4}\right) \tag{2.13}$$

where A, B, C and D are constants and is sometimes referred to as the cross elasticity equation because of the product term $e_{yy}e_{zz}$.

It is sometimes of interest to consider the total strain energy as made up of energy due to volume change and energy due to distortion. The energy due to volume changes is given by:

$$W_V = \tfrac{1}{2} p_m \Delta = \frac{(1-2\nu)}{6E}(p_{xx}+p_{yy}+p_{zz})^2 \tag{2.14}$$

The distortion energy is given by the difference with the total energy:

$$W_S = W - W_V = \frac{(1+\nu)}{6E}[(p_{xx}-p_{yy})^2+(p_{yy}-p_{zz})^2+(p_{zz}-p_{xx})^2$$
$$+6p_{xy}{}^2+6p_{yz}{}^2+6p_{zx}{}^2]$$
$$= -\frac{1+\nu}{E}J'_2 \qquad (2.15)$$

i.e. W_S is proportional to the second invariant of the deviatoric stress system. W_S is, of course, the energy associated with the deviatoric stress system as may be confirmed by substituting the deviatoric stresses into equation (2.7).

2.1.4. *Plane stress anisotropy*

The plane stress assumption, i.e. $p_{zz} = p_{yz} = p_{zx} = 0$, reduces the number of constants in the general case to six and the equations become:

$$\begin{aligned} e_{xx} &= a_{11}p_{xx}+a_{12}p_{yy}+a_{14}p_{xy} \\ e_{yy} &= a_{21}p_{xx}+a_{22}p_{yy}+a_{24}p_{xy} \\ e_{xy} &= a_{41}p_{xx}+a_{42}p_{yy}+a_{44}p_{xy} \end{aligned} \qquad (2.16)$$

where $a_{21} = a_{12}$, $a_{24} = a_{42}$ and $a_{14} = a_{41}$.
The elastic constants in other directions in the sheet may be found by the usual transformation relationships. Suppose we have a sheet, as shown in Figure 2.2(a) in which the constants are defined, in equation (2.16), in terms of the directions x and y as shown. If we require the elastic constants in some other direction, say x' at an angle θ to the x direction, then from the usual plane stress relationships we may write:

$$e'_{xx} = \frac{e_{xx}+e_{yy}}{2}+\frac{e_{xx}-e_{yy}}{2}\cos 2\theta+\frac{e_{xy}}{2}\sin 2\theta \qquad (2.17)$$

and also:

$$\begin{aligned} p_{xx} &= \frac{p'_{xx}+p'_{yy}}{2}+\frac{p'_{xx}-p'_{yy}}{2}\cos 2\theta+p'_{xy}\sin 2\theta \\ p_{yy} &= \frac{p'_{xx}+p'_{yy}}{2}-\frac{p'_{xx}-p'_{yy}}{2}\cos 2\theta-p'_{xy}\sin 2\theta \\ p_{xy} &= \frac{p'_{xx}-p'_{yy}}{2}\sin 2\theta-p'_{xy}\cos 2\theta \end{aligned} \qquad (2.18)$$

By substituting for e_{xx}, etc., from equation (2.16) in equation (2.17), e'_{xx} may be expressed in terms of p_{xx}, p_{yy} and p_{xy}. If these stresses are now replaced by p'_{xx}, p'_{yy} and p'_{xy} from equations (2.18) then there is a relationship between e'_{xx} and

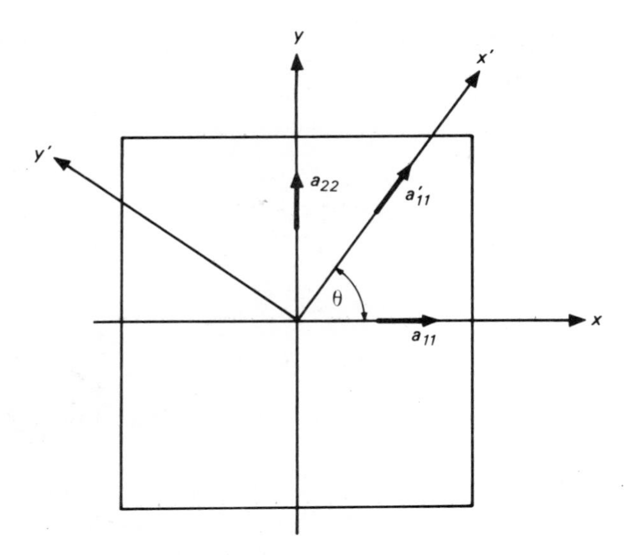

(a) Coordinates and moduli in an anisotropic sheet

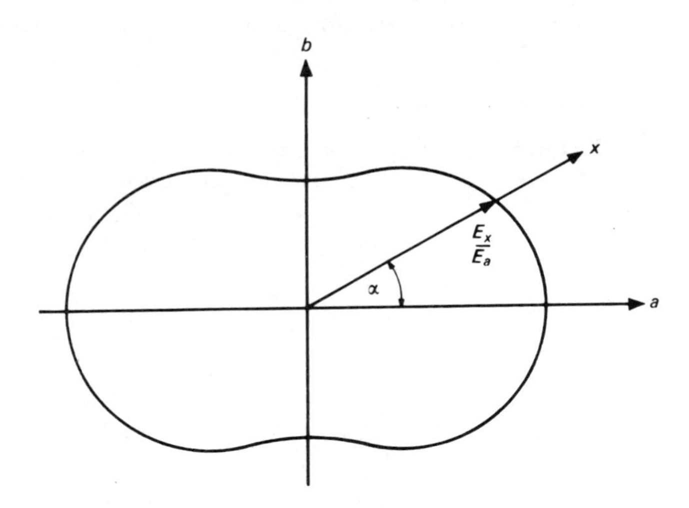

(b) Variation of modulus in an orthotropic sheet, with $E_b = \frac{1}{2} E_a$ and $E_a \left[\frac{1}{G_{ab}} - \frac{2\nu_{ab}}{E_b} \right] = 2$

Figure 2.2 Plane stress anisotropy

the component stresses of the form:

$$e'_{xx} = a'_{11}p'_{xx}+a'_{12}p'_{yy}+a'_{14}p'_{xy}$$

where a'_{11}, a'_{12} and a'_{14} are the elastic constants in the new direction x'. For example we have:

$$a'_{11} = a_{11}\cos^4\theta+a_{22}\sin^4\theta+2a_{12}\sin^2\theta\cos^2\theta$$
$$+2a_{14}\sin\theta\cos^3\theta+2a_{24}\cos\theta\sin^3\theta+a_{44}\sin^2\theta\cos^2\theta \qquad (2.19)$$

It will be noted that when:

$$\theta = 0, \qquad a'_{11} = a_{11} \qquad \text{and when} \quad \theta = \frac{\pi}{2}, \qquad a'_{11} = a_{22}$$

as expected. The limiting values of a'_{11} may be determined from $\mathrm{d}a'_{11}/\mathrm{d}\theta = 0$ and we have the condition:

$$(a_{22}-a_{11})\sin 2\theta+(a_{44}-a_{11}-a_{22}+2a_{12})\sin 2\theta\cos\theta$$
$$+(a_{14}+a_{24})\cos 2\theta+(a_{24}-a_{14})(1-2\cos^2 2\theta) = 0 \qquad (2.20)$$

The determination of the constant a'_{14} shows that this has the same form as equation (2.20) and the condition is equivalent to:

$$a'_{14} = 0$$

and similarly for a'_{22} the limiting condition is $a'_{24} = 0$.

A special case of interest is an orthotropic plate in which the two limiting conditions are perpendicular so that:

$$a_{14} = a_{24} = 0$$

This may be verified from equation (2.20) in which $a_{14} = 0$ is the condition for $\theta = 0$ and $a_{24} = 0$ is that for $\theta = \pi/2$. Orthotropic plates are encountered in fibre reinforced systems in which the fibres are laid in two perpendicular directions and the constants are defined in these two directions:

$$e_{aa} = \frac{1}{E_a}p_{aa}-\frac{\nu_{ab}}{E_b}p_{bb}$$
$$e_{bb} = \frac{1}{E_b}p_{bb}-\frac{\nu_{ba}}{E_a}p_{aa} \qquad (2.21)$$
$$e_{ab} = \frac{p_{ab}}{G_{ab}}$$

There is no interaction between shear and direct stress effects and from the general symmetry;

$$\frac{\nu_{ab}}{E_b} = \frac{\nu_{ba}}{E_a}$$

giving four basic constants.

Poisson's ratio is introduced to retain the normal definition in simple tension, e.g.

$$\frac{-e_{bb}}{e_{aa}} = \frac{\nu_{ba}}{E_a} p_{aa} \frac{E_a}{p_{aa}} = \nu_{ba}$$

The modulus in a direction x at an angle α to the a direction is therefore given by:

$$a_{11} = \frac{1}{E_x} = \frac{\cos^4\alpha}{E_a} + \frac{\sin^4\alpha}{E_b} + \frac{\sin^2 2\alpha}{4}\left(\frac{1}{G_{ab}} - \frac{2\nu_{ab}}{E_b}\right) \tag{2.22}$$

For the isotropic case, $E_x = E_a = E_b$, this reduces to equation (2.3)

$$G = \frac{E}{2(1+\nu)}$$

Figure 2.2(b) shows a typical form of the variation of E_x with α. The other relevant parameters are:

$$\begin{aligned} a_{12} &= \frac{-\nu_{ab}}{E_b} + \frac{\sin^2 2\alpha}{4}\left[\frac{1}{E_a} + \frac{1}{E_b} - \frac{1}{G_{ab}} + \frac{2\nu_{ab}}{E_b}\right] \\ a_{44} &= \frac{\cos^2 2\alpha}{G_{ab}} + \sin^2 2\alpha\left[\frac{1}{E_a} + \frac{1}{E_b} + \frac{2\nu_{ab}}{E_b}\right] \\ a_{14} &= \frac{\sin 2\alpha}{2}\left[\frac{1}{E_a} - \frac{1}{E_b}\right] + \frac{\sin 2\alpha \cos 2\alpha}{2}\left[\frac{1}{E_a} + \frac{1}{E_b} - \frac{1}{G_{ab}} + \frac{2\nu_{ab}}{E_b}\right] \end{aligned} \tag{2.23}$$

a_{24} and a_{22} are obtained by the substitution of $\alpha + \pi/2$ for α in the expressions for a_{14} and a_{11} respectively.

Another special case of practical importance is that of isotropic behaviour in one plane, say xy, and different properties in the z direction. A drawn fibre is a typical example of this system and if one of the directions of loading is taken as z then there are five elastic constants needed, i.e.:

$$a_{11} = a_{22},\ a_{33},\ a_{44} = a_{55} = a_{66},\ a_{23} = a_{31},\ a_{12}$$

with all other values zero.

2.2. Finite strain elasticity

2.2.1. *Definitions*

Polymers above their glass transition temperature exhibit rubber like properties in that they are essentially elastic up to strains of several hundreds of per cent. The assumptions of the infinitesimal theory are clearly invalid for this behaviour and a different constitutive relationship is required. The theory developed here will assume that the material is perfectly elastic so that all the work done will be stored in the form of strain energy for an isothermal deformation. Thus, as discussed in section (2.1.4), the strain energy per unit volume W may be expressed as a function of the three strain invariants:

$$W(I_1, I_2, I_3) \qquad (2.24)$$

since it must be an invariant quantity, and may be defined in terms of the external work performed.

It is possible to derive the theory for general deformations but it is extremely complex and a useful practical result is obtained by restricting our attention here to pure deformations, i.e. deformations involving no shearing action. Thus a unit cube of material, as shown in Figure 2.3, will deform to become a cuboid

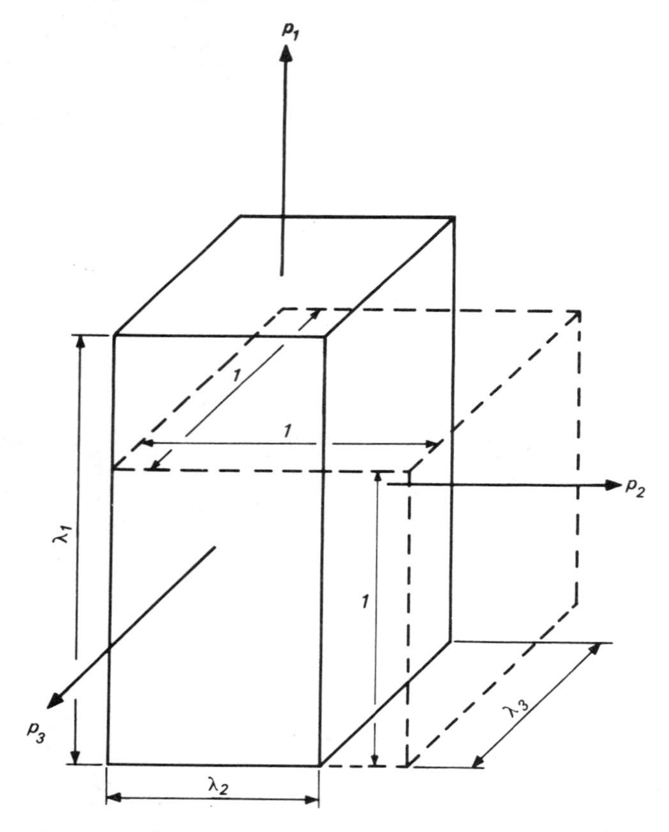

Figure 2.3 An element of unit side length undergoing a finite pure deformation.

of side lengths λ_1, λ_2 and λ_3 as shown. λ_1, λ_2 and λ_3 may be considered as the principal extension ratios since there is no shear and the stresses acting on the faces may be taken as the principal values p_1, p_2 and p_3. A change in the stored energy for a small change in the deformations may then be written as:

$$dW = p_1\lambda_2\lambda_3\, d\lambda_1 + p_2\lambda_1\lambda_3\, d\lambda_2 + p_3\lambda_1\lambda_2\, d\lambda_3 \qquad (2.25)$$

The term $p_1\lambda_2\lambda_3$, for example, is the force acting in direction one and $d\lambda_1$ is the movement of that force so $p_1\lambda_2\lambda_3\, d\lambda_1$ is the work done. The stresses are therefore given by expressions of the form:

$$p_1 = \frac{1}{\lambda_2\lambda_3}\frac{\partial W}{\partial \lambda_1} \tag{2.26}$$

which reduce to the infinitesimal case given in equation (2.9) when very small deformations are considered.

Attention here will be limited to the case of initially isotropic materials so that the rate of change of W may be expressed as:

$$\frac{\partial W}{\partial \lambda_1} = \frac{\partial W}{\partial I_1}\frac{\partial I_1}{\partial \lambda_1} + \frac{\partial W}{\partial I_2}\frac{\partial I_2}{\partial \lambda_1} + \frac{\partial W}{\partial I_3}\frac{\partial I_3}{\partial \lambda_1} \tag{2.27}$$

since W is a function of I_1, I_2 and I_3. If the invariants are now written in terms of the principal extension ratios we have, from section 1.3:

$$\begin{aligned} I_1 &= \lambda_1^2+\lambda_2^2+\lambda_3^2 \\ I_2 &= \lambda_1^2\lambda_2^2+\lambda_2^2\lambda_3^2+\lambda_3^2\lambda_1^2 \\ I_3 &= \lambda_1^2\lambda_2^2\lambda_3^2 \end{aligned} \tag{2.28}$$

so that:

$$\frac{\partial I_1}{\partial \lambda_1} = 2\lambda_1, \frac{\partial I_2}{\partial \lambda_1} = 2\lambda_1(\lambda_2^2+\lambda_3^2) \text{ and } \frac{\partial I_3}{\partial \lambda_1} = 2\lambda_1\, \lambda_2^2\lambda_3^2$$

Substituting these into equation (2.27) and combining them with equation (2.26) we have the result:

$$p_1 = \frac{2}{I_3^{\frac{1}{2}}}\left[\lambda_1^2\frac{\partial W}{\partial I_1} - \frac{I_3}{\lambda_1^2}\frac{\partial W}{\partial I_2} + I_2\frac{\partial W}{\partial I_2} + I_3\frac{\partial W}{\partial I_3}\right] \tag{2.29}$$

This may be regarded as the general expression for stress for a compressible, initially isotropic, material where $\partial W/\partial I_1$, $\partial W/\partial I_2$ and $\partial W/\partial I_3$ are the elastic property functions of the material. This general form will not be pursued further here since another simplification can be made without detracting significantly from the usefulness of the results. The analysis will be developed now for the case of an incompressible material which is a good approximation to the rubber-like behaviour of most polymers.

2.2.2. *Incompressible materials*

If it is assumed that the volume of the material remains constant during deformation then λ_1, λ_2 and λ_3 are not independent variables and are related by:

$$\lambda_1\lambda_2\lambda_3 = 1 \tag{2.30}$$

This restriction alters the form of equation (2.26) which now becomes:

$$p_1 = \frac{1}{\lambda_2\lambda_3}\frac{\partial W}{\partial \lambda_1}+p$$

where p is an arbitrary hydrostatic pressure. Clearly, energy considerations cannot define a hydrostatic stress acting on an incompressible material since it does no work, which accounts for the presence of the arbitrary term. Substituting for the constant volume condition we have the result:

$$p_1 = \lambda_1\frac{\partial W}{\partial \lambda_1}+p \qquad (2.31)$$

Any strain definition could be used to describe W and others will be considered in section (2.2.3) but it is usual to continue with the use of extension ratios. The invariants, equations (2.28), may now be written as:

$$\begin{aligned} I_1 &= \lambda_1{}^2+\lambda_2{}^2+\lambda_3{}^3 \\ I_2 &= \frac{1}{\lambda_1{}^2}+\frac{1}{\lambda_2{}^2}+\frac{1}{\lambda_3{}^2} \\ I_3 &= 1 \end{aligned} \qquad (2.32)$$

and the strain energy function is:

$$W(I_1, I_2)$$

Making the substitutions as before we have the results:

$$\left.\begin{aligned} p_1 &= 2\lambda_1{}^2\frac{\partial W}{\partial I_1}-\frac{2}{\lambda_1{}^2}\frac{\partial W}{\partial I_2}+p \\ p_2 &= 2\lambda_2{}^2\frac{\partial W}{\partial I_1}-\frac{2}{\lambda_2{}^2}\frac{\partial W}{\partial I_2}+p \\ p_3 &= 2\lambda_3{}^2\frac{\partial W}{\partial I_1}-\frac{2}{\lambda_3{}^2}\frac{\partial W}{\partial I_2}+p \end{aligned}\right\} \qquad (2.33)$$

The terms $\partial W/\partial I_1$ and $\partial W/\partial I_2$ can, in principle, be any functions of I_1 and I_2 providing they satisfy certain limitations. A very simple form which has been found to be a reasonable representation of behaviour is the Mooney-Rivlin function which may be regarded as a first order finite strain relationship:

$$W = \frac{\phi}{2}(I_1-3)+\frac{\psi}{2}(I_2-3)$$

where ϕ and ψ are constants. The derivatives now become:

$$2\frac{\partial W}{\partial I_1} = \phi \text{ and } 2\frac{\partial W}{\partial I_2} = \psi$$

giving equations of the form:

$$\left.\begin{aligned} p_1 &= \phi\lambda_1^2 - \frac{\psi}{\lambda_1^2} + p \\ p_2 &= \phi\lambda_2^2 - \frac{\psi}{\lambda_2^2} + p \\ p_3 &= \phi\lambda_3^2 - \frac{\psi}{\lambda_3^2} + p \end{aligned}\right\} \qquad (2.34)$$

For the special case of plane stress, say $p_3 = 0$, we have:

$$p = -\frac{\phi}{\lambda_1^2\lambda_2^2} + \psi\lambda_1^2\lambda_2^2$$

and hence:

$$\left.\begin{aligned} p_1 &= (\phi + \psi\lambda_2^2)\left(\lambda_1^2 - \frac{1}{\lambda_1^2\lambda_2^2}\right) \\ p_2 &= (\phi + \psi\lambda_1^2)\left(\lambda_2^2 - \frac{1}{\lambda_1^2\lambda_2^2}\right) \end{aligned}\right\} \qquad (2.35)$$

The form of the stress-strain curves predicted by the relationship may be illustrated by considering simple tension in which:

$$p_1 = p_T,\ p_2 = 0,\ \lambda_1 = \lambda,\ \lambda_2 = \lambda_3 = \frac{1}{\lambda^{\frac{1}{2}}}$$

and hence:

$$p_T = \lambda^2\left(\phi + \frac{\psi}{\lambda}\right)\left(1 - \frac{1}{\lambda^3}\right) \qquad (2.36)$$

Figure 2.4 shows p_T/ϕ versus λ and the non-linear form is apparent, particularly in compression. For most polymers ψ is rarely greater than $0{\cdot}2\phi$ and it can be seen that the difference with $\psi = 0$ is quite small in tension although significant in compression. A simplified version is sometimes used in which it is assumed that λ may be expressed as a small strain in equation (2.36) which gives the result:

$$p = 3(\phi + \psi)\, e \qquad (2.37)$$

where $\lambda = 1 + e$.

This line is shown in Figure 2.4 for $\psi = 0$ and it is clearly a reasonable approximation over the range $-50\% < e < 100\%$. Problems are solved using the usual Hooke's law expression with infinitesimal strain theory but including the true stress and considerable simplifications can be achieved.

It will be noted that initial isotropy was defined at the start of the analysis and the reason may be seen by considering the stiffness in two directions in plane stress, i.e.:

$$\frac{\partial p_1}{\partial \lambda_1} = (\phi + \psi\lambda_2^{\,2})\left(2\lambda_1 + \frac{2}{\lambda_2^{\,2}\lambda_1^{\,3}}\right)$$

and

$$\frac{\partial p_2}{\partial \lambda_2} = (\phi + \psi\lambda_1^{\,2})\left(2\lambda_2 + \frac{2}{\lambda_1^{\,2}\lambda_2^{\,3}}\right)$$

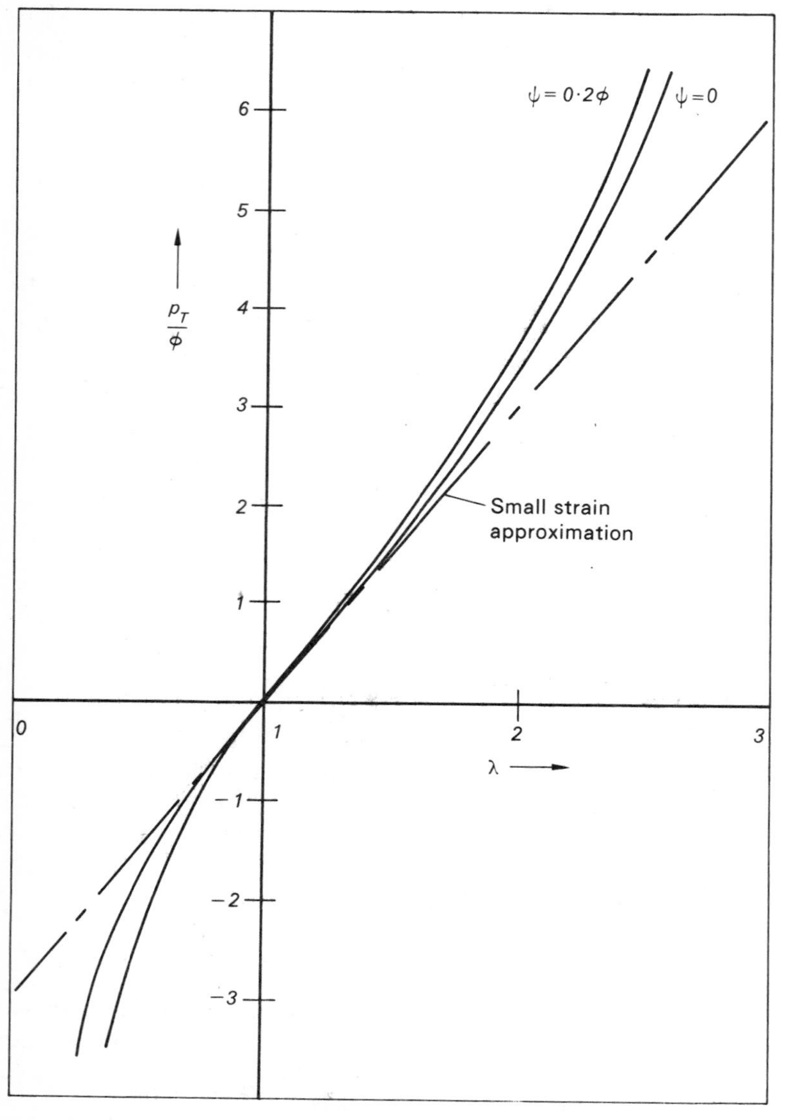

Figure 2.4 Mooney—Rivlin rubber elasticity in simple tension and compression.

Because of the non-linear nature of the behaviour these are different and the material may be regarded as anisotropic in the strained state.

2.2.3. *General strain formulation for incompressible materials*

The expressions for the stresses may be written in terms of a general strain definition:

$$\lambda = f(\xi)$$

so that we have:

$$p_1 = \frac{f(\xi_1)}{f'(\xi_1)} \frac{\partial W}{\partial \xi_1} + p \tag{2.38}$$

where

$$f'(\xi_1) = \frac{\mathrm{d}f(\xi_1)}{\mathrm{d}\xi_1}$$

From equations (1.43) we have the invariants:

$$\begin{aligned} I_1 &= \xi_1 + \xi_2 + \xi_3 \\ I_2 &= \xi_1\xi_2 + \xi_2\xi_3 + \xi_3\xi_1 \\ I_3 &= \xi_1\xi_2\xi_3 \end{aligned}$$

and there is a relationship between the invariants from the constant volume condition given by:

$$f(\xi_1)\,(f(\xi_2)\,f(\xi_3) = 1$$

The expression for the stress then becomes:

$$p_1 = \frac{f(\xi_1)}{f'(\xi_1)} \left[\frac{\partial W}{\partial I_1} + I_1 \frac{\partial W}{\partial I_2} - \frac{\partial W}{\partial I_2} \xi_1 + \frac{\partial W}{\partial I_3} \frac{I_3}{\xi_1} \right] + p \tag{2.39}$$

Any form of the function f may be used to give a convenient relationship and it is often dictated by the simplicity or otherwise of W in terms of the invariants of the strain used. Extension ratios clearly give a useable solution but other definitions give useful results.

2.2.4. *Natural strain formulation for incompressible materials*

Natural strain (see section 1.3.4) is defined as:

$$f(\varepsilon) = \exp \varepsilon$$

and the constant volume condition is:

$$\exp(\varepsilon_1 + \varepsilon_2 + \varepsilon_3) = 1$$

i.e.

$$\varepsilon_1 + \varepsilon_2 + \varepsilon_3 = 0$$

Thus the invariants are reduced to two since $I_1 = 0$ and further:

$$\frac{\mathrm{d}(\exp \varepsilon)}{\mathrm{d}\,\varepsilon} = \exp \varepsilon$$

so that:

$$\frac{f(\varepsilon)}{f'(\varepsilon)} = 1$$

The expression for the stress is therefore given from equation (2.39) as:

$$p_1 = -\frac{\partial W}{\partial I_2}\varepsilon_1 + \frac{\partial W}{\partial I_3}\frac{I_3}{\varepsilon_1} + p' \tag{2.40}$$

It is convenient to express the invariants in terms of simple tension strains, i.e.:

$$\varepsilon_1 = \bar{\varepsilon},\ \varepsilon_2 = \varepsilon_3 = -\frac{1}{2}\bar{\varepsilon}$$

so that:

$$I_2 = \varepsilon_1\varepsilon_2 + \varepsilon_2\varepsilon_3 + \varepsilon_3\varepsilon_1 = -\frac{3}{4}\bar{\varepsilon}^2$$

and similarly, $$I_3 = \varepsilon_1\varepsilon_2\varepsilon_3 = \frac{1}{4}\hat{\varepsilon}^3$$

Equation (2.40) then becomes:

$$p_1 = \frac{\partial W}{\partial \bar{\varepsilon}}\frac{2}{3}\frac{\varepsilon_1}{\bar{\varepsilon}} + \frac{\partial W}{\partial \hat{\varepsilon}}\frac{1}{3}\frac{\hat{\varepsilon}}{\varepsilon_1} + p$$

In general $\partial W/\partial\bar{\varepsilon}$ and $\partial W/\partial\hat{\varepsilon}$ may be functions of both $\bar{\varepsilon}$ and $\hat{\varepsilon}$ and since both have the dimensions of stress we will write:

$$\frac{\partial W}{\partial \bar{\varepsilon}} = p^* \quad\text{and}\quad \frac{\partial W}{\partial \hat{\varepsilon}} = p^{**}$$

p may be obtained by summation:

$$p = \frac{p_1 + p_2 + p_3}{3} + \frac{p^{**}}{3}\left(\frac{\bar{\varepsilon}}{\hat{\varepsilon}}\right)^2$$

giving final equations of the form:

$$\left.\begin{aligned}
\frac{p^{**}}{2}\left[\frac{\hat{\varepsilon}}{\varepsilon_1} + \left(\frac{\bar{\varepsilon}}{\hat{\varepsilon}}\right)^2\right] + p^*\frac{\varepsilon_1}{\bar{\varepsilon}} &= p_1 - \frac{1}{2}(p_2 + p_3)\\
\frac{p^{**}}{2}\left[\frac{\hat{\varepsilon}}{\varepsilon_2} + \left(\frac{\bar{\varepsilon}}{\hat{\varepsilon}}\right)^2\right] + p^*\frac{\varepsilon_2}{\bar{\varepsilon}} &= p_2 - \frac{1}{2}(p_1 + p_3)\\
\frac{p^{**}}{2}\left[\frac{\hat{\varepsilon}}{\varepsilon_3} + \left(\frac{\bar{\varepsilon}}{\hat{\varepsilon}}\right)^2\right] + p^*\frac{\varepsilon_3}{\bar{\varepsilon}} &= p_3 - \frac{1}{2}(p_1 - p_2)
\end{aligned}\right\} \tag{2.41}$$

It is of interest to note that the second invariant of the deviatoric stress system:

$$J_2' = p_1'p_2' + p_2'p_3' + p_3'p_1'$$

is usually written in terms of an equivalent simple tension value. Thus:

$$p_1 = \bar{p},\, p_2 = p_3 = 0 \text{ and } p_1' = \frac{2}{3}\bar{p},\, p_2' = p_3' = -\frac{\bar{p}}{3}$$

giving:

$$J_2' = -\frac{1}{3}\bar{p}^2$$

and from equations (2.41) we have:

$$\bar{p}^2 = p^{*2} + 2\left(\frac{\hat{\varepsilon}}{\bar{\varepsilon}}\right) p^*p^{**} + p^{**2}\left(\frac{\bar{\varepsilon}}{\hat{\varepsilon}}\right)^4 \tag{2.42}$$

If it is assumed that there is no dependence on $\hat{\varepsilon}$ then $p^{**} = 0$ and $p^* = \bar{p}$ giving equations of the form:

$$\varepsilon_1 \frac{\bar{p}}{\bar{\varepsilon}} = p_1 - \frac{1}{2}(p_2 + p_3)$$

which are identical to the Hencky total strain plasticity equations given in section 2.3.6.

The forms of the functions p^* and p^{**} depend on the strain dependence assumed in the strain energy function. A linear form would assume no terms of order higher than ε^2 so that:

$$W = \frac{a}{2}\bar{\varepsilon}^2$$

where a is a constant and:

$$p^* = a\bar{\varepsilon} \text{ and } p^{**} = 0$$

The stress equations now have the form:

$$\varepsilon_1 = \frac{1}{a}\left[p_1 - \frac{1}{2}(p_2 + p_3)\right]$$

and hence a may be regarded as a modulus. The inclusion of terms up to ε^4 gives an expression of the form:

$$W = \frac{a_1}{2}\bar{\varepsilon}^2 + \frac{a_2}{4}\bar{\varepsilon}^4 + \frac{a_3}{3}\hat{\varepsilon}^3$$

where a_1, a_2 and a_3 are constants and we have:

$$p^* = a_1\bar{\varepsilon} + a_2\bar{\varepsilon}^3,\; p^{**} = a_3\hat{\varepsilon}^2$$

The functions p^* and p^{**} may be determined experimentally by performing two simple tests. Firstly, from a simple tension test, which gives a curve of

p_T versus ε_T, we have:

$$\varepsilon_T = \bar{\varepsilon} = \hat{\varepsilon} \qquad \text{by definition}$$

and from equations (2.41):

$$p^{**} + p^{*} = p_T$$

The second test is a wide sheet clamped so that $\varepsilon_2 = 0$ ($\lambda_2 = 1$) and if the sheet is thin, $p_3 = 0$, and we may write:

$$p_1 = p_s,\; p_3 = 0,\; p_2 = \frac{1}{2} p_s,\; \varepsilon_2 = 0,\; \varepsilon_3 = -\varepsilon_1 = -\varepsilon_s$$

and hence:

$$\hat{\varepsilon} = 0,\; p^{**} = 0,\; p^{*} = \frac{\sqrt{3}}{2} p_s \text{ and } \varepsilon_s = \frac{\sqrt{3}}{2} \bar{\varepsilon}$$

From this test, therefore, we may obtain p^{*} as a function of $\bar{\varepsilon}$ from p_s versus ε_s. p^{**} may be obtained from the simple tension results since:

$$p^{**} = \left(p_T - \frac{\sqrt{3}}{2} p_s\right)$$

2.3. Plastic yielding

Plastic yielding is conventionally defined as the onset of permanent or irreversible deformation and the plastic deformation which occurs thereafter is taken as completely irreversible, as opposed to the perfectly reversible condition of the elastic theory. The theory of plasticity, which describes this type of deformation, has been developed for metals whose behaviour approximates closely to these assumptions. The behaviour of polymeric materials, on the other hand, does not generally approximate to them so closely because their deformations are almost always reversible in some sense. However, most polymers in the glassy state do undergo a process which can be equated with yielding and the concepts developed in plasticity theory are therefore of interest in connection with polymeric materials.

2.3.1. *Yield criteria*

In a simple tension test yielding is usually identified with a rapid decrease in the slope of the stress-strain curve. More rigorously, the onset of yield is that condition when permanent deformation first occurs, and can be identified, in an elastic-plastic material, by a process of loading and unloading. In polymeric materials the yielding is more difficult to define, because of the problem of deciding what are irrecoverable deformations in time-dependent materials. Thus, a rapid change in the slope of the stress-strain curve is often all that can be used in practice.

Once a polymer has yielded then its deformation usually takes place at constant volume. This means that, for most polymers, the yielding process can be regarded as occurring at constant volume, is thus unaffected by hydrostatic stress and may be described in terms of the deviatoric stresses. Moreover, the condition cannot be a function of any particular set of coordinates or components and must be governed by the invariants of the deviatoric stress system (see section 1.1.6) and may be written as:

$$\phi(J_2', J_3') = 0 \tag{2.43}$$

The first invariant $J_1' = 0$ for this system and the remaining two, as given in equations (1.18), are:

$$J_2' = p'_{xx}p'_{yy} + p'_{yy}p'_{zz} + p'_{zz}p'_{xx} - p_{xy}^2 - p_{yz}^2 - p_{xz}^2$$

and $$J_3' = p'_{xx}p'_{yy}p'_{zz} + 2p_{xy}p_{yz}p_{zx} - p'_{xx}p_{yz}^2 - p'_{yy}p_{zx}^2 - p'_{zz}p_{xy}^2$$

where $$p'_{xx} = p_{xx} - p_m, \text{ etc.}$$

and $$p_m = \frac{p_{xx} + p_{yy} + p_{zz}}{3} \text{ the hydrostatic stress}$$

J_3' is frequently omitted from the equations because good agreement with experimental data may be obtained by ignoring it. If this is done the yield condition may be written as:

$$\phi(J_2') = 0$$

and the simplest form for this function is when J_2' reaches some critical value, say J_0. When this happens yielding occurs, i.e.:

$$J_2' - J_0 = 0$$

which is the widely used Von Mises yield criterion. J_2' may be written in terms of the actual stress components as:

$$J_2' = -\tfrac{1}{6}\left[(p_{xx} - p_{yy})^2 + (p_{yy} - p_{zz})^2 + (p_{zz} - p_{xx})^2 + 6p_{xy}^2 + 6p_{zx}^2 + 6p_{yz}^2\right]$$

and J_0 is frequently changed for the value of stress p_Y at which yielding takes place in the simple tension test, i.e.:

$$J'_2 = -\tfrac{1}{6}(p_Y^2 + p_Y^2) = J_o$$

and finally:

$$2p_Y^2 = (p_{xx} - p_{yy})^2 + (p_{yy} - p_{zz})^2 + (p_{zz} - p_{xx})^2 + 6p_{xy}^2 + 6p_{zx}^2 + 6p_{yz}^2 \tag{2.44}$$

In terms of principal stresses this becomes:

$$2p_Y^2 = (p_1 - p_2)^2 + (p_2 - p_3)^2 + (p_3 - p_1)^2 \tag{2.45}$$

There are many interpretations of this criterion, including one in terms of a limiting value of shear strain energy (see equations (2.15) and (2.44)) and one which can be seen by rearranging equation (2.45) as:

$$2\left(\frac{p_Y-0}{2}\right)^2 = \left(\frac{p_1-p_2}{2}\right)^2+\left(\frac{p_2-p_3}{2}\right)^2+\left(\frac{p_3-p_1}{2}\right)^2$$

i.e. yield occurs when the root mean square of the maximum shear stresses reaches some critical value. Both are reasonable physical arguments since the process is at constant volume and is shear controlled. The fact that it is a shear process has produced another criterion of yielding which may be regarded as an approximation to that of Von Mises, namely the Tresca criterion. This states that yielding takes place when the maximum shear stress reaches some critical value:

$$\frac{p_Y}{2} = \frac{p_1-p_2}{2},\ \frac{p_2-p_3}{2},\ \frac{p_3-p_1}{2} \tag{2.46}$$

whichever is the greatest. This criterion frequently simplifies the mathematical manipulation and is used for this reason but is generally inferior to the Von Mises criterion as an interpretation of what actually happens physically.

2.3.2. *Graphical representation of the yield criteria in plane stress*

The plane stress condition is of special practical interest and corresponds to $p_{zz} = p_{yz} = p_{xz} = 0$. Hence the Von Mises criterion, i.e. equation (2.44), reduces to:

$$2p_Y^2 = (p_{xx}-p_{yy})^2+p_{yy}^2+p_{xx}^2+6p_{xy}^2$$

or

$$p_Y^2 = p_{xx}^2+p_{yy}^2-p_{xx}p_{yy}+3p_{xy}^2 \tag{2.47}$$

and in terms of principal stresses (equation (2.45)) it reduces to:

$$1 = \left(\frac{p_1}{p_Y}\right)^2+\left(\frac{p_2}{p_Y}\right)^2-\left(\frac{p_1}{p_Y}\right)\left(\frac{p_2}{p_Y}\right) \tag{2.48}$$

The form of this relationship may be illustrated by plotting p_1/p_Y versus p_2/p_Y as shown in Figure 2.5. The yield locus is an ellipse symmetrical about the $p_1/p_Y = p_2/p_Y$ line and simple tension corresponds to $p_2/p_Y = 0$ and $p_1/p_Y = 0$, i.e. $p_1/p_Y = \pm 1$ and $p_2/p_Y = \pm 1$. It can also be seen that the applied stress is greater than the yield stress p_Y for some stress states and the maximum may be determined by differentiating equation (2.48) and putting

$$\frac{d\left(\dfrac{p_1}{p_Y}\right)}{d\left(\dfrac{p_2}{p_Y}\right)} = 0$$

i.e.
$$0 = 2\left(\frac{p_2}{p_Y}\right) - \left(\frac{p_1}{p_Y}\right)$$

On substituting into equation (2.48) this gives the conditions

$$\frac{p_2}{p_Y} = \pm\frac{1}{\sqrt{3}} \quad \text{and} \quad \frac{p_1}{p_Y} = \pm\frac{2}{\sqrt{3}}$$

and differentiating with respect to p_1/p_Y gives similar conditions for p_2/p_Y. These lines for the maximum condition are shown in Figure 2.5. The Tresca

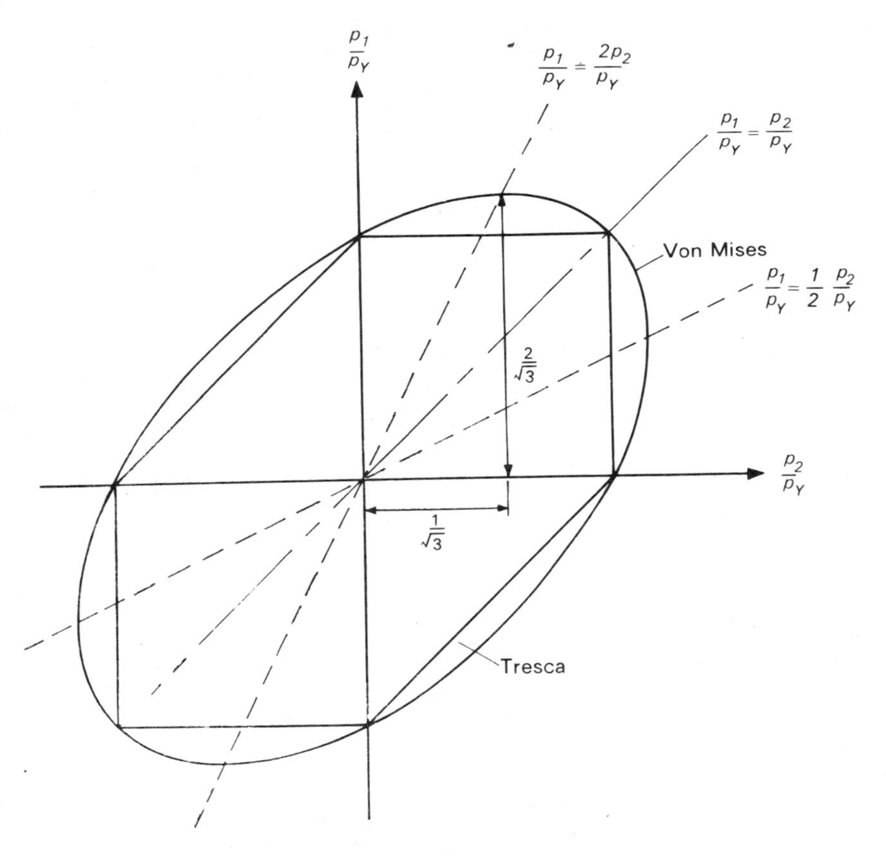

Figure 2.5 Yield criteria in plane stress.

condition is also shown in Figure 2.5 and clearly when p_1 and p_2 both have the same sign the maximum shear condition is given by:

$$p_1 = p_Y \quad \text{and} \quad p_2 = p_Y$$

as shown. When they have different signs we have:

$$p_1 - p_2 = p_Y$$

and it can be seen that the Tresca criterion inscribes the ellipse. The increase of applied stress above p_Y is not predicted.

2.3.3. *Graphical representation of the yield criteria in three dimensions*

The yield criteria can be illustrated graphically by considering the functions plotted in a three-dimensional stress space with cartesian coordinates p_1, p_2 and p_3 as shown in Figure 2.6. The Von Mises criterion becomes a circular cylinder equally inclined to the three coordinate directions, i.e. the axis of the cylinder is

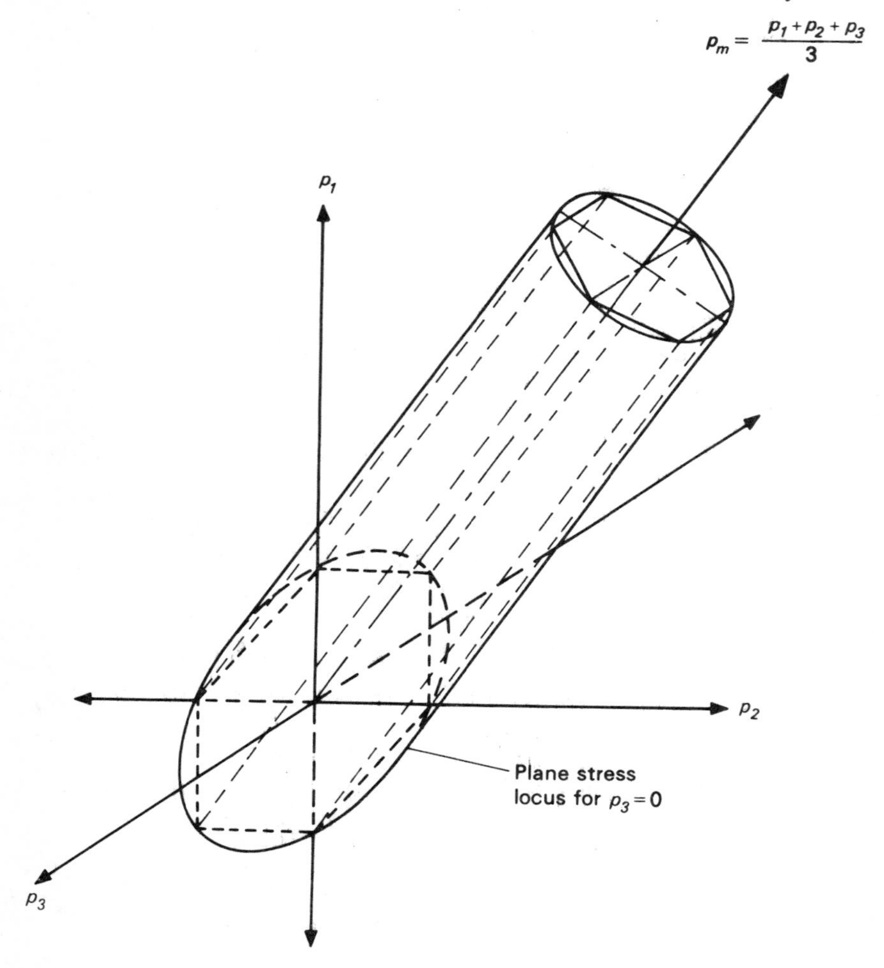

Figure 2.6 Three-dimensional yield criteria in stress space.

the direction of the hydrostatic stress p_m. A plane stress case, $p_3 = 0$, can be seen as the intercept of the cylinder on the p_1, p_2 plane. The Tresca criterion becomes a hexagonal prism. A more convenient way to represent this system is to imagine that the locus is viewed looking along the p_m direction, i.e. p_m acts normal to the paper. In this case the stress axes are at 120° intervals as shown in Figure 2.7 and the Von Mises criterion becomes a circle with the Tresca criterion

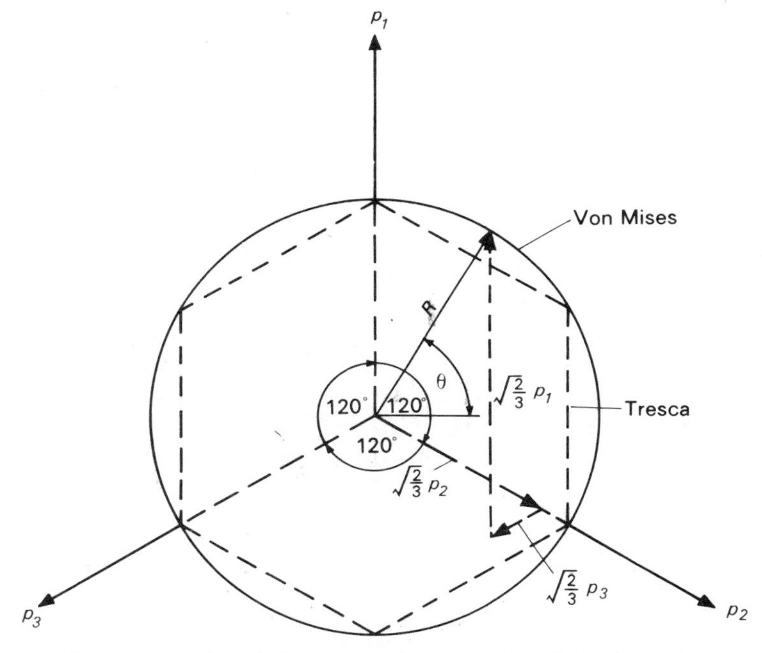

Figure 2.7 Stress space plot of criteria on a plane normal to the hydrostatic component of stress.

as an inscribed hexagon. The components of stress as seen on this plane are each multiplied by a factor due to the angle of incidence of the axis; i.e.

$$\sqrt{\frac{2}{3}}\,p_1, \qquad \sqrt{\frac{2}{3}}\,p_2 \qquad \text{and} \qquad \sqrt{\frac{2}{3}}\,p_3$$

The radius R of the circle may be calculated from the three components:

$$R^2 = \left(\frac{\sqrt{3}}{2}\sqrt{\frac{2p_2}{3}} - \frac{\sqrt{3}}{2}\sqrt{\frac{2p_3}{3}}\right)^2 + \left(\sqrt{\frac{2p_1}{3}} - \frac{1}{2}\sqrt{\frac{2p_2}{3}} - \frac{1}{2}\sqrt{\frac{2p_3}{3}}\right)^2$$

i.e. $(3\,R^2) = (p_1 - p_2)^2 + (p_2 - p_3)^2 + (p_3 - p_1)^2$

Hence R is on the yield locus and from equation (2.45) we have:

$$R = \sqrt{\frac{2}{3}}\,p_Y \tag{2.49}$$

and the radius of the circle is the projection of the yield stress p_y. Since all the components are multiplied by $\sqrt{\frac{2}{3}}$ it is usual to ignore this factor and plot the three components as p_1, p_2 and p_3 and the resulting radius becomes p_Y.

Any stress state at yield must lie on the yield locus and a parameter frequently used is the angle the radius makes with the horizontal direction. This may be written as:

$$\mu = \tan\theta = \frac{1}{\sqrt{3}}\left[\frac{(p_1-p_2)+(p_1-p_3)}{p_2-p_3}\right] \tag{2.50}$$

Clearly for any system in which $p_2 = p_3$, $\theta = \pi/2$, e.g. simple tension. The particular case where $\theta = 0$ is given by:

$$p_1 = \tfrac{1}{2}(p_2+p_3)$$

which is a state of plane strain under some circumstances (see section 1.1.8, (iv)). The main point of interest is that if the ratios between the stresses remain constant then θ remains constant and the stress state lies on a fixed radius.

A further point of interest is that the resultant stress, the radius, is limited to p_Y and hence, although individual components can be greater, they still give a value of p_Y at yield. In the extreme case of hydrostatic tension or compression p_m can have a value and not have any influence on yielding because of the constant volume condition. The result is important since it means that although a material has a finite yield stress there is no limit to the magnitude of the stress components which it can sustain.

2.3.4. *Yield criteria for materials sensitive to hydrostatic stress*

There is some evidence to suggest that the yield criteria of some polymers is not completely insensitive to the hydrostatic stress imposed. In Figure 2.6 this would imply, for example, that the radius of the Von Mises cylinder would vary with position along the p_m axis. The deformation could not be regarded as taking place at constant volume and the yield criteria would have to contain the hydrostatic stress. A simple form of this is to add the hydrostatic stress to the Von Mises criterion giving:

$$A(p_1+p_2+p_3)+B[(p_1-p_2)^2+(p_2-p_3)^2+(p_3-p_1)^2] = 1$$

in terms of the principal values, where A and B are constants. It should be noted that part of this expression changes sign from tension to compression and we may define A and B in terms of the yield stresses in simple tension and compression, p_{YT} and p_{YC} respectively.

i.e.
$$Ap_{YT}+2Bp_{YT}^2 = 1$$

and
$$-Ap_{YC}+2Bp_{YC}^2 = 1$$

i.e.
$$A = \frac{p_{YC}-p_{YT}}{p_{YC}p_{YT}} \quad \text{and} \quad B = \frac{1}{2p_{YC}p_{YT}}$$

and hence:

$$2p_{YC}p_{YT} = [(p_1-p_2)^2+(p_2-p_3)^2+(p_3-p_1)^2] \\ +2(p_{YC}-p_{YT})[p_1+p_2+p_3] \quad (2.51)$$

If the yield stress is the same in tension and compression then the usual Von Mises condition is recovered. The yield locus as shown in Figure 2.6 is now a cone and the apex describes the condition for yielding under a hydrostatic stress alone:

i.e. $\quad p_1 = p_2 = p_3 = p \quad$ and $\quad p = \dfrac{p_{YC}p_{YT}}{3(p_{YC}-p_{YT})}$

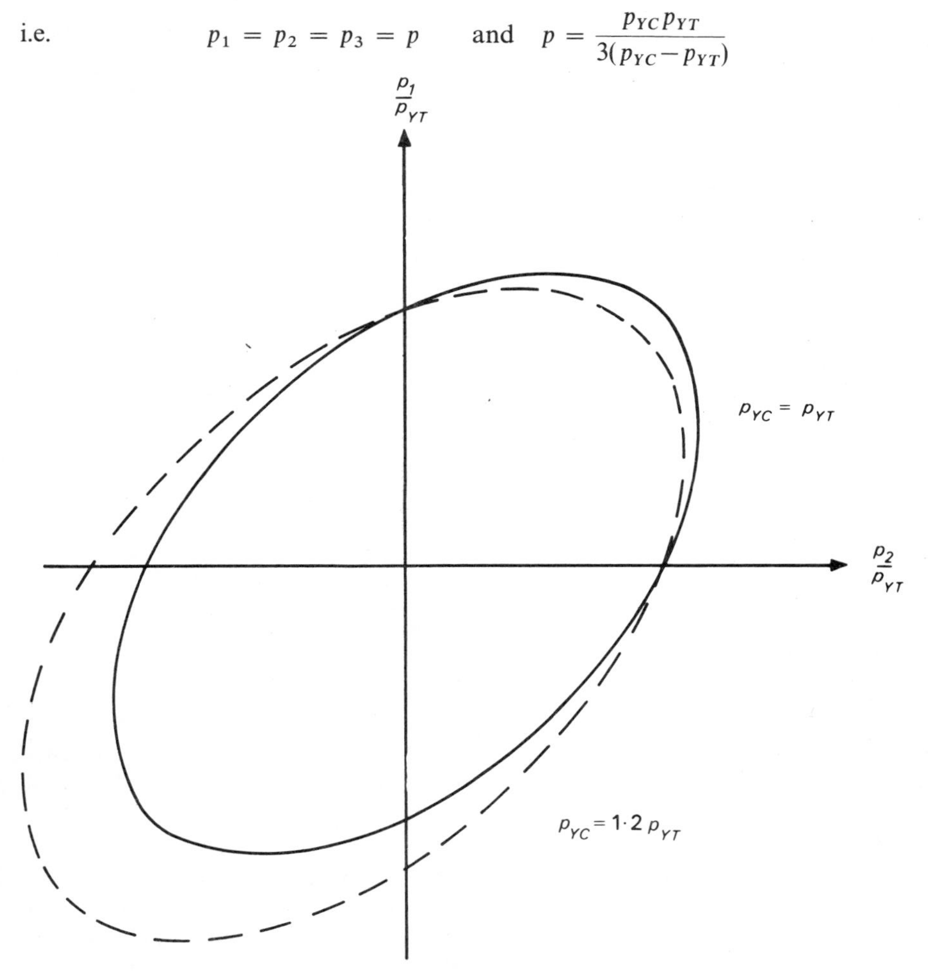

Figure 2.8 Von Mises yield criterion including a hydrostatic stress term.

If $p_{YC} > p_{YT}$ then p is a hydrostatic tension but for $p_{YC} < p_{YT}$ it is a hydrostatic compression.

The plane stress locus ($p_3 = 0$) is shown in Figure 2.8 for the case of

$p_{YC} = 1{\cdot}2\,p_{YT}$ and the increase in the yield stress in compression is apparent. The maximum stress component is given by:

$$\frac{p_1}{p_{YT}} = \frac{2}{\sqrt{3}}\sqrt{\left(\frac{p_{YC}}{p_{YT}}-1\right)^2 + \frac{p_{YC}}{p_{YT}} - \left(\frac{p_{YC}}{p_{YT}}-1\right)}$$

2.3.5. *Perfectly plastic materials*

This is a special class of materials which have a stress–strain curve in simple tension as shown in Figure 2.9. It is a property of this material that once it has yielded the stress remains constant at p_Y for increasing strain. Two forms are considered; the elastic perfectly plastic material as shown in which there is

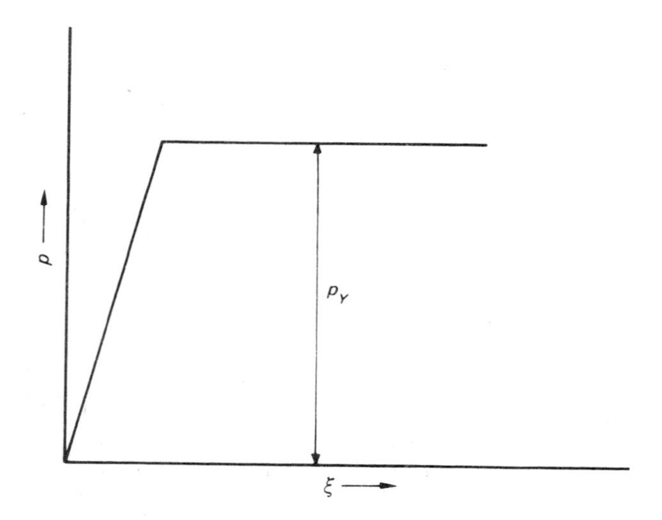

Figure 2.9 An elastic perfectly plastic material.

elastic behaviour prior to yielding and a rigid perfectly plastic material which undergoes no strain prior to yielding, i.e. is infinitely stiff. Clearly the elastic assumption is more realistic and the rigid model is usually only employed for mathematical convenience.

For many polymeric materials the yielding behaviour can be represented quite adequately in this simple way. The assumption of a constant p_Y will give stress distributions when combined with the equilibrium equations and some solutions will be discussed in later sections. Such solutions will not provide information about the state of strain since the assumed constitutive equation implies that stress is independent of strain. It is only when the yield stress varies with strain that it becomes possible to relate the stress and strain states and Figure 2.10 shows curves for (a) a work hardening material where p_Y

increases with strain, and (b) a work softening material which exhibits the reverse effect.

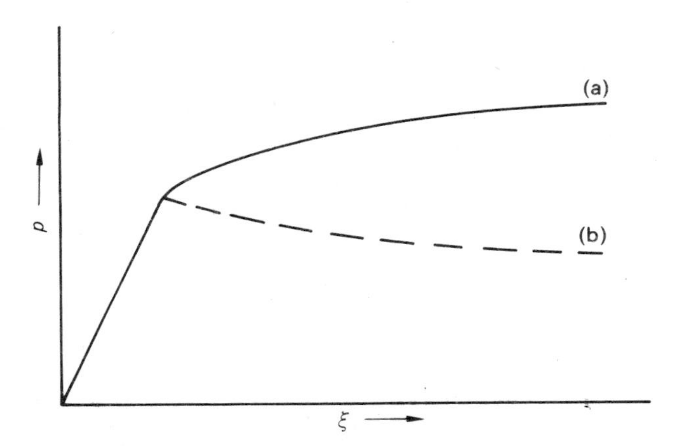

Figure 2.10 Work hardening (a) and work softening materials (b).

2.3.6. *The plastic flow equations*

If we assume that the plastic deformation is to take place at constant volume then by an analysis similar to that used for rubber elasticity we may write down the expression for the increment of plastic work done by small changes in strain in terms of the principal components as:

$$\mathrm{d}W_p = (p_1 - p)\frac{\mathrm{d}\lambda_1}{\lambda_1} + (p_2 - p)\frac{\mathrm{d}\lambda_2}{\lambda_2} + (p_3 - p)\frac{\mathrm{d}\lambda_3}{\lambda_3}$$

where p is an arbitrary hydrostatic stress. Introducing the natural strain definition:

$$\mathrm{d}\varepsilon = \frac{\mathrm{d}\lambda}{\lambda}$$

we have:

$$p_1 - p = \frac{\partial W_p}{\partial \varepsilon_1}$$

The fundamental difference between elastic and plastic behaviour is that the plastic work cannot be defined by the current state of strain since it is an irreversible process and is thus history dependent. All that can be said is that the change in plastic work can be written as a function of the change in the strains. The increment of plastic work must therefore be a function of the invariants of the change in strain and we have:

$$\mathrm{d}W_p\,(I_1, I_2, I_3)$$

where $I_1 = d\varepsilon_1 + d\varepsilon_2 + d\varepsilon_3$, $I_2 = d\varepsilon_1 d\varepsilon_2 + d\varepsilon_2 d\varepsilon_3 + d\varepsilon_3 d\varepsilon_1$ and $I_3 = d\varepsilon_1 d\varepsilon_2 d\varepsilon_3$. Since the deformation is assumed to be at constant volume $I_1 = 0$ and if terms higher than order $d\varepsilon^2$ are ignored we have:

$$dW_p(I_2)$$

It is usual to write I_2 in terms of its value in simple tension:

i.e. $$d\varepsilon_1 = d\bar{\varepsilon}, \qquad d\varepsilon_2 = d\varepsilon_3 = -\frac{d\bar{\varepsilon}}{2}$$

so that:

$$I_2 = -\tfrac{3}{4}\, d\bar{\varepsilon}^2 = d\varepsilon_1\, d\varepsilon_2 + d\varepsilon_2\, d\varepsilon_3 + d\varepsilon_3\, d\varepsilon_1$$

where $d\bar{\varepsilon}$ is called the equivalent strain increment.
We may now write:

$$\frac{\partial W_p}{\partial \varepsilon_1} = \frac{\partial W_p}{\partial I_2}\frac{\partial I_2}{\partial \varepsilon_1} = \frac{\partial W_p}{\partial \bar{\varepsilon}} \cdot \frac{2}{3}\frac{d\varepsilon_1}{d\bar{\varepsilon}}$$

and hence:

$$p_1 - p = \frac{2}{3}\frac{d\varepsilon_1}{d\bar{\varepsilon}}\frac{\partial W_p}{\partial \bar{\varepsilon}} \tag{2.52}$$

Summing this and the two similar equations for p_2 and p_3 we have:

$$p_1 + p_2 + p_3 - 3p = \frac{2}{3}\frac{\partial W_p}{\partial \bar{\varepsilon}}\frac{d\varepsilon_1 + d\varepsilon_2 + d\varepsilon_3}{d\bar{\varepsilon}} = 0$$

i.e. $$p = \frac{p_1 + p_2 + p_3}{3}$$

so that the stress equations have the form:

$$p_1' = \frac{2}{3}\left[p_1 - \frac{1}{2}(p_2 + p_3)\right] = \frac{2}{3}\frac{\partial W_p}{\partial \bar{\varepsilon}}\frac{d\varepsilon_1}{d\bar{\varepsilon}} \tag{2.53}$$

It is usual to define $\partial W_p/\partial \bar{\varepsilon}$ in terms of the second invariant of the deviatoric stress state, i.e.:

$$J_2' = p_1' p_2' + p_2' p_3' + p_3' p_1'$$

and substituting from equation (2.53) and similar equations for $d\varepsilon_2$ and $d\varepsilon_3$ we have:

$$J_2' = \frac{4}{9}\left[\frac{\partial W_p}{\partial \bar{\varepsilon}}\right]^2 \frac{d\varepsilon_1\, d\varepsilon_2 + d\varepsilon_2\, d\varepsilon_3 + d\varepsilon_3\, d\varepsilon_1}{d\bar{\varepsilon}^2} = -\frac{1}{3}\left[\frac{\partial W_p}{\partial \bar{\varepsilon}}\right]^2$$

J'_2 may be written in terms of its value in simple tension, the equivalent stress $\bar{p}$, so that:

$$p_1 = \bar{p}, \qquad p_2 = p_3 = 0 \qquad \text{and} \quad p_1' = \frac{2}{3}\bar{p}, \qquad p_2' = p_3' = -\frac{\bar{p}}{3}$$

giving:

$$J'_2 = -\frac{2}{3}\bar{p}\frac{\bar{p}}{3}+\frac{\bar{p}}{3}\frac{\bar{p}}{3}-\frac{\bar{p}}{3}\frac{2}{3}\bar{p} = -\frac{\bar{p}^2}{3}$$

and hence:

$$\bar{p} = \frac{\partial W_p}{\partial \bar{\varepsilon}} \tag{2.54}$$

The final form of the stress–strain relationships is therefore:

$$\mathrm{d}\varepsilon_1 = \frac{\mathrm{d}\bar{\varepsilon}}{\bar{p}}[p_1-\tfrac{1}{2}(p_2+p_3)] \tag{2.55}$$

and similar expressions for $\mathrm{d}\varepsilon_2$ and $\mathrm{d}\varepsilon_3$. These are the Lévy–Mises equations of plasticity and describe the most general form of plastic deformation. The equivalent stresses and strains are usually written in the forms:

$$\begin{aligned} 2\bar{p}^2 &= (p_1-p_2)^2+(p_2-p_3)^2+(p_3-p_1)^2 \\ \tfrac{9}{2}\,\mathrm{d}\bar{\varepsilon}^2 &= (\mathrm{d}\varepsilon_1-\mathrm{d}\varepsilon_2)^2+(\mathrm{d}\varepsilon_2-\mathrm{d}\varepsilon_3)^2+(\mathrm{d}\varepsilon_3-\mathrm{d}\varepsilon_1)^2 \end{aligned} \tag{2.56}$$

and from equation (2.45) it can be seen that $\bar{p}$ is the current yield stress of the material and that the assumption of only $\mathrm{d}\varepsilon^2$ terms in the energy equation results in the Von Mises criterion for yielding. Work hardening problems may be solved by using a curve of $\bar{p}$ versus $\bar{\varepsilon}$ to define the material properties. If, at any instant, the stress state is known then $\bar{p}$ may be calculated and for any chosen value of $\mathrm{d}\bar{\varepsilon}$ the values of the component strain increments determined from equations (2.55). The next value of $\bar{p}$ may then be determined from the $\bar{p}$ versus $\bar{\varepsilon}$ curve and the necessary stress components to give yielding determined. The next set of strain increments may then be found and so on. The process is equivalent to integrating equation (2.55) to give the final strain:

$$\varepsilon_1 = \int_o^{\bar{\varepsilon}} [p_1-\tfrac{1}{2}(p_2+p_3)]\,\frac{\mathrm{d}\bar{\varepsilon}}{\bar{p}}$$

where p_1, p_2, p_3 vary with the stress history and $\mathrm{d}\bar{\varepsilon}$ must be positive.

It is clear from equation (2.56) that a strain space diagram may be constructed in a similar manner to that for stress space and both are shown in Figure 2.11. A strain history may be regarded as a strain path drawn in this space and to construct it ε_1, ε_2 and ε_3 may all be varied. At any point A the equivalent strain $\bar{\varepsilon}$ is given by the radius as shown and $\mathrm{d}\bar{\varepsilon}$ is along the strain path as shown. The stress space is also shown and at the same point A′ we have $\bar{p}$ and $\mathrm{d}\bar{p}$ and the stress path corresponding to the strain path. Now A′ lies on the current yield circle and the flow rule, equation (2.55), requires that $\mathrm{d}\bar{\varepsilon}$ be in the same direction as $\bar{p}$. This may be seen from the diagram since the angle $\mathrm{d}\bar{\varepsilon}$ makes with the horizontal is given by:

$$\tan \mathrm{d}\psi = \frac{1}{\sqrt{3}}\left[\frac{(\mathrm{d}\varepsilon_1-\mathrm{d}\varepsilon_2)+(\mathrm{d}\varepsilon_1-\mathrm{d}\varepsilon_3)}{\mathrm{d}\varepsilon_2-\mathrm{d}\varepsilon_3}\right]$$

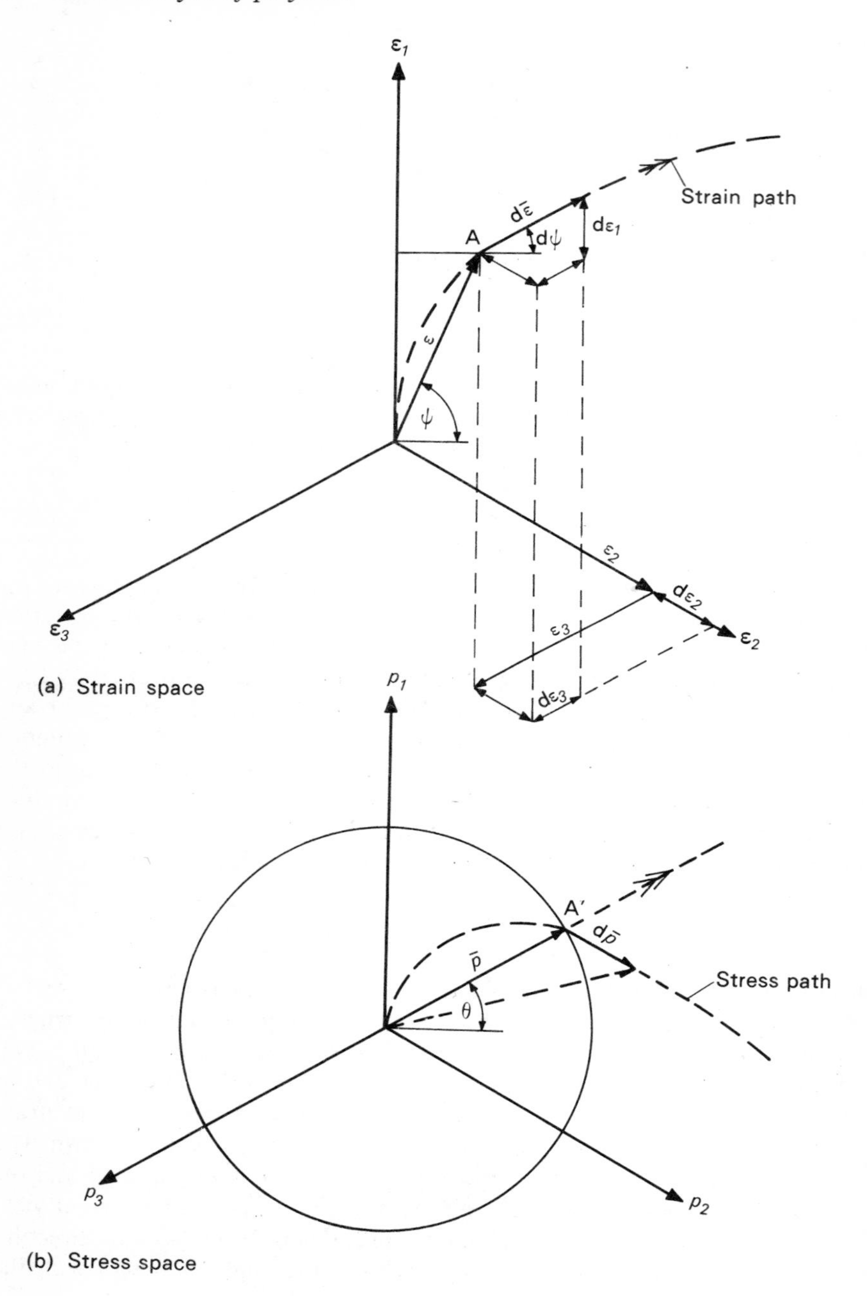

Figure 2.11 Strain and stress paths.

and substituting from equation (2.55) we have:

$$\tan d\psi = \frac{1}{\sqrt{3}}\left[\frac{(p_1-p_2)+(p_1-p_3)}{p_2-p_3}\right] = \tan\theta$$

the angle which $\bar{p}$, which is now the yield stress, makes with the horizontal. This is known as the normality rule, i.e. the strain increment is in the direction of the current yield stress and normal to the yield locus.

As $\bar{p}$ moves along the stress path the direction of $d\bar{\varepsilon}$ is determined and the strain path may be drawn. This representation is of particular interest when there are elastic strains involved as well since the elastic strain component is in the $d\bar{p}$ direction (from elastic theory) and a vector sum of strain components must be made. Thus for these general stress and strain paths it is vital to know if the material is elastic or undergoing flow since the strain paths will differ. In polymers it is not usually possible to lay down a simple definition of plastic flow and thus one can only use the ideas described here with certainty when it does not matter which is operating. The obvious condition is when the stress path is a radial line which must result in a radial line strain path since both $\bar{p}$ and $d\bar{p}$ are in the same direction and the strain increments can be simply summed. This condition is known as 'constant stress ratios' which results in constant strain ratios. A further condition which must be satisfied is that only loading must be considered since unloading will require differentiation between reversible and irreversible deformation.

Therefore, here we will consider only problems of loading with constant stress and strain ratios, in which case equations (2.55) may be integrated and give an alternative form of a non-linear representation:

i.e.
$$\varepsilon_1 = \frac{\bar{\varepsilon}}{\bar{p}}\left[p_1-\tfrac{1}{2}(p_2+p_3)\right] \tag{2.57}$$

sometimes known as the Hencky equations.

Solutions are produced by defining the $\bar{p}$ versus $\bar{\varepsilon}$ curve and then, at any instant, if we know the stress state then, for example, $\bar{p}$ and hence $\bar{\varepsilon}$ may be found. Substitution gives the corresponding strain states. Precisely the same result may be derived from an elastic analysis (see section 2.2.5) and if no unloading is considered the two forms may be taken as identical since elastic and plastic deformation are indistinguishable under such circumstances.

The reduction of three-dimensional stress and strain systems to a single relationship using equivalent stress and strain is of great value. It has been shown to be applicable to a wide range of behaviour and is not confined to plastic flow. This stems from the fact that it is apparently the most general form of averaging for stress and strain states. Many phenomena are controlled by invariants and it would appear that the second is usually dominant.

Bibliography

1. Timoshenko, S. P. and Goodier, J. N. *Theory of Elasticity*, Third Edition McGraw-Hill, New York (1970).
2. Hearmon, R. F. S. *An Introduction to Applied Anisotropic Elasticity*, Oxford University Press (1961).
3. Rivlin, R. S. 'Large Elastic Deformations', Chap. 10 in *Rheology*, Vol. 1, Ed. Frederick R. Eirich, Academic Press Inc., New York (1956).
4. Green A. E. and Adkins, J. E. *Large Elastic Deformations*, Oxford University Press (1960).
5. Hill, R. *Mathematical Theory of Plasticity*, Oxford University Press (1950).
6. Johnson, W. and Mellor, P. B. *Plasticity for Mechanical Engineers*, D. Van Nostrand, London (1962).

3 Time dependent behaviour

3.1. Definitions

In the previous chapter we discussed elastic behaviour, which is defined as completely and instantaneously reversible, and plastic behaviour which is completely irreversible, so that both are independent of any time factors. The elastic strain which results from any stress system is completely determined by the current state of that stress system and is not influenced by how long it has been applied or in what manner. Most metals conform to this type of behaviour at low stresses but it is a feature of the behaviour of polymeric materials that they do not. They are all time dependent or visco-elastic (although some rubbers are only so to a slight extent) and it is this aspect of their mechanical properties which is of major interest in design calculations.

Any successful constitutive relationship must be capable of describing a number of distinct features of the time dependence which are illustrated in Figure 3.1 for a simple tension stress system of stress p and strain e.

(a) *Creep*

If a stress p is applied at zero time and held constant the strain e increases with time t at a decreasing rate and the amount of this strain increases with stress. The behaviour of an elastic material is shown as a broken line for comparison purposes.

(b) *Relaxation*

When a strain is applied at zero time and held constant then the stress decreases from its value at $t = 0$ as time passes.

(c) *Recovery*

If the stress is removed, either partially or entirely, the strain decreases or 'recovers' as a function of time, in other words, there is delayed recovery.

(d) *Constant rate stressing*

Constant rates of stress application result in a non-linear increase of strain with time. A linearly elastic material would give linear strain increases. If stress–strain lines are drawn for different stress rates then the curve rises more steeply as stress rate increases (it is the same for all rates for an elastic material).

(e) *Constant rate straining*

The same behaviour obtains in that with increasing strain rate the stress–strain curve rises more steeply.

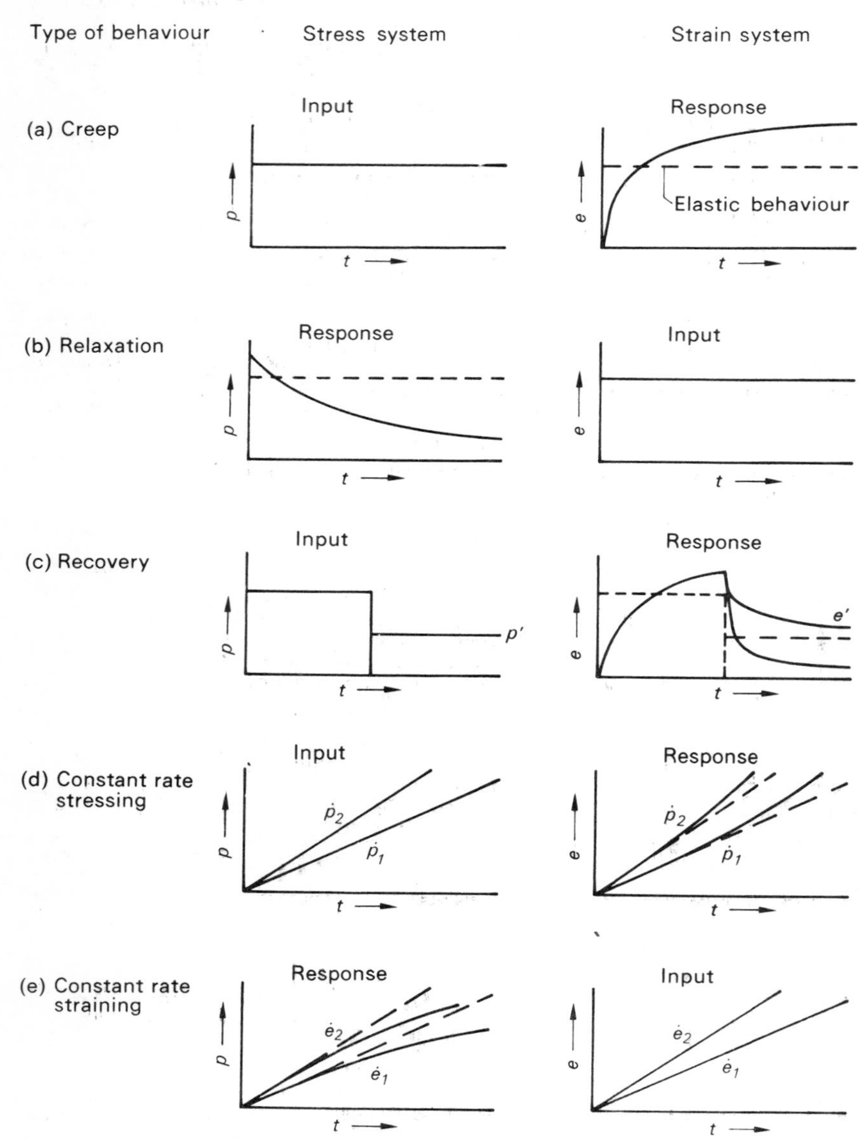

Figure 3.1 The main features of polymer time dependence.

Any constitutive relationship must account for these five forms of behaviour, at least qualitatively, and it is then a matter of experimental observation to determine if any such relationship is a good quantitative description of behaviour. We shall now examine various theoretical relationships which are available and comment on their value in engineering stress analysis. In all cases we shall assume that the temperature remains constant.

3.2. Equation of state theory

The simplest possible form of constitutive equation which includes time as a variable is obtained by postulating that there exists a unique relationship between stress, strain and time. We shall define zero time as when the first event happens such that:

$$p = e = 0 \qquad \text{for} \quad t = 0$$

and we have

$$e = \phi\,(p, t) \tag{3.1}$$

The strain e will always be taken as infinitesimal in this chapter. In effect we are saying that there exists a three-dimensional surface which is unique for a given material and may be drawn as shown in Figure 3.2. Sections drawn through the

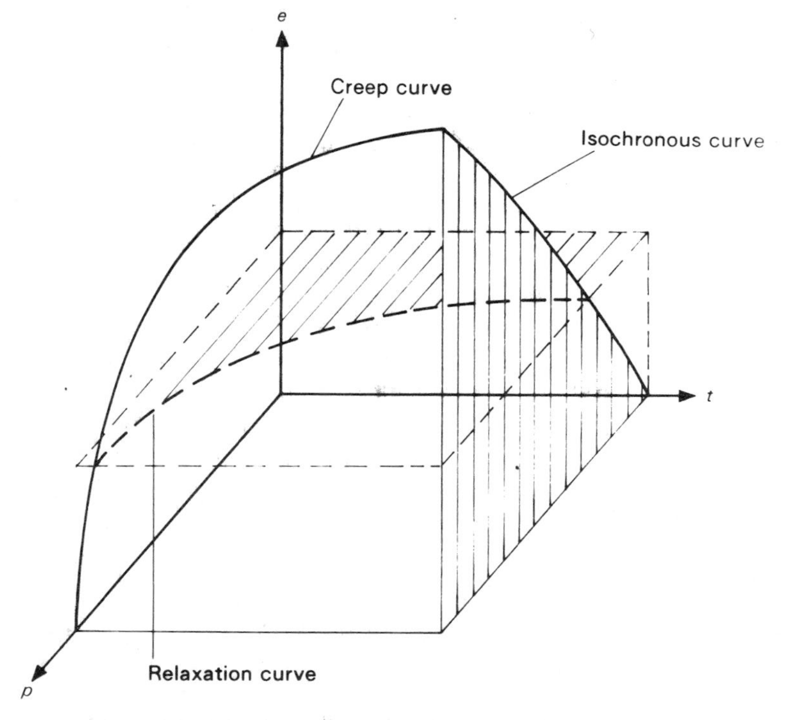

Figure 3.2 The three-dimensional stress-strain-time surface.

surface at constant stress levels give creep curves and sections at constant strain give relaxation curves as shown. Another important cross-plot which is also shown is a section for a constant time, or isochronous curve. This results in a stress–strain curve for each time. The cross-plot procedure is further illustrated in Figure 3.3 and there is a reasonable representation of creep and relaxation.

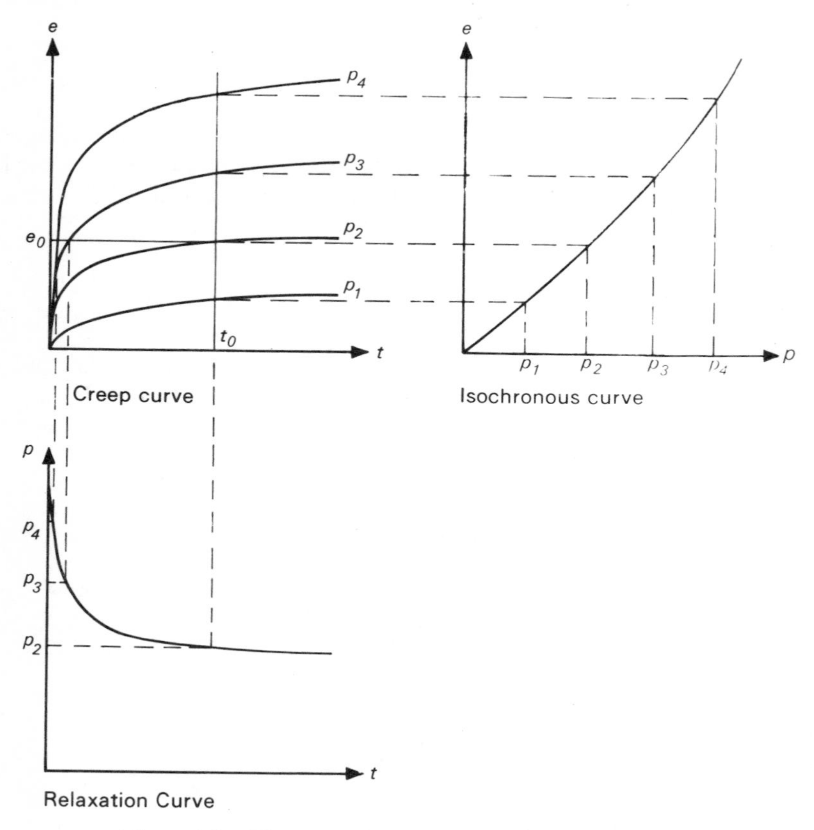

Figure 3.3. The cross-plot method for an equation of state.

Lines of constant strain rate may also be drawn on the creep curves and it is possible to determine the appropriate p versus t curves. Similarly constant stress rate behaviour can be shown to be given in a reasonable form.

An important result from this approach is that each isochronous curve is independent of the loading system used for its determination, i.e. at a time t_o for a stress of p, we have a strain e, whether we apply that stress instantaneously at $t = 0$ and then hold it constant or if we apply it at a uniform rate over the whole time span t_0 or indeed arrive at the final stress in any manner. Thus a test of the hypothesis of a unique surface is to derive isochronous curves for various loading systems (i.e. creep, relaxation, etc.) and compare them.

It is usual to distinguish three types of function in equation (3.1).

(a) Is the completely general form:

$$e = \phi(p, t)$$

which may be expressed as a polynomial:

$$e = C_1(t)\,p + C_2(t)\,p^2 + C_3(t)p^3 + \cdots \tag{3.2}$$

(b) Variables separable. In this case it is assumed that ϕ may be represented as the product of two functions:

$$e = C(t) f(p) \tag{3.3}$$

(c) Linear form. If it is assumed that:

$$f(p) = p$$

then we have the linear form:

$$e = C(t)\,p \tag{3.4}$$

A linear material therefore has a linear isochronous stress–strain relationship which provides a simple check on this property.

From the preceding discussion it is clear that for continuously increasing or constant loading systems the equation of state theory gives a reasonable qualitative representation. However, if the case of recovery is considered a basic weakness becomes apparent. For example, if the stress is removed completely at some time t_0 the surface requires that $e = 0$ also and hence the strain would return to zero immediately. Polymers clearly do not do this but exhibit delayed recovery. Similar anomalies become apparent if any other form of unloading, reversal or abrupt change of load is examined and we are thus obliged to consider, in detail, behaviour under changing loads.

3.3. Superposition theory

The concept of a superposition theory may be explained by considering the system shown in Figure 3.4 in which a constant stress p_1 is increased to a value p_2 at a time τ. We have two basic pieces of information available in that the strain response for each stress level is known,

i.e. $$e_1 = \phi(p_1, t) \text{ and } e_2 = \phi(p_2, t)$$

If we now invoke the simple superposition principle that the strain at time t is given by the algebraic sum of the three components shown in Figure 3.4 then

$$e(t) = e_1(t) - e_1(t-\tau) + e_2(t-\tau)$$

and hence: $$e(t) = \phi(p_1, t) - \phi(p_1, t-\tau) + \phi(p_2, t-\tau) \tag{3.5}$$

This kind of superposition is applicable to any form of the function ϕ but is based on the assumption that the various components are independent of the history effect, e.g. $e_2(t-\tau)$ is not influenced by e_1.

If $p_2 - p_1 = \delta p$, a small change in stress, we may write:

$$e(t) = \phi(p_1, t) + \frac{\partial \phi}{\partial p}(p_1, t-\tau)\delta p \tag{3.6}$$

If any particular history is taken as made up of a series of steps δp applied at times τ then $e(t)$ may be summed to give:

$$e(t) = \sum_{n=0}^{n} \frac{\partial \phi}{\partial p}(p, t-\tau)\, \delta p_n \tag{3.7}$$

which in the limit reduces to:

$$e(t) = \int_0^p \frac{\partial \phi}{\partial p}(p, t-\tau)\, \mathrm{d}p \tag{3.8}$$

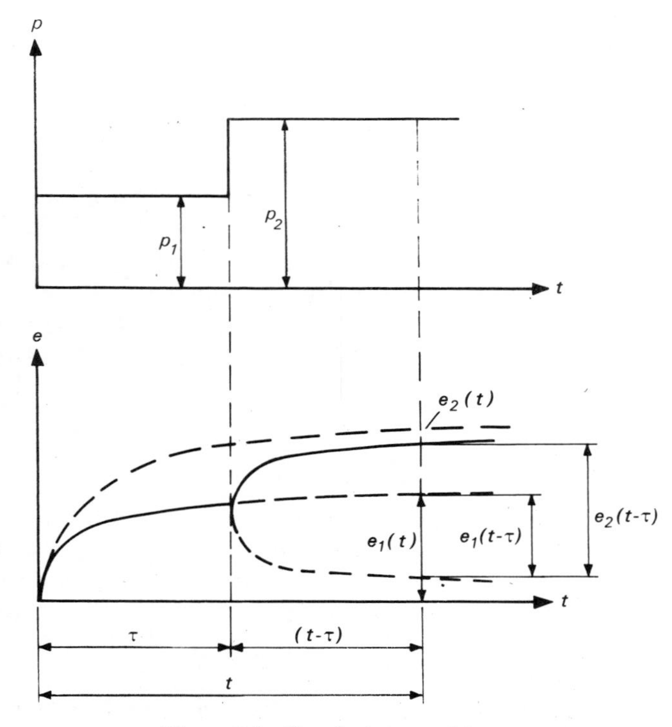

Figure 3.4 Simple superposition.

This equation, called a convolution integral, is fundamental to all visco-elastic analysis and incorporates the mechanism by which the time variable τ is integrated from 0 to t. This expresses the history idea in which the result at time t is the summation of the whole history, as expressed in dp, combined with the material properties, expressed in $\partial\phi/\partial p\,(p, t)$. It is usually more convenient to write equation (3.8) as a function of τ giving:

$$e(t) = \int_0^t \frac{\partial\phi}{\partial p}(p, t-\tau)\frac{\mathrm{d}p(\tau)}{\mathrm{d}\tau}\,\mathrm{d}\tau \tag{3.9}$$

Note that p is a function of τ and the term $\mathrm{d}p(\tau)/\mathrm{d}\tau$ is a complete time history function and that $\partial\phi/\partial p\,(p, t-\tau)$ represents the contribution of the material behaviour from the time τ to t, i.e. $t-\tau$.

For a step input at $t = 0$ this reduces to the basic creep expression:

$$e(t) = \phi(p, t)$$

Special forms of this equation may be written in particular cases. For example, when the stress and time variables are separable we have:

$$e(t) = \int_0^t C(t-\tau)\frac{\partial f(p)}{\partial p}\frac{\mathrm{d}p}{\mathrm{d}\tau}\,\mathrm{d}\tau$$

which gives:

$$e(t) = \int_0^t C(t-\tau)\frac{\mathrm{d}f(p)}{\mathrm{d}\tau}\,\mathrm{d}\tau \tag{3.10}$$

This is called the Leaderman form of the convolution integral. For the special case of a linear material we have:

$$e(t) = \int_0^t C(t-\tau)\frac{\mathrm{d}p}{\mathrm{d}\tau}\,\mathrm{d}\tau \tag{3.11}$$

A more general type of superposition may be used in which all possible interactions are allowed between stress changes. In this case we have:

$$e(t) = \int_0^t C_1(t-\tau)\frac{\mathrm{d}p}{\mathrm{d}\tau}\,\mathrm{d}\tau + \int_0^t\int_0^t C_2(t-\tau_1, t-\tau_2)\frac{\mathrm{d}p}{\mathrm{d}\tau_1}\frac{\mathrm{d}p}{\mathrm{d}\tau_2}\,\mathrm{d}\tau_1\mathrm{d}\tau_2 + \ldots \tag{3.12}$$

This multiple integral contains all those previously mentioned as special cases and in principle is capable of describing a wide range of behaviour. However, its practical application is still the subject of research and its validity is not sufficiently proven to warrant further consideration here.

Returning to the single integrals, equations (3.9), (3.10) and (3.11) are a formal representation of the idea of superimposing the responses to abrupt changes in stress and they are capable of giving a consistent picture of behaviour of unloading. This is easily seen in Figure 3.4 where, when $p_2 = 0$, the recovery curve is certainly of the required form. Although more complex than the equation of state theory, the convolution integral is more satisfactory. To explain in more detail the significance of the convolution integral the linear form will be discussed in detail as the linear theory forms the basis of most of the analytical work connected with the visco-elastic behaviour of polymeric materials.

3.4. Linear visco-elasticity theory

3.4.1. *Introduction*

Linear visco-elasticity theory conforms to the definition previously given so far as the adjective 'linear' is concerned in that the ratio of stress to strain for any history is a function of time only. The term visco-elasticity appears because time dependent behaviour can be viewed as a combination of elastic and viscous effects. If we consider the response under the conditions of a creep test for a linear material (equation (3.4)) we may write:

$$e(t) = C(t)\,p$$

where $C(t) =$ the creep compliance

and for a general stressing history (equation (3.11)) this can be expressed as:

$$e(t) = \int_0^t C(t-\tau)\,\frac{\mathrm{d}p(\tau)}{\mathrm{d}\tau}\,\mathrm{d}\tau$$

This is the well known Boltzmann superposition integral and has been widely studied because of its importance in other time dependent systems in electricity and dynamics. The most convenient analytical method for studying its behaviour is to transform the variables contained in it using Laplace transforms. A complete description of the properties of these will be found in mathematical texts but a résumé will be given here as an understanding is essential for subsequent sections.

3.4.2. *Laplace transforms*

If y is a function of t then it is transformed to a function of a new variable s through the following expression:

$$\bar{y} = \int_0^\infty y\,e^{-st}\,\mathrm{d}t \tag{3.13}$$

where $\bar{y}$ is the Laplace transform of y.
The process of transformation may be illustrated with the function:

$$y = t^2$$

and

$$\bar{y} = \int_0^\infty t^2\,\mathrm{e}^{-st}\,\mathrm{d}t$$

Integrating by parts we have:

$$\bar{y} = \left[t^2\,\frac{\mathrm{e}^{-st}}{-s}\right]_0^\infty - \int_0^\infty \frac{2t}{-s}\,\mathrm{e}^{-st}\,\mathrm{d}t$$

i.e. $$\bar{y} = \left[t^2 \frac{e^{-st}}{-s} \right]_0^\infty - \left[\frac{2t}{-s} \frac{e^{-st}}{-s} \right]_0^\infty + \left[\frac{2e^{-st}}{-s.s.s} \right]_0^\infty$$

and $$\bar{y} = \frac{2}{s^3}$$

Thus we can say that the Laplace transform of t^2 is $2/s^3$. Other functions may be found in standard tables and a selection of those which are of interest in this work is given in table 3.1.

Table 3.1 *Laplace transforms*

$f(t)$	$\bar{f}(t)$	
t^n	$\frac{n}{s^{n+1}}$	(n an integer)
t^n	$\frac{\Gamma(1+n)}{s^{1+n}}$	n not an integer but > 0. Γ is the gamma function.
$\sin \omega t$	$\frac{\omega}{\omega^2+s^2}$	
$\cos \omega t$	$\frac{s}{\omega^2+s^2}$	
e^{at}	$\frac{1}{s-a}$	

A further important property of Laplace transforms is in the transformation of derivatives.

For example:

$$\overline{\frac{dy}{dt}} = \int_0^\infty \frac{dy}{dt} e^{-st} \, dt$$

$$= \left[e^{-st} y \right]_0^\infty + s \int_0^\infty y \, e^{-st} \, dt$$

and $$\overline{\frac{dy}{dt}} = -y_0 + s\bar{y} \tag{3.14}$$

where y_0 is the value of y at $t = 0$. The second derivative becomes:

$$\overline{\frac{d^2y}{dt^2}} = -\left[\left(\frac{dy}{dt} \right)_0 + sy_0 \right] + s^2\bar{y}$$

D

and thus the nth derivative is:

$$\overline{\frac{d^n y}{dt^n}} = -\left[\left(\frac{d^{n-1}y}{dt^{n-1}}\right)_0 + s\left(\frac{d^{n-2}y}{dt^{n-2}}\right)_0 + \ldots + s^{n-1}y_0\right] + s^n\bar{y} \tag{3.15}$$

In the discussions given here $t = 0$ is taken as the starting point of all histories and we take the derivatives just before anything has happened. This amounts to the concept of -0 and $+0$ such that at -0 all the derivatives are zero but at $+0$ they take values. Thus for the visco-elastic analysis we may write:

$$\overline{\frac{d^n y}{dt^n}} = s^n\,\bar{y} \tag{3.16}$$

Borel's theorem states that a convolution integral may be transformed:

$$\overline{\int_0^t f(\tau)\,g(t-\tau)\,d\tau} = \overline{\int_0^t g(\tau)f(t-\tau)\,d\tau} = g(s)f(s)$$

so that equation (3.11) may be transformed to give:

$$\bar{e} = \bar{C}(s)\,\overline{\frac{dp(t)}{dt}}$$

and hence

$$\bar{e} = \bar{C}(s)\,\bar{p}\,s \tag{3.17}$$

This equation in the transformed variables $\bar{e}$, $\bar{p}$ and s can be manipulated according to the usual algebraic rules and is therefore much simpler to deal with than the corresponding convolution integral.

3.4.3. *Relaxation*

Using the same approach as that adopted previously for the result of a creep test (p = constant) the relaxation behaviour of a linear material can be described in the following form:

$$p(t) = M(t)\,e \tag{3.18}$$

where $M(t)$ = the relaxation modulus

and is a function of time only. Clearly $M(t)$ and $C(t)$ are not simply related as $M(t)$ describes a constant strain history and $C(t)$ a constant stress history. For a linear elastic material the relaxation and creep relationships are:

$$p = Me \quad \text{and} \quad e = Cp$$

where M and C are constants so that:

$$M = \frac{1}{C} = E, \text{ the Young's modulus} \tag{3.19}$$

The strain history effects may be described using the Boltzmann superposition principle which leads to the convolution integral:

$$p(t) = \int_0^t M(t-\tau)\frac{\mathrm{d}e(\tau)}{\mathrm{d}\tau}\,\mathrm{d}\tau \tag{3.20}$$

Any strain history may be included in the derivative term and $M(t)$ defines the material behaviour. The relationship between $C(t)$ can be seen by taking transforms of equation (3.20) giving:

$$\bar{p} = M(s)\overline{\frac{\mathrm{d}e(t)}{\mathrm{d}t}} = \overline{M}(s)\,\bar{e}\,s \tag{3.21}$$

Combining equations (3.21) and (3.17) gives the result:

$$s\,\overline{M}(s) = \frac{1}{s\,\overline{C}(s)} \tag{3.22}$$

Thus $M(t)$ and $C(t)$ are related in a similar way to that of the constants of linear elastic materials given in equation (3.19), except that the elastic constants are replaced by s times the transformed variables.

Equation (3.22) can be rewritten as:

$$\overline{M}(s)\,\overline{C}(s) = \frac{1}{s^2}$$

which, when inverted back to the variable t, becomes:

$$\begin{aligned}\int_0^t M(t-\tau)\,C(\tau)\,\mathrm{d}\tau &= t\\ \int_0^t C(t-\tau)\,M(\tau)\,\mathrm{d}\tau &= t\end{aligned} \tag{3.23}$$

Thus if $M(t)$ or $C(t)$ is known the other may be derived.

3.4.4. *The differential equation representation of time dependence*

The discussion so far has been in terms of convolution integrals but the type of behaviour described by them can also be expressed in other ways. For example, in equation (3.17):

$$\bar{e} = s\,\overline{C}(s)\,\bar{p}$$

and the function of s, $s\ C(s)$, could be written as the ratio of two polynomials in s:

i.e.
$$s\,C(s) = \frac{Q(s)}{P(s)}$$

where
$$P(s) = a_o + a_1 s + a_2 s^2 + \cdots + a_n s^n$$

and $$Q(s) = b_o + b_1 s + b_2 s^2 + \cdots + b_n s^n$$

Equation (3.17) now becomes:

$$P(s)\,\bar{e} = Q(s)\,\bar{p}$$

and since

$$\overline{\frac{d^n y}{dt^n}} = s^n\,\bar{y}$$

we have:

$$a_n \frac{d^n e}{dt^n} + \cdots + a_2 \frac{d^2 e}{dt^2} + a_1 \frac{de}{dt} + a_o e = b_o p + b_1 \frac{dp}{dt} + \cdots + b_n \frac{d^n p}{dt^n} \qquad (3.24)$$

Thus the behaviour of a linear visco-elastic material may be represented by a linear differential equation.

3.4.5. *Sinusoidal loading*

Another useful equation can be derived from (3.24) by considering a sinusoidal stress input:

$$p = p_0 \sin \omega t$$

Clearly the right-hand side of the equation becomes:

$$b_0 p_0 \sin \omega t + b_1 p_0 \omega \cos \omega t - b_2 p_0 \omega^2 \sin \omega t + \cdots, \text{ etc.}$$

which can be written as:

$$p_0 A_1(\omega) \sin \omega t + p_0 B_1(\omega) \cos \omega t$$

where $A_1(\omega)$ and $B_1(\omega)$ are material property functions of the frequency ω.

The steady state solution of the linear differential equation with this as the right-hand side is of the same form;

$$e = p_0 A_2(\omega) \sin \omega t + p_0 B_2(\omega) \cos \omega t$$

where $A_2(\omega)$ and $B_2(\omega)$ are again material property functions of ω. This solution can be interpreted as made up of an in-phase component $p_0 A_2(\omega)$ and an out-of-phase component $p_0 B_2(\omega)$. For problems involving sinusoidal inputs of either stress or strain it is usual to define the material properties, not as functions of time, but as functions of ω and separate an in-phase and out-of-phase modulus or compliance. This representation is considered further in section 3.5.3.

3.5. Special linear visco-elastic materials and models

The special forms of the constitutive equation frequently used for linear visco-elastic materials arose out of the early studies which considered that visco-elastic behaviour was, indeed, a combination of viscous and elastic components. Two basic elements were proposed:

(1) The linear spring shown in Figure 3.5(a) in which:

$$e = \frac{p}{E}$$

It will be assumed that the spring has a unit cross-sectional area and unit length so that the stress p is the load and the strain e the deflection. The stiffness is then the modulus E.

(2) The viscous element shown in Figure 3.5(b) which obeys Newton's viscosity relationship:

$$\frac{\mathrm{d}e}{\mathrm{d}t} = \frac{p}{\mu}$$

where μ is the viscosity of the element. Again we assume a unit cross-sectional area and length.

Various visco-elastic models are made up of different combinations of these two elements.

3.5.1. *The Maxwell model*

A simple model, known as the Maxwell model, consists of a spring and a viscous element in series as shown in Figure 3.5(c). Now if e_1 is the strain in the spring and e_2 is the strain in the viscous element then:

$$e = e_1 + e_2 \tag{3.25}$$

where e is the overall strain. The stress in each element is the same since these are loads for the unit elements and they are in series. Thus we may write:

$$e_1 = \frac{p}{E} \quad \text{and} \quad \frac{\mathrm{d}e_2}{\mathrm{d}t} = \frac{p}{\mu}$$

and differentiating equation (3.25) we have:

$$\frac{\mathrm{d}e}{\mathrm{d}t} = \frac{\mathrm{d}e_1}{\mathrm{d}t} + \frac{\mathrm{d}e_2}{\mathrm{d}t}$$

giving:

$$\frac{\mathrm{d}e}{\mathrm{d}t} = \frac{1}{E}\frac{\mathrm{d}p}{\mathrm{d}t} + \frac{p}{\mu} \tag{3.26}$$

This is the constitutive equation for a material described by the Maxwell model and a particular form of equation (3.24) in which all the coefficients are zero, except:

$$a_1 = 1, \qquad b_0 = \frac{1}{\mu} \qquad \text{and} \quad b_1 = \frac{1}{E}$$

It is of interest to examine the response of such a material to various stress and strain histories.

(a) *Creep*
Here $p = p_0$ constant and hence $\mathrm{d}p/\mathrm{d}t = 0$. Thus, from equation (3.26), we have:

$$\frac{\mathrm{d}e}{\mathrm{d}t} = \frac{p_0}{\mu}$$

i.e.

$$e = \frac{p_0}{\mu}t + A$$

where A is a constant.

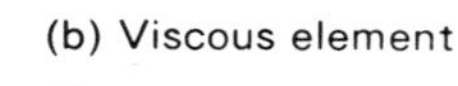

(a) Linear spring

(b) Viscous element

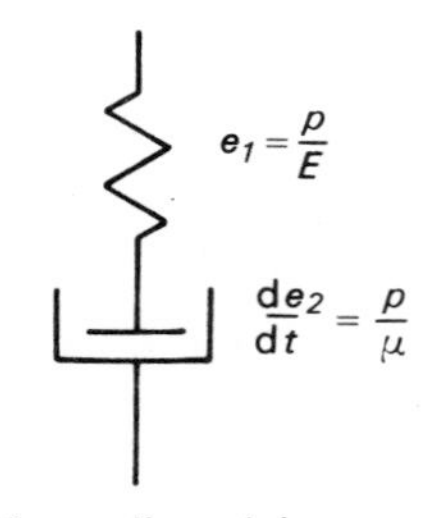

(c) Maxwell model

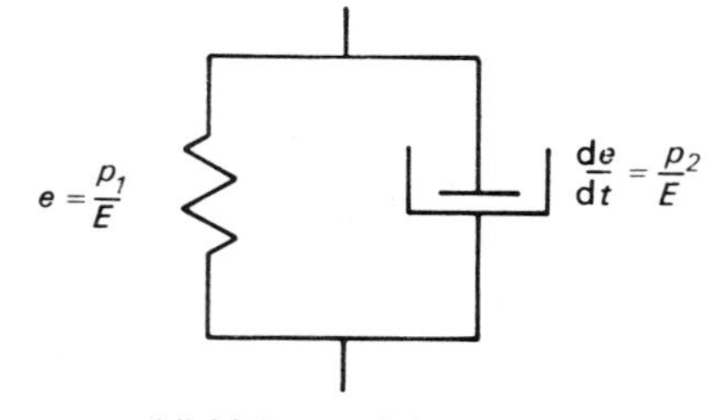

(d) Voigt model

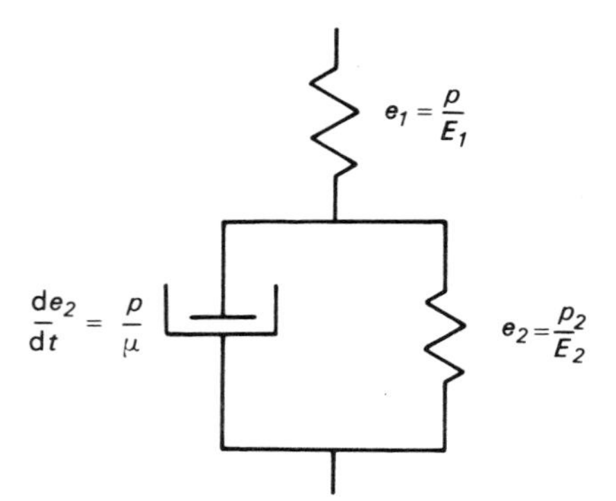

(e) Standard linear solid

Figure 3.5 Special linear visco-elastic material models.

This corresponds to the response of the viscous element as inspection of the model confirms (see Figure 3.6). The applied stress p_0 is carried directly by the

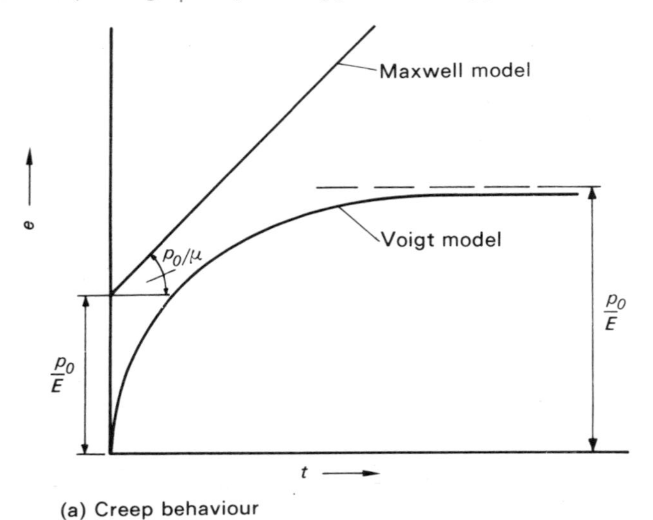

(a) Creep behaviour

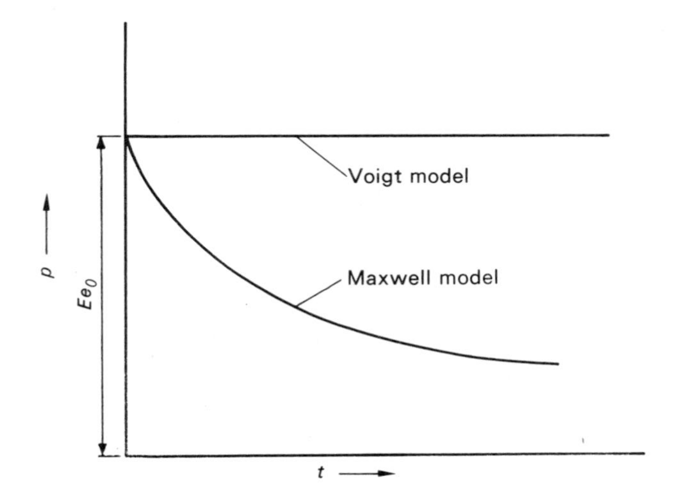

(b) Relaxation behaviour

Figure 3.6 The behaviour of Maxwell and Voigt models.

viscous element giving this viscous response and the spring gives an elastic strain so that $A = p_0/E$ and the creep compliance function is:

$$C(t) = \frac{e}{p_0} = \frac{t}{\mu} + \frac{1}{E}$$

This is not a very realistic form of behaviour because the strain increases linearly with time and we require the strain rate to decrease with time.

(b) *Relaxation*

In this case $e = e_0$ and $\mathrm{d}e/\mathrm{d}t = 0$ and equation (3.26) becomes:

$$0 = \frac{1}{E}\frac{\mathrm{d}p}{\mathrm{d}t} + \frac{p}{\mu}$$

Integrating gives:

$$\ln p = -\frac{E}{\mu}t + A$$

where A is a constant. If $p = p_0$ at $t = 0$ then we have:

$$p = p_0 \exp\left(-\frac{Et}{\mu}\right)$$

p_0 may be determined by the fact that in the initial straining (equation (3.26)), $\mathrm{d}e/\mathrm{d}t$ is very large and hence so is $\mathrm{d}p/\mathrm{d}t$ which dominates the equation and we have:

$$\frac{\mathrm{d}e}{\mathrm{d}t} = \frac{1}{E}\frac{\mathrm{d}p}{\mathrm{d}t}$$

i.e. $p = E\,e$, elastic behaviour giving a final result of:

$$p = E\,e_0 \exp\left(-\frac{Et}{\mu}\right)$$

The mechanism resulting in this behaviour can be seen from the model because the viscous element is initially 'frozen out' by the rapid initial increase in the strain and the response is the elastic behaviour of the spring. With the strain held constant the viscous element comes into play and the load decreases as the element extends and the spring contracts. The form of the curve (see Figure 3.6(b)) is as required in that the stress falls at an ever decreasing rate and tends to zero at long times.

One other definition is frequently used for the term μ/E. This is written τ_0 (it has the units of time) and called the relaxation time. The relaxation modulus may be written as:

$$M(t) = \frac{p}{e_0} = E \exp\left(-\frac{t}{\tau_0}\right) \qquad (3.27)$$

(c) *Constant stress rate*

For the same reason as in the case of creep this simply produces a viscous response:

i.e.
$$\frac{dp}{dt} = R \text{ a constant}$$

and hence
$$\frac{de}{dt} = \frac{R}{E} + \frac{Rt}{\mu}$$

(d) *Constant strain rate*
Because the relaxation behaviour was satisfactory the constant strain rate is also of the required form. It is of interest to solve the equation in two ways:
(i) Differential equation approach: $de/dt = T$ a constant and hence from equation (3.26) we have:

$$T = \frac{1}{E}\frac{dp}{dt} + \frac{p}{\mu}$$

Integrating we have:

$$-\frac{t}{\tau_0} = \ln(\mu T - p) + A$$

and where A is a constant and since $p = 0$ at $t = 0$,
$$p(t) = \mu T\left[1 - \exp\left(-\frac{t}{\tau_0}\right)\right]$$

(ii) Convolution integral approach: in general the integral is:

$$p(t) = \int_0^t M(t-\tau)\frac{de}{d\tau}\,d\tau$$

and hence:

$$p(t) = \int_0^t E\exp\left(-\frac{t-\tau}{\tau_0}\right) T\,d\tau$$

from equation (3.27),

and
$$p(t) = ET\left[\tau_0 \exp\left(-\frac{t-\tau}{\tau_0}\right)\right]_0^t$$

i.e.
$$p(t) = \mu T\left[1 - \exp\left(-\frac{t}{\tau_0}\right)\right]$$

as before.

3.5.2. *The Voigt model*

Another simple model is the Voigt, or Kelvin, model which consists of the same two elements as the Maxwell model but placed in parallel (see Figure 3.5(d)). In this case the stress p is made up of that in the elastic and viscous elements such that:

$$p = p_1 + p_2$$

and as:

$$e = \frac{p_1}{E} \text{ and } \frac{de}{dt} = \frac{p_2}{\mu}$$

we have:

$$\frac{p}{\mu} = \frac{E}{\mu} e + \frac{de}{dt} \tag{3.28}$$

Again all the coefficients in equation (3.24) are zero except:

$$b_0 = \frac{1}{\mu}, \; a_0 = \frac{E}{\mu} \text{ and } a_1 = 1$$

(a) *Creep*
If we have $p = p_0$ a constant, then equation (3.28) becomes:

$$\frac{p_0}{\mu} = \frac{E}{\mu} e + \frac{de}{dt}$$

which gives:

$$e = \frac{p_0}{E}\left[1 - \exp\left(-\frac{t}{\tau_0}\right)\right]$$

when $e = 0$ at $t = 0$. In this case τ_0 is often referred to as the retardation time. This is a satisfactory form as $e \to p_0/E$ as $t \to \infty$ at a decreasing rate, shown in Figure 3.6(a). Reference to the model shows that this is the expected behaviour because the spring is restrained by the viscous element during the initial loading and then is allowed to extend as the viscous element extends. This goes on until the spring is fully extended at an elastic strain of $e = p_0/E$. The creep compliance function is:

$$C(t) = \frac{1}{E}\left[1 - \exp\left(-\frac{t}{\tau_0}\right)\right] \tag{3.29}$$

(b) *Relaxation*
If $e = e_0$ a constant, then no relaxation can take place because the spring is held at a constant strain. This is shown by the equation:

$$\frac{p}{\mu} = \frac{Ee_0}{\mu}$$

i.e.

$$p = Ee_0,$$

which represents linear elastic behaviour.

(c) *Constant stress rate*
The convolution integral in this case becomes:

$$e(t) = \int_0^t \frac{1}{E}\left[1-\exp\left(-\frac{t-\tau}{\tau_0}\right)\right] R \, d\tau$$

using the creep compliance function from equation (3.29) and hence:

$$e(t) = \frac{R}{E}\left[t-\tau_0\left\{1-\exp\left(-\frac{t}{\tau_0}\right)\right\}\right]$$

(d) *Constant strain rate*
As in relaxation we obtain an elastic reponse with an additional viscous term:

$$p = ETt+\mu T$$

It is clear that the Maxwell model describes relaxation but not creep while the Voigt model describes creep but not relaxation. Their individual deficiencies may be overcome by combining them.

3.5.3. *The standard linear solid*
(a) *Creep and relaxation*
The simplest combination is shown in Figure 3.5(e) and consists of a Voigt model with a spring in series. If we let the modulus of the additional spring be E_1 and of the spring in the Voigt model be E_2 as shown then the differential equation is:

$$\frac{de}{dt}+\frac{E_2 e}{\mu} = \frac{E_1+E_2}{\mu}\frac{1}{E_1}p+\frac{1}{E_1}\frac{dp}{dt} \qquad (3.30)$$

from which it can be seen that both creep and relaxation result in satisfactory relationships. The creep compliance function is:

$$C(t) = \frac{1}{E_1}+\frac{1}{E_2}\left[1-\exp\left(-\frac{t}{\tau_2}\right)\right] \qquad (3.31)$$

where $\tau_2 = \mu/E_2$ and the relaxation modulus becomes:

$$M(t) = E_1-\frac{E_1{}^2}{E_1+E_2}\left[1-\exp\left(\frac{t}{\tau_1}\right)\right] \qquad (3.32)$$

where $\tau_1 = \mu/(E_1+E_2)$.

Thus adequate qualitative representations of both relaxation and creep behaviour are obtained in a single model and the time parameters for the two responses, τ_1 and τ_2, are different. The forms of the functions are of interest and they are shown schematically in Figure 3.7. Both $M(t)$ and $1/C(t)$ tend to E_1 as $t \to 0$ and to $E_1E_2/(E_1+E_2)$ as $t \to \infty$ which is apparent from an inspection of the model. The curves are sigmoidal in form and their shape depends on τ_1 and τ_2. The bulk of the change takes place over the time range where $t \simeq \tau_1$ or τ_2

and thus, as $\tau_1 < \tau_2$ in this case, the $M(t)$ curve will always be the lower of the two. This shape of curve gives rise to the concept of the short time modulus E_1 and the long time modulus $E_1E_2/(E_1+E_2)$ which is sometimes used.

If any attempt is to be made to fit these curves to real data then it is apparent that the fixed shape allows little flexibility. It is therefore usual to consider

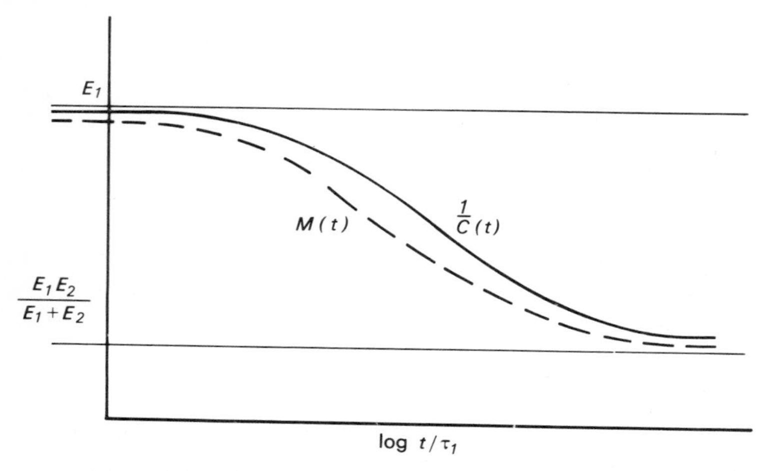

Figure 3.7 Modulus functions for the standard linear solid.

models with very many elements to reproduce some shape and in general a large number of Maxwell models in parallel is used to predict relaxation behaviour giving a function of the form:

$$M(t) = \sum_{i=1}^{n} E_i \exp\left(-\frac{t}{\tau_i}\right)$$

and in the limit:

$$M(t) = \int_0^\infty E(\tau) \exp\left(-\frac{t}{\tau}\right) d\tau$$

This is the relaxation spectrum concept in which changes in $M(t)$ may be linked with relaxation times within the spectrum. The long time modulus for this system tends to zero. Similarly, a set of Voigt models in series gives an equation of the form:

$$C(t) = \int_0^\infty \frac{1}{E(\tau)}\left[1-\exp\left(-\frac{t}{\tau}\right)\right] d\tau$$

which gives a spectrum of times for creep behaviour. This approach has been commonly used in attempts to relate molecular structure to macroscopic properties, and in particular to relate the various relaxation times to identifiable molecular motions within the structure.

Returning to the standard linear solid, we find that by using $C(t)$ and $M(t)$ the expressions for constant stress rate and constant strain rate may be derived. For constant stress rate this may be written in the form of a modulus:

$$E_{\text{stress}} = \frac{E_1 E_2}{E_1 + E_2}\left[\frac{1}{1 - \frac{\tau_1}{t}\left(1 - \exp\left(-\frac{t}{\tau_2}\right)\right)}\right]$$

and for constant strain rate:

$$E_{\text{stress}} = \frac{E_1 E_2}{E_1 + E_2}\left[1 + \frac{E_1}{E_2}\frac{\tau_1}{t}\left(1 - \exp\left(-\frac{t}{\tau_1}\right)\right)\right]$$

Equations (3.31) and (3.32) may be written to give comparable forms:

$$E_{\text{creep}} = \frac{E_1 E_2}{E_1 + E_2}\frac{1}{\left[1 - \frac{E_1}{E_1 + E_2}\exp\left(-\frac{t}{\tau_2}\right)\right]}$$

and

$$E_{\text{relax}} = \frac{E_1 E_2}{E_1 + E_2}\left[1 + \frac{E_1}{E_2}\exp\left(-\frac{t}{\tau_1}\right)\right]$$

These four expressions give the slopes of the isochronous curves for the four systems of loading for any given value of t. All tend to the long time modulus for $t \to \infty$ but for other values of t they are in a fixed order namely:

$$E_{\text{relax}} < E_{\text{creep}} < E_{\text{strain}} < E_{\text{stress}}$$

For example if we let $E_1 = E_2 = E, t = \tau_2$ and noting that $\tau_1 = \tau_2 (E_2/E_1 + E_2)$ then:

$$E_{\text{stress}} = E\frac{1}{1 + \exp(-1)}, \quad E_{\text{strain}} = E\frac{3 - \exp(-2)}{4}$$

$$E_{\text{creep}} = E\frac{1}{2 - \exp(-1)}, \quad E_{\text{relax}} = E\frac{1 + \exp(-2)}{2}$$

giving the values:

$$E_{\text{relax}} = 0{\cdot}568E,\ E_{\text{creep}} = 0{\cdot}612E,\ E_{\text{strain}} = 0{\cdot}717E, \text{ and } E_{\text{stress}} = 0{\cdot}732E.$$

As previously stated these differences are the major manifestations of time dependent behaviour and the order predicted is that for all visco-elastic materials.

(b) *Sinusoidal loading*

Sinusoidal loading is of a considerable practical interest and it is convenient to consider it in relation to the standard linear solid. If we consider first a cyclic stress input:

$$p = p_0 \sin \omega t$$

where ω = the angular frequency then this function can be used together with $C(t)$ from equation (3.31) to give the convolution integral:

$$e(t) = \int_0^t \left[\frac{1}{E_1} + \frac{1}{E_2}\left\{1 - \exp\left(-\frac{t-\tau}{\tau_2}\right)\right\}\right] p_0 \omega \cos \omega\tau \, d\tau$$

which may be evaluated to give:

$$e(t) = p_0\left(\frac{1}{E_1} + \frac{1}{E_2}\frac{1}{1+\omega^2\tau^2{}_2}\right)\sin \omega t - \frac{p_0}{E_2}\frac{\omega\tau_2}{1+\omega^2\tau^2{}_2}\cos \omega t$$
$$+\frac{p_0}{E_2}\frac{\omega\tau_2}{1+\omega^2\tau^2{}_2}\exp\left(-\frac{t}{\tau_2}\right)$$

It can be seen that the first term may be regarded as in-phase with the sinusoidal input and the second term as out-of-phase. The third term is a transitory effect which tends to zero for long times and is usually ignored. The results are conventionally written in terms of in-phase and out-of-phase moduli designated E' and E'' respectively which in this case are:

$$\left.\begin{aligned} \frac{1}{E'} &= \frac{1}{E_1} + \frac{1}{E_2}\frac{1}{1+\omega^2\tau^2{}_2} \\ \text{and} \qquad \frac{1}{E''} &= \frac{1}{E_2}\frac{\omega\tau_2}{1+\omega^2\tau^2{}_2} \end{aligned}\right\} \tag{3.33}$$

For high frequencies ($\omega\tau_2 \gg 1$), $E' \to E_1$ and $E'' \to \infty$ while for very low frequencies, ($\omega\tau_2 \ll 1$), $E' \to E_1E_2/(E_1+E_2)$ and $E'' \to \infty$. Again an inspection of the model confirms these results.

A further parameter of practical importance is the loss factor written as:

$$\tan \delta = \frac{\text{Out-of-phase component}}{\text{In-phase component}}$$

which in this case becomes:

$$\tan \delta_1 = \frac{E_1\omega\tau_2}{(E_1+E_2)+E_2\omega^2\tau^2{}_2} \tag{3.34}$$

Two special cases of interest are the Maxwell model with $E_2 = 0$ and hence:

$$E' = E_1, \; E'' = \omega\mu$$

and the Voigt model with $E_1 = \infty$ giving:

$$E' = E_2\,(1+\omega^2\tau_2{}^2), \; E'' = E_2\frac{1+\omega^2\tau_2{}^2}{\omega\tau_2}$$

If the input is of the form:

$$e = e_0 \sin \omega t$$

which may be regarded as strain cycling compared with stress cycling used previously, then $M(t)$, equation (3.32), must be used in the convolution integral and the results become:

$$p(t) = e_0 \frac{E_1 E_2}{E_1+E_2}\left(1+\frac{E_1}{E_2}\frac{\omega^2\tau_1{}^2}{1+\omega^2\tau_1{}^2}\right)\sin \omega t +$$

$$e_0 \frac{E_1{}^2}{E_1+E_2}\frac{\omega\tau_1}{1+\omega^2\tau_1{}^2}\cos \omega t + \frac{e_0\omega\tau_1}{1+\omega^2\tau_1{}^2}\frac{E_1{}^2}{E_1+E_2}\exp\left(-\frac{t}{\tau_1}\right)$$

As before, for long times the third term tends to zero and we have:

$$\left.\begin{aligned} E' &= \frac{E_1 E_2}{E_1+E_2}\left[1+\frac{\omega^2\tau_1{}^2}{1+\omega^2\tau_1{}^2}\frac{E_1}{E_2}\right] \\ E'' &= \frac{E_1{}^2}{E_1+E_2}\frac{\omega\tau_1}{1+\omega^2\tau_1{}^2}\end{aligned}\right\} \qquad (3.35)$$

As would be expected from the difference in creep and relaxation moduli, the moduli from stress cycling and from strain cycling are different. For example, if both expressions for E' are written in terms of τ_1, we have, from equations (3.35):

$$E'_{\text{strain}} = \frac{E_1 E_2}{E_1+E_2}\left[1+\frac{\omega^2\tau_1{}^2}{1+\omega^2\tau_1{}^2}\frac{E_1}{E_2}\right]$$

and from equations (3.33):

$$E'_{\text{stress}} = \frac{E_1 E_2}{E_1+E_2}\left[1+\frac{\omega^2\tau_1{}^2}{\dfrac{E_2}{E_1+E_2}+\omega^2\tau_1{}^2}\frac{E_1}{E_2}\right]$$

The only difference is in the term $E_2/(E_1+E_2)$ and clearly $E'_{\text{strain}} < E'_{\text{stress}}$ as expected. The two expressions are shown plotted in Figure 3.8 as a function of $\log 1/\omega\tau_1$ and for low frequencies, $\omega\tau_1 \ll 1$, both tend to a value of $E_1E_2/(E_1+E_2)$ which corresponds to the long time response as shown in Figure 3.7 for other histories. Similarly at high frequencies, i.e. short times, both tend to a value of E_1. The loss factor is given by:

$$\tan\delta_2 = \frac{E_1\omega\tau_1}{E_2+(E_1+E_2)\,\omega^2\tau_1{}^2} = \frac{E_1\omega\tau_2\dfrac{E_2}{E_1+E_2}}{E_2+(E_1+E_2)\,\omega^2\tau_2{}^2\dfrac{E_2{}^2}{(E_1+E_2)^2}} = \tan\delta_1$$

Thus the loss factor expression is the same for both types of loading and is shown in Figure 3.8. At both high and low frequencies it tends to zero and there is a maximum of:

$$\tan\hat{\delta} = \frac{1}{2}\frac{E_1}{E_2}\sqrt{\frac{E_2}{E_1+E_2}}$$

for
$$\omega\tau_1 = \sqrt{\frac{E_2}{E_1+E_2}}$$

The special cases for strain cycling are: for the Maxwell model ($E_2 = 0$)

$$E' = E_1 \frac{\omega^2\tau_1{}^2}{1+\omega^2\tau_1{}^2}, \quad E'' = E_1 \frac{\omega\tau_1}{1+\omega^2\tau_1{}^2}$$

and for the Voigt model ($E_1 = \infty$)

$$E' = E_2, \quad E'' = \omega\mu$$

The result for the Maxwell model is used to define relaxation spectra from E' and E'' as functions of ω in a similar manner to creep and relaxation data so that:

$$E' = \int_0^\infty E(\tau) \frac{\omega^2\tau^2}{1+\omega^2\tau^2} \,d\tau \quad \text{and} \quad E'' = \int_0^\infty E(\tau) \frac{\omega\tau}{1+\omega^2\tau^2} \,d\tau$$

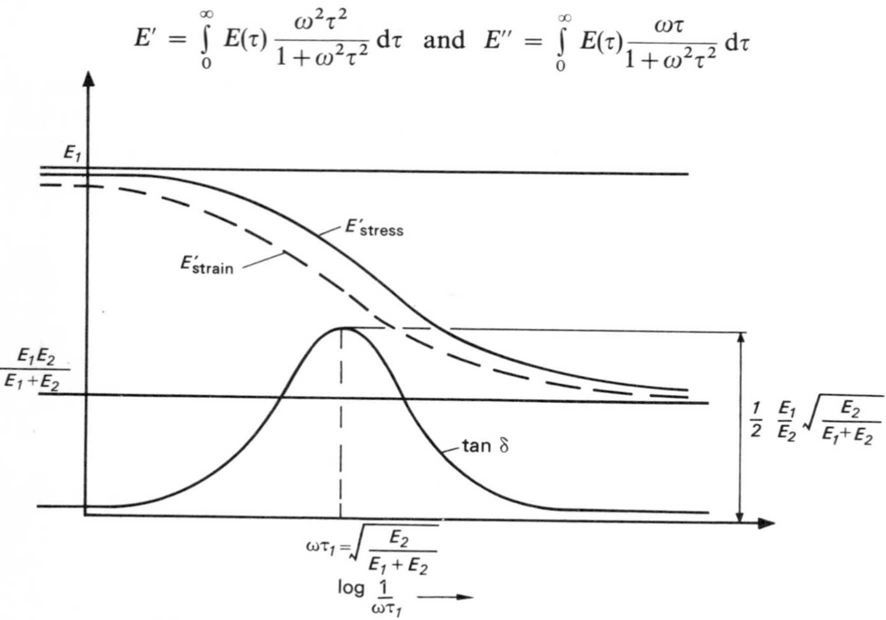

Figure 3.8 Sinusoidal loading of a standard linear solid.

A clearer view of the significance of the terms may be derived by considering the energy absorbed in a full cycle ($t = 2\pi/\omega$) which is given by:

$$W = \int_0^{\frac{2\pi}{\omega}} p \frac{de}{dt} \,dt$$

For stress cycling ($p = p_0 \sin \omega t$) we have:

$$e(t) = \frac{p_0}{E'} \sin \omega t - \frac{p_0}{E''} \cos \omega t$$

ignoring the transient effects.

Thus the energy is given by:

$$W_{\text{stress}} = \int_0^{2\pi} \left[\frac{p_0}{E_1} \cos \omega t + \frac{p_0}{E''} \sin \omega t \right] p_0 \sin \omega t \, \mathrm{d}(\omega t)$$

Evaluating this integral shows that the in-phase term goes to zero as expected, because it is the elastic form of response, while the out-of-phase gives an absorbed value of:

$$W_{\text{stress}} = \frac{\pi p_0{}^2}{E''} = \frac{\pi p_0{}^2}{E_2} \frac{\omega \tau_2}{1+\omega^2 \tau_2{}^2}$$

Similarly for strain cycling we have:

$$W_{\text{strain}} = \pi e_0^2 \, E'' = \pi e_0^2 \frac{E_1{}^2}{E_1+E_2} \frac{\omega \tau_1}{1+\omega^2 \tau_1{}^2}$$

Clearly the energy absorption has a similar dependence on ω although the values will be different.

3.5.4. *Power law materials*

There are many polymers for which, when either the stress in a relaxation test or the strain in creep test is plotted against time on a log-log basis, a good linear fit to the data is obtained. This implies a general time dependence of the form t^n where n is a number generally less than unity. Thus for creep we may write a general relationship of the form:

$$C(t) = \frac{t^n}{E_0} \tag{3.36}$$

for the creep compliance where E_0 is a constant. This form is, of course, only of limited applicability, but is practically very useful.

To determine the expression for $M(t)$ for this material we must consider the transform relationship equation (3.22):

$$\overline{M} = \frac{1}{s^2 \overline{C}}$$

for which it is necessary to determine $\overline{C}$. This involves the use of the gamma function which we define as:

$$\Gamma(m) = \int_0^\infty x^{m-1} \, \mathrm{e}^{-x} \, \mathrm{d}x$$

The transform of C is:

$$\overline{C} = \frac{1}{E_0} \int_0^\infty t^n \, \mathrm{e}^{-st} \, \mathrm{d}t$$

and if $x = st$:

$$\overline{C} = \frac{1}{E_0}\frac{1}{s^{1+n}}\int_0^\infty x^n e^{-x}\,dx$$

$$\overline{C} = \frac{\Gamma(1+n)}{E_0\, s^{(1+n)}}$$

and hence:

$$\overline{M} = \frac{E_0}{s^{(1-n)}}\frac{1}{\Gamma(1+n)}$$

Now if we consider the function:

$$\int_0^\infty t^{-n} e^{-st}\,dt = \frac{1}{s^{(1-n)}}\Gamma(1-n)$$

and hence we may invert $\overline{M}$ to give:

$$M(t) = \frac{E_0\, t^{-n}}{\Gamma(1+n)\,\Gamma(1-n)}$$

Reference to the properties of the gamma function gives the result:

$$\Gamma(1+n)\,\Gamma(1-n) = \frac{n\pi}{\sin n\pi}$$

and hence

$$M(t) = \left(\frac{\sin n\pi}{n\pi}\right) E_0\, t^{-n} \tag{3.37}$$

Clearly this representation is not valid at $t \to 0$ but is quite good over limited ranges. The isochronous moduli may be written as:

$$E_{\text{creep}} = E_0\, t^{-n}$$

$$E_{\text{relax}} = \left(\frac{\sin n\pi}{n\pi}\right) E_0\, t^{-n}$$

Their difference is due to the factor $(\sin n\pi/n\pi)$ and it is useful to consider numerical values in the range $0 < n < 0{\cdot}2$ since n is rarely greater than 0·2.

n	$\dfrac{\sin n\pi}{n\pi}$
0	1·000
0·05	0·995
0·10	0·984
0·15	0·964
0·20	0·936

For typical values of n around 0·10 the error in assuming $E_{\text{creep}} = E_{\text{relax}}$ is less than 2%. This fact has important practical consequences and will be discussed later (section 3.7.1).

Constant rate stressing may be analysed using the convolution integral:

$$e(t) = \int_0^t \frac{(t-\tau)^n}{E_0} \frac{\mathrm{d}p}{\mathrm{d}\tau} \mathrm{d}\tau$$

which gives the result:

$$e(t) = \frac{pt^n}{E_0(1+n)}$$

and

$$E_{\text{stress}} = (1+n) E_0 t^{-n}$$

Constant rate straining gives:

$$p(t) = \left(\frac{\sin n\pi}{n\pi}\right) \int_0^t E_0 (t-\tau)^{-n} \frac{\mathrm{d}e}{\mathrm{d}\tau} \mathrm{d}\tau$$

which reduces to:

$$E_{\text{strain}} = \left(\frac{\sin n\pi}{n\pi}\right) \frac{E_0}{(1-n)} t^{-n}$$

For $n = 0{\cdot}1$ we have the values
$E_{\text{relax}} = 0{\cdot}984\ E_0\, t^{-n}$, $E_{\text{creep}} = 1{\cdot}000\ E_0\, t^{-n}$, $E_{\text{strain}} = 1{\cdot}092\ E_0\, t^{-n}$ $E_{\text{stress}} = 1{\cdot}100\ E_0\, t^{-n}$, which is the same order as observed previously for the standard linear solid. It should be noted, however, that with this realistic representation the total variation is within $\pm 6\%$ of the mean.

This type of material also provides a useful example of the use of the convolution integral method of predicting responses to different stress inputs. For example we may consider a ramp input at constant rate up to a stress of p_0 at a time t_0 and the stress remaining constant thereafter as shown in Figure 3.9(a). The response for a time $t > t_0$ is given by:

$$e(t) = \frac{p_0}{t_0} \int_0^{t_0} \frac{(t-\tau)^n}{E_0} \mathrm{d}\tau = \frac{p_0[t^{1+n} - (t-t_0)^{1+n}]}{t_0(1+n)E_0}$$

i.e.

$$e(t) = \frac{p_0 t_0^{\,n}}{(1+n)E_0} \left(\frac{t}{t_0}\right)^{1+n} \left[1 - \left(1 - \frac{t_0}{t}\right)^{1+n}\right]$$

For $t \gg t_0$ this may be expanded as a series:

$$e(t) = \frac{p_0 t^n}{E_0} \left[1 - \frac{n}{2}\left(\frac{t_0}{t}\right) + \ldots\right]$$

showing, as expected, that for this condition the result tends to become a creep response.

The response to a sawtooth input, Figure 3.9(b), may be obtained by subtracting the creep response from t_0 to t:

$$e(t) = \frac{p_0 t_0^n}{(1+n)E_0}\left(\frac{t}{t_0}\right)^{1+n}\left[1-\left(1-\frac{t_0}{t}\right)^{1+n}\right] - \frac{p_0(t-t_0)^n}{E_0}$$

$$= \frac{p_0 t^n}{(1+n)E_0}\left[\frac{t}{t_0} - \frac{t}{t_0}\left(1-\frac{t_0}{t}\right)^{1+n} - (1+n)\left(1-\frac{t_0}{t}\right)^n\right]$$

i.e.

$$e(t) = \frac{p_0 t^n}{E_0}\left[\frac{n}{2}\frac{t_0}{t} + \frac{5}{6}n(1-n)\left(\frac{t_0}{t}\right)^2 + \cdots\right]$$

so that $e(t) \to 0$ for long times as expected.

The response to a ramp input followed by ramp unloading, Figure 3.9(c), may be deduced from:

$$e(t) = \frac{p_0}{t_0}\int_0^{t_0} \frac{(t-\tau)^n}{E_0}\,\mathrm{d}\tau - \frac{p_0}{t_0}\int_{t_0}^{2t_0} \frac{(t-\tau)^n}{E_0}\,\mathrm{d}\tau$$

$$= \frac{p_0}{t_0}\,\frac{1}{(1+n)E_0}\left[t^{1+n} - (t-t_0)^{1+n} + (t-t_0)^{1+n} - (t-2t_0)^{1+n}\right]$$

$$= \frac{p_0 t^n}{(1+n)E_0}\left[\frac{t}{t_0} - \frac{t}{t_0}\left(1-2\frac{t_0}{t}\right)^{1+n}\right]$$

3.6. General linear visco-elasticity equations

We shall now consider the equations for describing general three-dimensional systems of stress and strain in linear visco-elasticity. The discussion will be limited to isotropic materials for which, for example, the elasticity equation for the x direction has the form:

$$e_{xx} = \frac{1}{E}\left[p_{xx} - \nu(p_{yy} + p_{zz})\right]$$

and for one dimension:

$$e_{xx} = \frac{p_{xx}}{E}$$

The equivalent linear visco-elasticity form is:

$$e_{xx}(t) = \int_0^t C(t-\tau)\frac{\mathrm{d}p_{xx}}{\mathrm{d}\tau}\,\mathrm{d}\tau$$

and the general equation becomes:

$$e_{xx}(t) = \int_0^t C(t-\tau)\frac{\mathrm{d}p_{xx}}{\mathrm{d}\tau}\,\mathrm{d}\tau - \int_0^t R(t-\tau)\frac{\mathrm{d}}{\mathrm{d}\tau}(p_{yy} + p_{zz})\,\mathrm{d}\tau$$

where $R(t)$ is a time dependent material property. Clearly the equations are derived with the condition that:

$$C(t) \equiv \frac{1}{E} \qquad \text{and} \quad R(t) \equiv \frac{\nu}{E}$$

By taking transforms we have:

$$\begin{aligned}
\bar{e}_{xx} &= \overline{C}(s)\, \bar{p}_{xx}\, s - \overline{R}(s)\, (\bar{p}_{yy} + \bar{p}_{zz})\, s \\
\bar{e}_{yy} &= \overline{C}(s)\, \bar{p}_{yy}\, s - \overline{R}(s)\, (\bar{p}_{xx} + \bar{p}_{zz})\, s \\
\bar{e}_{zz} &= \overline{C}(s)\, \bar{p}_{zz}\, s - \overline{R}(s)\, (\bar{p}_{xx} + \bar{p}_{yy})\, s
\end{aligned}$$

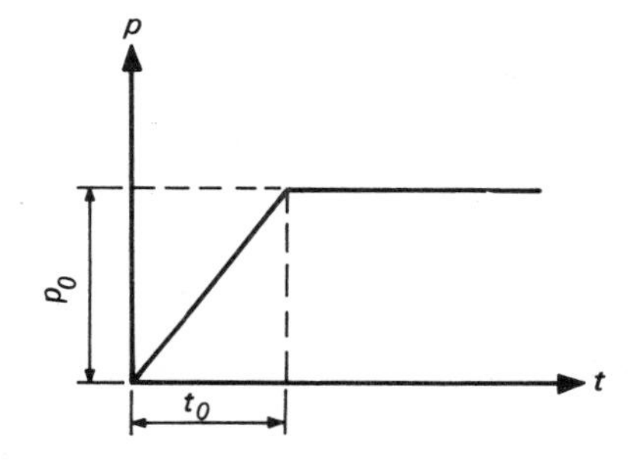

(a) Ramp loading

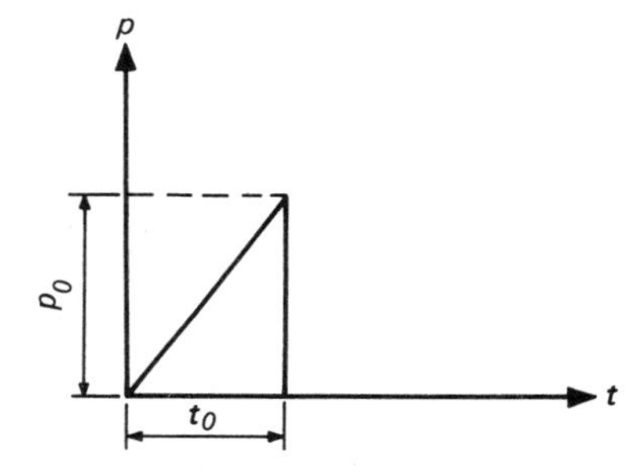

(b) Sawtooth loading

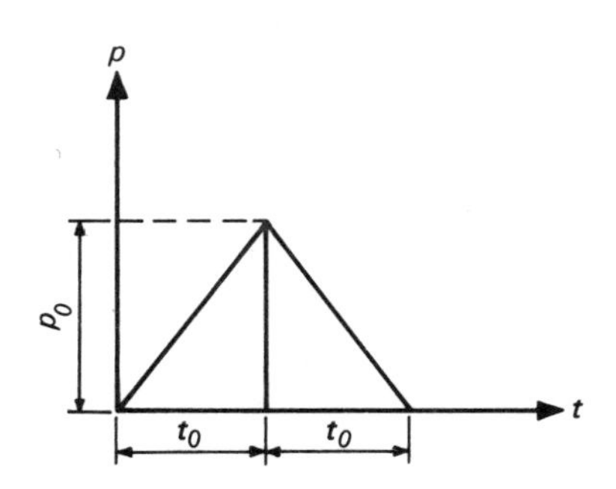

(c) Ramp loading - unloading

Figure 3.9 Loading histories.

Thus the visco-elasticity equations are formed from the elasticity equations by replacing the terms in the elasticity equations with the transforms multiplied by the variable s. The function $R(t)$ can be regarded as:

$$s\overline{R}(s) = \frac{s\bar{\nu}}{\overline{\overline{E}}}$$

Similarly the relationships between moduli now hold between the transforms and we have:

$$s\overline{G} = \frac{s\overline{E}}{2(1+s\overline{v})}$$

and
$$s\overline{E} = 3s\overline{K}(1-2s\overline{v})$$

There is no problem with creep and relaxation behaviour in this form because:

$$s\overline{C} = \frac{1}{s\overline{M}} = \frac{1}{s\overline{E}}$$

i.e. the modulus remains the same for both, providing it is considered as a transformed variable.

This property is useful in giving a general method of solving linear viscoelasticity problems in that, as the stress–strain relationships are written as transforms, so we may write the equilibrium, compatibility and boundary conditions as transforms and obtain a solution as if for an elastic case, in terms of transforms. This must then be inverted to give the solution in terms of the true variables. For example we may write down the equilibrium equations as:

$$\frac{\partial \overline{p}_{xx}}{\partial x} + \frac{\partial \overline{p}_{xy}}{\partial y} + \frac{\partial \overline{p}_{xz}}{\partial z} + \overline{X} = 0, \text{ etc.}$$

with the strains in the form:

$$\overline{e}_{xx} = \frac{\partial \overline{u}}{\partial x}, \text{ etc.}$$

to give a set of equations in terms of transforms.

A special case of set conditions, called proportional loading, is when the spatial distributions are not functions of time, i.e. when the time dependence with respect to x, y and z is identical. For example an applied force F in the general case must have the form:

$$F = \alpha(x, y, z, t)$$

but the special case becomes:

$$F = \phi(x, y, z).\psi(t)$$

Taking transforms gives:

$$\overline{F}(s) = \phi.\overline{\psi}(s)$$

as ϕ is not a function of t. The solution for this system is the product of that for ϕ and that for $\overline{\psi}(s)$ which means, in effect, that the elasticity solution is simply multiplied by some function of t. The stress and strain distributions remain the same but their magnitude is a function of time.

This method of solution, via transforms, provides a general method for any system and is useful in that, providing the material functions are known, it is simply a matter of solving an elasticity problem and then inverting the solution. Both processes can be performed numerically which is useful now that computers are widely available.

3.7. Practical simplifications

The application of visco-elasticity analysis to the solution of practical problems presents several difficulties. In whatever form the theory is used the equations tend to be complicated and in addition the data required is often difficult to obtain experimentally with any accuracy. The general problem of defining the applied loads with any degree of precision must also be overcome before a solution is obtained. In the light of these practical constraints, therefore, it is often useful to consider simplifications in the analysis which will give answers commensurate with the precision of the definition of the problem.

Most problems may be classified as either long or short time in nature and simplifications made accordingly. In long time problems the loads are usually constant, change very slowly or at infrequent intervals and may be regarded as modifications of creep behaviour. Most problems involving short time effects are concerned with cyclic loading such as vibrations and can be analysed using the concept of in-phase and out-of-phase moduli. There are a few problems in which load changes in short times and their effects are of interest, e.g. solid fuel rocket motors, in which case a full visco-elasticity solution must be attempted. Their number is not large, however, and if we exclude the vibration problems almost all engineering applications are long time in nature. Some useful simplifications are possible in long time problems and they are worthy of further consideration.

3.7.1. *Long-time problems*

In many polymers both creep and relaxation behaviour may be represented with a power law time dependence of the form t^n where n is about 0·1. The analysis of such materials was considered in section 3.5.4 and it was noted that the manifestation of visco-elastic effects in the differences of isochronous moduli from different loading histories was such that variations of about n were predicted. For example for $n = 0{\cdot}1$ the constant rate of stress modulus was 1·1 times the creep modulus and the relaxation modulus was 0·984 times the creep modulus. The reason for the minor nature of these effects is that the parameters vary very slowly with time and, since the effects arise from change rather than absolute values, they are small.

For many practical long time problems accuracies of about 10% are more than adequate, particularly since most loads cannot be defined to much better than 10%, and it is therefore reasonable to ignore the time history effects and simply treat the creep modulus as a time dependent elastic modulus. In this case conventional linear-elasticity solutions may then be used with the time dependent modulus replacing Young's modulus in the equations and time dependent versions of the other elastic constants also being used. The practical advantages of this are obvious since no additional analysis of problems is necessary and it is only required to examine the effect of changing elastic constants.

The general elasticity equations become:

$$e_{xx}(t) = \frac{1}{E(t)}\left[p_{xx} - \nu(t)\,(p_{yy} + p_{zz})\right], \text{ etc.}$$

It is difficult to obtain $\nu(t)$ experimentally and two assumptions are often made.

(1) *Constant* ν Poisson's ratio does vary with time but it often does not affect results appreciably. For this reason a value in the region of 0·3–0·4 is used. It is possible to derive a time dependent shear modulus on this basis using:

$$G(t) = \frac{E(t)}{2(1+\nu)}$$

(2) *Constant* K Perhaps more realistic physically is the assumption of constant bulk modulus since the changes in K are generally much less than those in E. A varying ν may be derived:

i.e.

$$\nu(t) = \frac{1}{2}\left[1 - \frac{E(t)}{3K}\right]$$

3.8. Non-linear theory

At the time of writing this topic presents a difficult problem. It is possible to write down a large number of plausible non-linear theories and some were mentioned in section 3.3 but it is not within the scope of this book to discuss the many published versions of the theory. Engineering stress analysis requires a workable theory which gives a reasonable representation of non-linear effects and as yet none of those proposed has been shown to be applicable in other than very restricted conditions. Some have great generality, in particular the multiple integral form, but unless some simplifications can be introduced they are not of great practical interest. A useable theory will be taken here to mean one which may be manipulated without undue complication and gives an improvement on the linear theory where necessary.

Anything other than a single integral form is unlikely to be easy to use and for this reason some modified form of the single integral seems most promising. The one which has had the greatest success so far is termed the Leaderman form which was derived earlier, equation (3.10), in the form:

$$e(t) = \int_0^t C(t-\tau)\,\frac{\mathrm{d}f(p)}{\mathrm{d}\tau}\,\mathrm{d}\tau$$

This form is only applicable to materials whose behaviour is such that time and stress or strain effects are separable, i.e. in creep:

$$e(t) = C(t)f(p)$$

For relaxation:

$$p(t) = M(t)\,g(e)$$

and hence by a similar analysis we have:

$$p(t) = \int_0^t M(t-\tau)\frac{\mathrm{d}g(e)}{\mathrm{d}\tau}\,\mathrm{d}\tau$$

Simple superposition is assumed in this context and the equations may be regarded as describing a linear material in which stress is defined as $f(p)$ and strain as $g(e)$. However the equations are not truly linear because to achieve this it would be necessary to have f and g satisfying the conditions:

$$g(t) = C(t)f \qquad \text{and} \quad f(t) = M(t)\,g$$

in which case the previously defined transform relationship between M and C would be applicable. In general, it is simpler to regard stress and strain history systems as separate and determine the corresponding functions independently.

The evaluations of the integrals for any stressing or straining history may be performed numerically or by various approximate methods. For example if we consider a strain history and take a relaxation modulus:

$$M(t) = E_0 t^{-n}$$

then

$$p(t) = \int_0^t E_0(t-\tau)^{-n}\frac{\mathrm{d}g}{\mathrm{d}\tau}\,\mathrm{d}\tau$$

Considering a typical form of g:

$$g = e - \alpha e^2$$

where α = constant, we may solve for the case of constant rate straining.

$$e = Rt$$

i.e.

$$p(t) = \int_0^t E_0(t-\tau)^{-n}(R - 2R^2\alpha\tau)\,\mathrm{d}\tau$$

which, on integrating by parts, gives:

$$p(t) = E_0 t^{-n}\left[\frac{e}{1-n} - \frac{2\alpha e^2}{(1-n)(2-n)}\right]$$

The first term is the linear solution and the effect of the non-linear part may be illustrated by considering typical values; i.e. $n = 0{\cdot}1$, $\alpha = 5$ for the two functions of g:

$$g_{\text{relax}} = e - \alpha e^2 = e - 5e^2$$

$$g_{\text{strain}} = \frac{e}{1-n} - \frac{2\alpha e^2}{(1-n)(2-n)} = 1{\cdot}11e - 5{\cdot}85e^2$$

which are compared below for strains up to 0·10.

e	g_{relax}	g_{strain}
0·005	0·0049	0·0054
0·010	0·0095	0·0105
0·020	0·0180	0·0199
0·050	0·0375	0·0409
0·100	0·0500	0·0525

The effect can be seen to be similar to the linear case in that g_{strain} is higher than g_{relax} and the differences are of the same order as the linear theory.

Bibliography

1. Faupel, J. H. *Engineering Design*, Wiley, New York (1964).
2. Baer, E. (editor) *Engineering Design for Plastics*, Reinhold, New York (1964).
3. Christensen, R. M. *Theory of Visco-elasticity*, Academic Press, New York (1971).
4. Churchill, R. V. *Operational Mathematics*, Second Edition, McGraw-Hill, New York (1956).
5. Goldman, S. *Transformations Calculus and Electrical Transients*, Prentice-Hall Inc., New York (1949).

4 Problems involving bending

4.1. Introduction

This chapter will discuss the analysis of structures which support the loads to which they are subjected by bending. The basis of the analysis will be the study of the stresses and deflections in slender beams which is usually classified as simple beam theory. The subject is covered in great detail in all elementary texts on strength of materials and so no attempt will be made here to cover the standard theory in depth. Instead an outline will be given followed by a selection of topics which are of particular relevance to designing with polymers.

4.2. Summary of simple elastic beam theory

The theory is best considered in terms of the usual steps of a continuum solution.

4.2.1. *Equilibrium*

An element of a beam of length δx is shown in Figure 4.1 and is subjected to a uniformly distributed load w per unit length together with a shear force Q on one face which increases to $Q + \delta Q$ on the opposite face. In addition there is a bending moment M which increases to $M + \delta M$ as shown. The sign convention employed here is as follows:

(i) Direction x along the beam positive from left to right.

(ii) A positive bending moment produces 'hogging', i.e. the one shown in Figure 4.1 is positive.

(iii) A positive shear force is one which produces anticlockwise rotation, i.e. as shown in Figure 4.1.

The equilibrium of the element may be considered first in terms of moments about AD:

$$M - (M + \delta M) + (Q + \delta Q)\,\delta x - w\,\delta x\,\frac{\delta x}{2} = 0$$

and ignoring terms of higher than the first order we have:

$$\frac{dM}{dx} = Q \tag{4.1}$$

Equilibrium of vertical forces gives:

$$Q-(Q+\delta Q)+w\,\delta x = 0$$

and hence:

$$\frac{\mathrm{d}Q}{\mathrm{d}x} = w \tag{4.2}$$

which may be combined with equation (4.1) to give:

$$\frac{\mathrm{d}^2M}{\mathrm{d}x^2} = \frac{\mathrm{d}Q}{\mathrm{d}x} = w \tag{4.3}$$

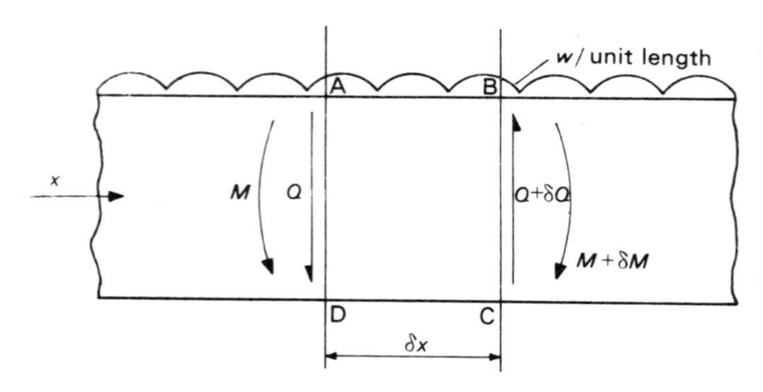

Figure 4.1 Section of a beam.

The direct stress in the x direction, p_{xx}, as shown in Figure 4.2, must satisfy two equilibrium conditions:

(i) Axial load—for beams with an axial load F we have:

$$F = \int_0^d p_{xx}b\,\mathrm{d}y \tag{4.4}$$

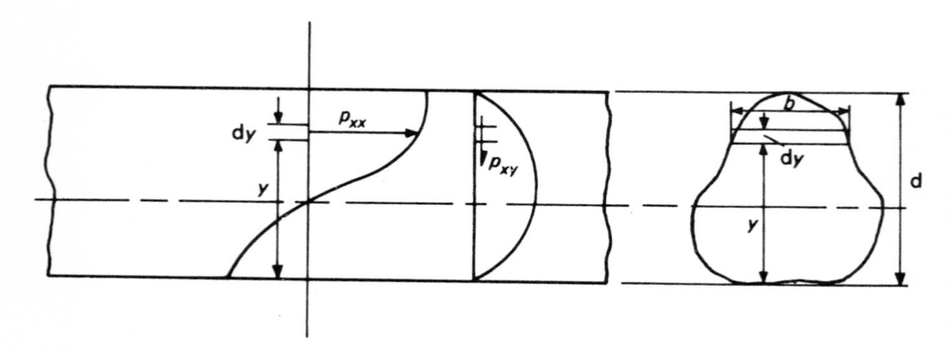

Figure 4.2 The stresses in a beam.

and for zero axial load, $F = 0$.

(ii) Bending moment:

$$M = \int_0^d p_{xx} b y \, \mathrm{d}y \tag{4.5}$$

(iii) Shear force:

$$Q = \int_0^d p_{xy} b \, \mathrm{d}y \tag{4.6}$$

The stress distributions will be considered further in later sections but the above relationships provide a sufficient basis for the following discussion.

4.2.2. *Strain and deflection*

The bending of the element ABCD is shown in Figure 4.3. where it becomes A′B′C′D′. Since, in a bent section, the outer fibres go from tension on one side to compression on the other then there must be a section on which there is zero strain, say $O–O$, called the neutral axis. Suppose the element is bent such that

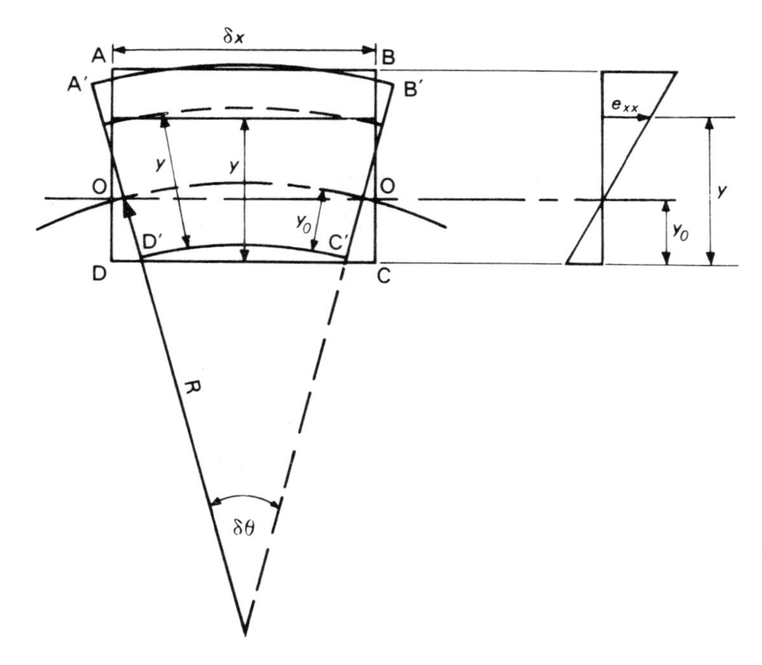

Figure 4.3 The strain in a beam.

the local radius of curvature is R then, if plane sections are assumed to remain plane, an element at a distance y from DC has a strained length given by:

$$\delta x' = [R + (y - y_0)] \, \delta\theta$$

where y_0 = the distance to the neutral axis.

Since O–O is not strained, by definition, we may write:

$$\delta x = R\,\delta\theta$$

The engineer's strain in the x direction is thus given by:

$$e_{xx} = \frac{\delta x' - \delta x}{\delta x}$$

and
$$e_{xx} = \frac{y - y_0}{R} \tag{4.7}$$

i.e. the strain is directly proportional to the distance from the neutral axis giving a linear distribution as shown in Figure 4.3.

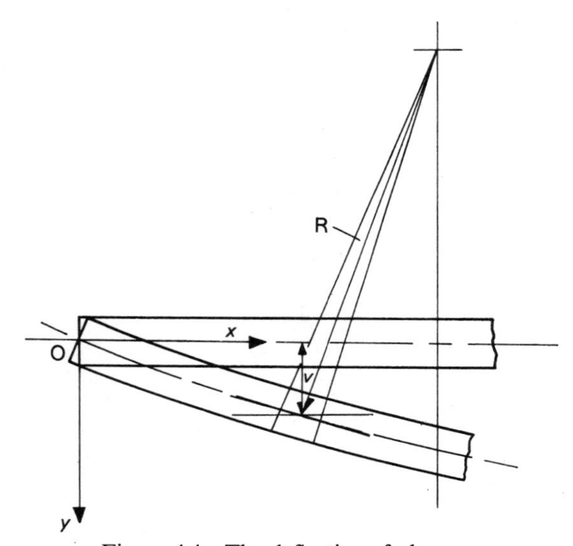

Figure 4.4 The deflection of a beam.

The deflection of a beam may be deduced from Figure 4.4 in which the deflection in the y direction, v, is taken as positive downwards. The radius of curvature R may be written as:

$$R = \frac{\left[1 + \left(\frac{\mathrm{d}v}{\mathrm{d}x}\right)^2\right]^{3/2}}{\frac{\mathrm{d}^2 v}{\mathrm{d}x^2}} \tag{4.8}$$

If it is assumed that the slopes are small then this takes the familiar form:

$$\frac{\mathrm{d}^2 v}{\mathrm{d}x^2} = \frac{1}{R} \tag{4.9}$$

In considering slender beams it is usual to assume that any deflection due to shearing will be negligible and thus is all due to bending. This question will be discussed again in section 4.5.

4.2.3. *Constitutive equations*

Simple theory assumes that the material is isotropic and Hookean. Further it is assumed that the stress p_{yy} may be ignored (see section 4.5) and hence e_{xx} may be deduced from the simple tension relationship:

$$e_{xx} = p_{xx}/E \tag{4.10}$$

4.2.4. *The beam equations*

The results may now be combined to give the familiar beam equations. For example the combination of equations (4.10) and (4.7) gives a linear stress distribution of the form:

$$p_{xx} = \frac{E}{R}(y - y_0) \tag{4.11}$$

and y_0 may be determined from equation (4.4) which gives:

$$\int_0^d b(y - y_0)\,\mathrm{d}y = 0 \tag{4.12}$$

for zero axial load.

The bending moment is deduced from equation (4.5):

$$M = \int_0^d \frac{E}{R}(y - y_0)by\,\mathrm{d}y$$

i.e.

$$M = \frac{E}{R}\int_0^d by(y - y_0)\,\mathrm{d}y \tag{4.13}$$

The integral in this expression will be recognised as the second moment of area about the neutral axis

$$I = \int_0^d by(y - y_0)\,\mathrm{d}y \tag{4.14}$$

and hence:

$$\frac{M}{I} = \frac{E}{R}$$

or

$$\frac{1}{R} = \frac{\mathrm{d}^2 v}{\mathrm{d}x^2} = \frac{M}{EI} \tag{4.15}$$

from equation (4.9).

4.2.5. *Centrally loaded simply supported rectangular beams*

The nature of the assumptions used so far is best illustrated by considering the simple example of a beam shown in Figure 4.5. The beam is simply supported at

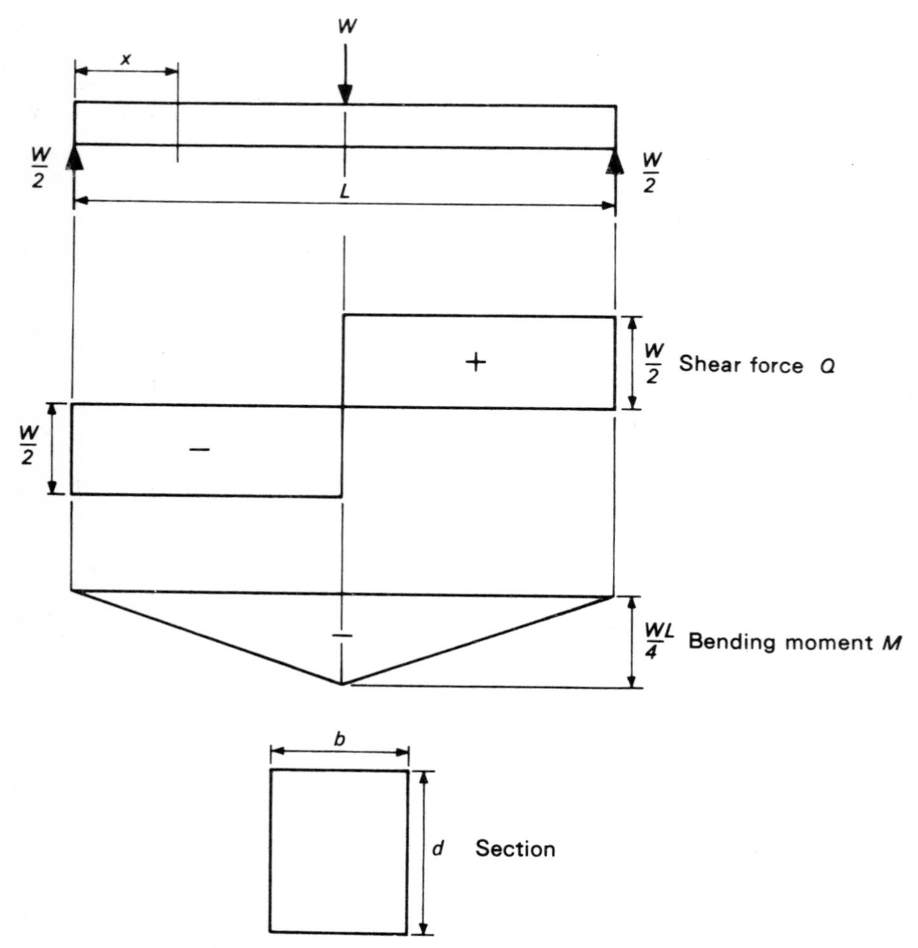

Figure 4.5 A centrally loaded, simply supported beam.

each end, i.e. a zero bending moment at each end, and the shear force and bending moment diagrams which are shown conform to the definitions given previously. Clearly both Q and M are discontinuous under the load but we may write down the expression for Q on the left-hand side of W as:

$$Q = -\frac{W}{2}$$

from vertical equilibrium and from equation (4.1) we have:

$$\frac{\mathrm{d}M}{\mathrm{d}x} = -\frac{W}{2}$$

giving:

$$M = -\frac{W}{2}x \tag{4.16}$$

since $M = 0$ at $x = 0$.

For a rectangular section b is constant and hence from equation (4.12) the position of the neutral axis may be determined:

$$b\int_0^d (y-y_0)\,\mathrm{d}y = 0$$

i.e.

$$\frac{d^2}{2} - y_0 d = 0$$

$$y_0 = d/2 \tag{4.17}$$

Thus the neutral axis is at the mid section as expected. From equation (4.14) we may determine I:

$$I = b\int_0^b y(y-d/2)\,\mathrm{d}y = \frac{bd^3}{12} \tag{4.18}$$

which is the standard result.

The deflection may be derived from equation (4.15) by substituting from equation (4.16) to give:

$$\frac{\mathrm{d}^2 v}{\mathrm{d}x^2} = -\frac{1}{EI}\frac{W}{2}x$$

Integrating twice we have:

$$v = -\frac{W}{EI}\frac{x^3}{12} + C_1 x + C_2$$

where C_1 and C_2 are constants. Taking the boundary conditions:

$$v = 0 \text{ at } x = 0 \text{ and } \frac{\mathrm{d}v}{\mathrm{d}x} = 0 \text{ at } x = L/2$$

gives $C_1 = \dfrac{WL^2}{16EI}$ and $C_2 = 0$

and hence:

$$v = \frac{WL^2}{16EI}x - \frac{W}{12EI}x^3$$

The deflection at the centre δ where $x = L/2$ is thus:

$$\delta = \frac{WL^3}{48EI} \tag{4.19}$$

For the case of a uniformly distributed load w we have:

$$Q = -\frac{wL}{2} + wx$$

i.e.

$$M = -\frac{wLx}{2} + \frac{wx^2}{2}$$

and by the same analysis, the deflection at the centre is:

$$\delta = \frac{5}{384} \frac{wL^4}{EI} \tag{4.20}$$

This result may be used to compute the deflection of a beam due to its own weight since, for a beam of uniform section:

$$w = \frac{W_B}{L}$$

where W_B = the weight of the beam. The deflection due to the weight is thus:

$$\delta_B = \frac{5}{384} \frac{W_B L^3}{EI}$$

and the total deflection may be written as:

$$\delta = \delta + \delta_B = \left(W + \frac{5}{8} W_B\right) \frac{L^3}{48EI} \tag{4.21}$$

i.e. it is necessary to add $\frac{5}{8} W_B$ to the load at the centre to correct for the weight of the beam.

4.2.6. *Shear stress distribution*

Combining equations (4.11) and (4.15) gives an expression for p_{xx} of the form:

$$p_{xx} = \frac{M}{I}(y - y_0)$$

Differentiating with respect to x we have:

$$\frac{\partial p_{xx}}{\partial x} = \left(\frac{y - y_0}{I}\right) \frac{\partial M}{\partial x} = \left(\frac{y - y_0}{I}\right) Q \tag{4.22}$$

and the two-dimensional equilibrium equations in plane stress (equations

(1.27)) give:

$$\frac{\partial p_{xx}}{\partial x}+\frac{\partial p_{xy}}{\partial y} = 0$$

and hence

$$\frac{\partial p_{xy}}{\partial y} = -\left(\frac{y-y_0}{I}\right)Q$$

Now Q is not a function of y and hence the expression may be integrated to give:

$$p_{xy} = -\frac{Q}{I}\left[\frac{y^2}{2}-y_0y+C_1(x)\right] \tag{4.23}$$

Thus for a rectangular section in which $y_0 = d/2$ we have:

$$p_{xy} = -\frac{Q}{I}\left[\frac{y^2}{2}-\frac{d}{2}y+C_1(x)\right]$$

and since $p_{xy} = 0$ at $y = 0$ and d, $C_1(x) = 0$ and hence:

$$p_{xy} = -\frac{Q}{I}\frac{y}{2}(y-d) \tag{4.24}$$

It is of interest to evaluate the integral,

$$b\int_0^d p_{xy}\,\mathrm{d}y = -\frac{Qb}{2I}\int_0^d (y^2-dy)\,\mathrm{d}y = \frac{Q}{I}\frac{bd^3}{12} = Q$$

as in equation (4.6). This equation arises since the general equilibrium equations include equation (4.6) as a special case. The distribution given by equation (4.24) is parabolic and the maximum value is at $y = d/2$,

i.e. $$(p_{xy})_{\max} = \frac{Q}{I}\frac{\mathrm{d}^2}{8}$$

and for the centrally loaded beam we have:

$$Q = -\frac{W}{2}$$

giving $$(p_{xy})_{\max} = -\frac{Wd^2}{16I}$$

The maximum direct stress is given when $y = d$ or $y = 0$, i.e.

$$(p_{xx})_{\max} = \frac{M}{I}\frac{d}{2} = -\frac{Wd}{4I}L$$

and hence the ratio is given by:

$$\frac{(p_{xy})_{\max}}{(p_{xx})_{\max}} = \frac{d}{4L} \tag{4.25}$$

Thus for slender beams the shear stress is very much less than p_{xx} although near the supports, where M is small, it is the dominating factor but its influence on the total deflection is small. However for thick section beams allowance must be made for the shear stress when considering the deflection analysis (see section 4.5).

From the second plane stress equilibrium equation,

$$\frac{\partial p_{yy}}{\partial y} + \frac{\partial p_{xy}}{\partial x} = 0$$

and hence for a rectangular beam,

$$\frac{\partial p_{yy}}{\partial y} = +\frac{\partial}{\partial x}\left[\frac{Q}{I}\frac{y}{2}(y-d)\right]$$

i.e.

$$\frac{\partial p_{yy}}{\partial y} = \frac{y}{2I}(y-d)\frac{\partial Q}{\partial x}$$

Substituting from equation (4.3) we have,

$$\frac{\partial p_{yy}}{\partial y} = \frac{wy}{2I}(y-d)$$

and integrating gives:

$$p_{yy} = \frac{w}{2I}\left[\frac{y^3}{3} - \frac{dy^2}{2} + C_2(x)\right]$$

Clearly p_{yy} only exists if there is a distributed load and the boundary conditions may be taken as:

$$p_{yy} = 0 \quad \text{at} \quad y = 0$$

and hence

$$C_2(x) = 0$$

i.e.

$$p_{yy} = \frac{w}{2I}y^2(y/3 - d/2) \tag{4.26}$$

Note that when $y = d$,

$$p_{yy} = \frac{wd^2}{2I}(d/3 - d/2) = -\frac{wd^3}{12I} = -\frac{w}{b}$$

i.e. equal and opposite to the applied load on the surface. The maximum value of p_{xx} for uniformly distributed loading is:

$$(p_{xx})_{\max} = \frac{M}{I}\frac{d}{2} = -\frac{3}{16}\frac{wL^2d}{I}$$

and hence:

$$\frac{(p_{yy})_{max}}{(p_{xx})_{max}} = \frac{4}{9}\left(\frac{d}{L}\right)^2 \tag{4.27}$$

Again the ratio is small for most of the beam and p_{yy} can be ignored when determining strains in slender beams as previously assumed and is rarely necessary in any analysis.

4.3. Three point bending as a test method

The simply supported beam discussed in section 4.2.5 is a useful test method for determining small strain behaviour in polymers. Its value may be illustrated by considering the deflection at the centre of the beam:

$$\delta = \frac{WL^3}{48EI}$$

and the maximum strain in the beam:

$$\hat{e} = \frac{d}{2R} = \frac{d}{2}\frac{M}{EI}$$

and since the maximum moment is given by:

$$M = \frac{WL}{4}$$

we have,

$$\hat{e} = \frac{d}{8}\frac{WL}{EI}$$

and

$$\delta = \frac{1}{6}\frac{\hat{e}}{d}L^2$$

i.e.

$$\hat{e} = 6\left(\frac{d}{L}\right)\left(\frac{\delta}{L}\right) \tag{4.28}$$

To achieve the same strain in tensile bar of length L the deflection would be:

$$\delta' = \hat{e}L = 6\frac{d}{L}\delta$$

i.e. for the same strain and specimen size there is a magnification factor in bending given by:

$$\frac{\delta}{\delta'} = \frac{L}{6d}$$

Thus, provided $L/d > 6$, there is a larger deflection in bending for the same strain. As the basic theory requires slender beams this is generally achieved with

L/d ratios of up to 75, i.e. magnifications of about 12.

Further, the applied load in tension is given by:

$$W_t = bd\, p_{xx} = E\hat{e}bd$$

and in bending

$$W_b = E\hat{e}bd\left(\frac{2}{3}\frac{d}{L}\right)$$

i.e.

$$\frac{W_b}{W_t} = \frac{2}{3}\frac{d}{L}$$

Thus the applied loads may be lower by a factor of about 100 with consequent simplification in test apparatus design.

As a test method it has its limitations in that each deflection gives the average response to strains ranging from $-\hat{e}$ to $+\hat{e}$ at each section over the whole length of the beam. If the material is known to be linear and the same in tension as in compression then there is no problem but in many polymers this is not certain. However, in practice, the data obtained is frequently used in bending calculations and would be self-correcting. In addition, providing the strains are kept below about $\frac{1}{2}\%$ the behaviour of most polymers can be regarded as linear and the differences between tension and compression are not large. This being so then a 'creep bending' modulus may be determined by loading a beam with several loads and determining δ as a function of t. If cross-plots are made of W versus δ at various values of t then a check on the assumption of linearity can be made. Assuming it to be true then the modulus at any time can be determined from the slope of the graph since:

$$E(t) = \frac{L^3}{48I}\frac{W'}{\delta(t)}$$

where

$$W' = W + \tfrac{5}{8}W_B$$

and a graph of $E(t)$ versus t may be constructed.

4.4. Some deviations from simple linearity

In previous sections reference has been made to the assumptions inherent in the bending formulae. If the dimensions of the beam are known then it is possible to estimate errors due to ignoring such factors as shear deflection, friction at the supports, etc. Specialist elasticity texts will provide most of the necessary analysis but it is worth including here some special cases of particular interest when dealing with polymers.

4.4.1. *Different moduli in tension and compression*

If it is assumed that the material is linear but that the moduli are different in tension and compression, say E_T and E_C, then we must return to the section on constitutive equations. We now have:

$$e_T = \frac{p_{xxT}}{E_T} \quad \text{and} \quad e_C = \frac{p_{xxC}}{E_C}$$

Clearly all the analysis of the preceding sections still applies and we have:

$$e_{xx} = \frac{y-y_0}{R} \quad \text{from equation (4.7)}$$

hence

$$p_{xx} = \frac{E_T}{R}(y-y_0), \quad y < y_0$$

and

$$p_{xx} = \frac{E_C}{R}(y-y_0), \quad y > y_0$$

Substituting in equation (4.4) for zero axial load for a rectangular beam we have:

$$\int_0^{y_0} \frac{E_T}{R}(y-y_0)\,\mathrm{d}y + \int_{y_0}^{d} \frac{E_C}{R}(y-y_0)\,\mathrm{d}y = 0$$

$$\therefore \qquad E_T\left(\frac{{y_0}^2}{2} - {y_0}^2\right) + E_C\left(\frac{d^2}{2} - y_0 d - \frac{{y_0}^2}{2} + {y_0}^2\right) = 0$$

and

$$\frac{y_0}{d} = \frac{1}{1+\sqrt{E_T/E_C}} \tag{4.29}$$

As $E_T/E_C \to 0$ then $y_0/d \to 1$, i.e. the neutral axis approaches the upper edge and similarly as $E_T/E_C \to \infty$, $y_0/d \to 0$, the lower surface. The bending moment may be deduced from equation (4.5) giving,

$$M = \int_0^{y_0} \frac{b}{R} E_T y(y-y_0)\,\mathrm{d}y + \int_{y_0}^{d} \frac{bE_C}{R} y(y-y_0)\,\mathrm{d}y$$

which on evaluating in conjunction with equation (4.29) gives:

$$M = \frac{E_T}{R}\left(\frac{bd^3}{12}\right)\left(\frac{2\sqrt{E_T/E_C}}{1+\sqrt{E_T/E_C}}\right)^2 \tag{4.30}$$

For $E_T = E_C = E$ we have equation (4.15):

$$M = \frac{E}{R}\left(\frac{bd^3}{12}\right)$$

and we may define a modulus in bending as:

$$E_B = E_T\left(\frac{2\sqrt{E_T/E_C}}{1+\sqrt{E_T/E_C}}\right)^2 \tag{4.31}$$

It is of interest to note that if the bending and tensile moduli are different at a given set of conditions, say some time, then the compression modulus can be estimated from:

$$E_C = E_T\left(\frac{2-\sqrt{E_B/E_T}}{\sqrt{E_B/E_T}}\right)^2 \tag{4.32}$$

4.4.2. *Non-linear stress–strain relationships*

Let us suppose that the behaviour of a material in simple tension may be represented as:

$$p_{xx} = E_0 e_{xx}^n$$

where E_0 and n are constants.
If we restrict our attention to a rectangular beam the neutral axis remains at the centre since the stress–strain curve is assumed to be the same in tension and compression, i.e. p is taken as the same sign as e and evaluated using $|e|$. The moment expression (equation (4.5)) may now be written as:

$$M = \int_0^d b\frac{E_0}{R^n}(y-d/2)^n y\,\mathrm{d}y \tag{4.33}$$

i.e.

$$M = \frac{bE_0}{R^n}\int_0^d y(y-\mathrm{d}/2)^n\,\mathrm{d}y$$

$$= \frac{bE_0}{R^n}\left[\frac{y(y-d/2)^{n+1}}{n+1} - \frac{(y-d/2)^{n+2}}{(n+1)(n+2)}\right]_0^d$$

$\therefore$

$$M = \frac{E_0}{R^n}\frac{bd^{n+2}}{2^{n+1}(n+2)} \tag{4.34}$$

Assuming small deflections as before we may write:

$$\frac{\mathrm{d}^2 v}{\mathrm{d}x^2} = \left[\frac{n+2}{2}\left(\frac{2}{d}\right)^{n+2}\frac{1}{bE_0}\right]^{1/n} M^{1/n} \tag{4.35}$$

and any beam problem may be solved by determining M and integrating as before. For example, the simply supported, centrally loaded beam may be solved since,

$$M = -\frac{W}{2}x$$

Integrating and using the same boundary conditions as before we have:

$$\delta = \frac{n}{1+2n}\frac{L^2}{2d}\left[\frac{n+2}{2}\frac{WL}{bd^2E_0}\right]^{1/n} \tag{4.36}$$

Note that for $n = 1$ equation (4.19) is recovered.

The most convenient representation for plotting experimental results may be obtained by rearranging equation (4.36) to give:

$$\left[6\frac{\delta d}{L^2}\right] = \left[\frac{3n}{1+2n}\right]\left[\frac{n+2}{24}\frac{d}{E_0}\right]^{1/n}\left[\frac{WL}{I}\right]^{1/n} \tag{4.37}$$

Thus if $6\,\delta d/L^2$ is plotted versus WL/I as logarithms the slope will give the power n and E_0 may be found from the intercept. Typical values of n for polymers are around 0·9. Solutions may be obtained for different values of E_0 and n in tension and compression but they are quite complicated and of limited practical value.

4.5. Shear deflections in beams

In simple bending theory it was shown that any deflection due to the shear stress p_{xy} could be ignored for slender beams but when deep beams are considered considerable shear deflections are possible. It was shown previously that the shear stress has a parabolic distribution and in a rectangular section, equation (4.24) gives:

$$p_{xy} = -\frac{Q}{I}\frac{y}{2}(y-d)$$

An examination of the shear strain is of interest now since it is this which gives the additional deflection. For small strains, from section 1.3.6, we have:

$$e_{xy} = \frac{\partial u}{\partial y} + \frac{\partial v}{\partial x}$$

and

$$e_{xx} = \frac{\partial u}{\partial x}$$

and hence the radius of curvature of the beam, R, is given by:

$$\frac{1}{R} = \frac{\partial^2 v}{\partial x^2} = -\frac{\partial^2 u}{\partial x\,\partial y} + \frac{\partial e_{xy}}{\partial x} = -\frac{\partial e_{xx}}{\partial y} + \frac{\partial e_{xy}}{\partial x}$$

Now $\qquad p_{xx} = \frac{M}{I}\left(y - \frac{d}{2}\right)$ and we may write:

$$e_{xx} = \frac{M}{EI}\left(y-\frac{d}{2}\right) - \frac{v}{E}p_{yy}$$

and

$$\frac{\partial e_{xx}}{\partial y} = \frac{M}{EI} - \frac{v}{E}\frac{\partial p_{yy}}{\partial y} = \frac{M}{EI} + \frac{v}{E}\frac{\partial p_{xy}}{\partial x}$$

from equations (1.27).

Similarly:

$$e_{xy} = \frac{p_{xy}}{G} = -\frac{Q}{IG}\frac{y}{2}(y-d)$$

i.e.
$$\frac{\partial e_{xy}}{\partial x} = -\frac{y}{2}\frac{(y-d)}{GI}\frac{\partial Q}{\partial x}$$

and
$$\frac{v}{E}\frac{\partial p_{xy}}{\partial x} = -\frac{y}{2}\frac{(y-d)}{I}\frac{v}{E}\frac{\partial Q}{\partial x}$$

$\therefore$
$$\frac{\partial^2 v}{\partial x^2} = -\left[\frac{M}{EI}+\frac{y}{2}\frac{(y-d)}{I}\left(\frac{2+v}{E}\right)\frac{\partial Q}{\partial x}\right] \tag{4.38}$$

This expression has the same first terms as equation (4.15) except for the sign which should be changed to conform to the bending system convention. Now $\partial^2 v/\partial x^2$ is assumed to be constant over the whole section of the beam so that plane sections remain plane but reference to equation (4.38) shows that the shear stresses produce distortions described by the function of y. If we confine our attention to the neutral axis, $y = d/2$, at which the deflection is determined then:

$$\frac{d^2 v}{dx^2} = \frac{M}{EI}-\frac{3}{2}\frac{1}{bd}\left(\frac{2+v}{E}\right)\frac{dQ}{dx}$$

A further difficulty with this expressions is illustrated in Figure 4.6 where the three point loading system is shown. Clearly Q is constant here and does not appear in the above equation since it describes changes in curvature which the constant shear does not give. The simplest method of avoiding this problem and associated problems with boundary conditions (e.g. $dv/dx = 0$ at the centre for bending) is to consider the bending and shear deflections separately and then sum them:

i.e.
$$\frac{d^2 v_B}{dx^2} = \frac{M}{EI}$$

and
$$\frac{dv_S}{dx} = -\frac{3}{2bd}\left(\frac{2+v}{E}\right)Q \tag{4.39}$$

giving
$$v = v_B + v_S$$

For the case of the simply supported beam in Figure 4.6 the central bending deflection is given by

$$\delta_B = \frac{WL^3}{48EI}$$

and for shear,
$$Q = -\frac{W}{2}$$

and
$$\frac{dv_S}{dx} = \frac{3}{2bd}\frac{W}{2}\left(\frac{2+v}{E}\right)$$

and since $v_S = 0$ at $x = 0$ we have, at the beam centre,

$$\delta_S = \frac{3}{8}\frac{WL}{bd}\left(\frac{2+v}{E}\right) = \frac{WL}{16EI}d^2(1+v/2)$$

and
$$\delta = \frac{WL^3}{48EI}\left[1+3(1+v/2)\frac{d^2}{L^2}\right] \tag{4.40}$$

Clearly for slender beams with $d \ll L$ the additional term has a small effect but in deeper sections it can be important. Two further examples are of interest:

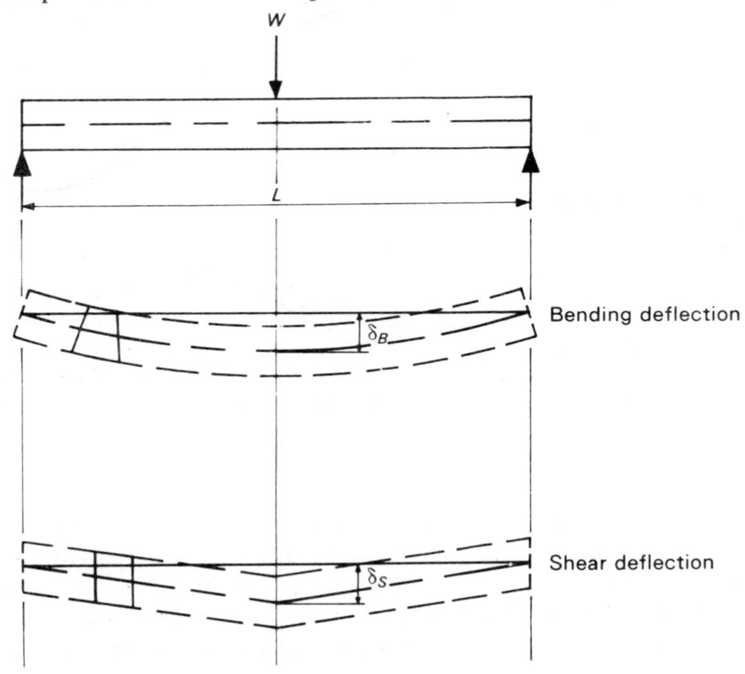

Figure 4.6 Shear deflections in a beam.

(i) A simply supported beam with a uniformly distributed load w/unit length:

$$Q = -\frac{wL}{2}+wx$$

and hence
$$\delta_S = \frac{3}{16bd}\left(\frac{2+v}{E}\right)wL^2$$

and since
$$\delta_B = \frac{5wL^4}{385EI}$$

we have:

$$\delta = \frac{5}{384}\frac{wL^4}{EI}\left[1+\frac{12}{5}\left(1+\frac{v}{2}\right)\frac{d^2}{L^2}\right] \tag{4.41}$$

(ii) For a cantilever of length L built in at one end with a load W at the other the total deflection becomes:

$$\delta = \frac{WL^3}{3EI}\left[1+\frac{3}{4}\left(1+\frac{v}{2}\right)\frac{d^2}{L^2}\right] \tag{4.42}$$

Exact solutions derived from stress functions which include the distortion of sections give the following forms for the additional terms:

$$\text{(i)}\ \frac{12}{5}\left(\frac{4}{5}+\frac{v}{2}\right)\frac{d^2}{L^2} \text{ and (ii) } \frac{3}{4}(1+v)\frac{d^2}{L^2}$$

The differences are thus quite small and this simple analysis gives a good representation of the deflection due to shear.

4.6. Sandwich beams

Composite beams made with two skins of material with a high modulus, e.g. a metal, separated by a core of a low density, often foam, material are frequently encountered in plastics designs. By a suitable choice of materials and dimensions composite beams may be produced with prescribed performance characteristics. In this section we shall deal with a simple 'sandwich' and show how its elastic performance may be predicted.

4.6.1. *Determination of the neutral axis*

Consider a sandwich section shown in Figure 4.7 of width b and depth d with a bottom skin of thickness h_1, and a top skin of thickness h_2. The moduli are taken to be E_S for the skin and E_C for the core. The general theory is used in that the linear strain distribution of the simple theory is assumed as well as the zero axial load condition. The stress distribution is discontinuous and equation (4.4) now becomes:

$$\int_0^{h_1} p_{xx}{}^S\,dy + \int_{h_1}^{d-h_2} p_{xx}{}^C\,dy + \int_{d-h_2}^{d} p_{xx}{}^S\,dy = 0 \tag{4.43}$$

and the stresses are given by:

$$p_{xx}{}^S = E_S\frac{y-y_0}{R},\ p_{xx}{}^C = E_C\frac{y-y_0}{R}$$

Substituting in equation (4.43) and assuming R is the same for all three sections (i.e. the linear strain assumption) we have:

$$E_S\left[\int_0^{h_1}(y-y_0)\,dy + \int_{d-h_2}^{d}(y-y_0)\,dy\right] + E_C\int_{h_1}^{d-h_2}(y-y_0)\,dy = 0$$

and solving for y_0 gives

$$y_0 = \frac{1}{2}\left[\frac{E_S d^2 - (E_S - E_C)\{(d-h_2)^2 - h_1{}^2\}}{E_S d - (E_S - E_C)\{d - (h_1 + h_2)\}}\right] \tag{4.44}$$

Clearly $y_0 = d/2$ for the special cases of $E_S = E_C$ and for $h_1 = h_2$.

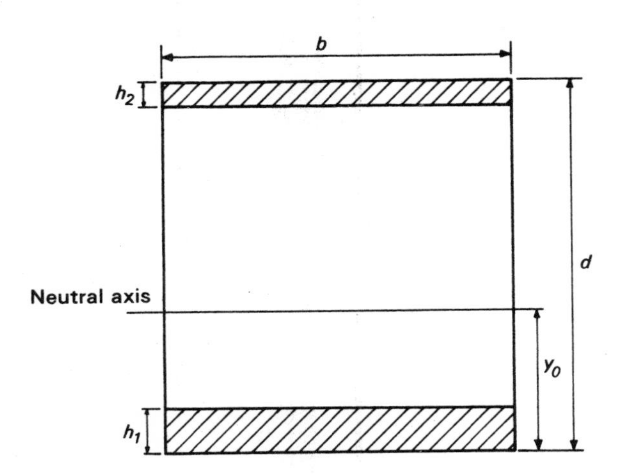

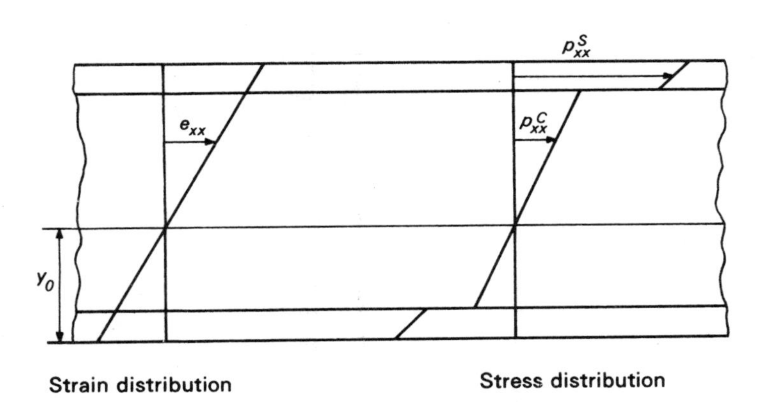

Figure 4.7 Cross section of a sandwich beam.

4.6.2. *Bending moments*

By a similar analysis equation (4.5) now becomes:

$$M = \frac{bE_S}{R}\int_0^{h_1}(y-y_0)\,y\,\mathrm{d}y + \frac{bE_C}{R}\int_{h_1}^{d-h_2}(y-y_0)\,y\,\mathrm{d}y + \frac{bE_S}{R}\int_{d-h_2}^{d}(y-y_0)\,y\,\mathrm{d}y$$

giving,

$$M = \frac{b}{R}\frac{1}{3}[E_S d^3 - (E_S - E_C)((d-h_2)^3 - h_1{}^3)]$$

$$- \frac{1}{4}\frac{[E_S d^2 - (E_S - E_C)((d-h_2)^2 - h_1{}^2)]^2}{[E_S d - (E_S - E_C)(d - (h_1 + h_2))]} \quad (4.45)$$

after the substitution for y_0 from equation (4.44). Comparison with the simple case, equation (4.15):

$$M = \frac{IE}{R} = \frac{bd^3}{12}\frac{E}{R}$$

suggests that an equivalent stiffness to EI be defined as S:

i.e.

$$S = \frac{bd^3}{12} E_S \left\{ 4 - 4\left(\frac{E_S - E_C}{E_S}\right)\left(\left(1 - \frac{h_2}{d}\right)^3 - \left(\frac{h_1}{d}\right)^3\right) - 3\frac{\left[1 - \left(\frac{E_S - E_C}{E_S}\right)\left(\left(1 - \frac{h_2}{d}\right)^2 - \left(\frac{h_1}{d}\right)^2\right)\right]^2}{\left[1 - \left(\frac{E_S - E_C}{E_S}\right)\left(1 - \left(\frac{h_1}{d} + \frac{h_2}{d}\right)\right)\right]} \right\} \quad (4.46)$$

Thus any beam bending problem may be solved in the usual fashion except that EI will be replaced by the function S computed from the values for the beam.

A special case of interest is when $h_1 = h_2 = h$ as this is frequently used in practice. This gives:

$$S = \frac{bd^3}{12} E_S \left[1 - \left(\frac{E_S - E_C}{E_S}\right)\left(1 - 2\left(\frac{h}{d}\right)\right)^3 \right] \quad (4.47)$$

Further if $E_C \ll E_S$ and $h/d \ll 1$ this reduces to:

$$S = E_S \frac{bd^2 h}{2} \quad (4.48)$$

4.6.3. *Optimisation*

In certain designs it is useful to make sandwich walls and it is desirable to achieve the most efficient combination of components. As an example of the method consider the case when we have ρ, either the weight or cost per unit volume of material. Let the values be ρ_S for the skin and ρ_C for the core. In this case the weight (or cost) per unit length of beam, for equal skin thicknesses, becomes

$$w = 2bh\rho_S + b(d - 2h)\rho_C$$

i.e.

$$w = bd\rho_C + 2bh(\rho_S - \rho_C) \quad (4.49)$$

Considering now the case of a beam of known b and d, for which we are to find the optimum skin thicknesses for a maximum stiffness to weight ratio

$$B = \frac{S}{w}$$

Clearly w rises as a linear function of h for $\rho_S > \rho_C$ which is the usual condition. An examination of equation (4.47) shows that S also increases with h and therefore the existence of an optimum condition will depend upon the nature of the dependence of S upon h. The optimum condition is when:

$$\frac{dB}{dh} = 0, \text{ i.e. } S\frac{dw}{dh} = w\frac{dS}{dh} \tag{4.50}$$

Now by differentiating equation (4.49) we have:

$$\frac{dw}{dh} = 2b(\rho_S - \rho_C)$$

and from equation (4.47):

$$\frac{dS}{dh} = \left(\frac{bd^2}{2}\right)(E_S - E_C)\left(1 - \frac{2h}{d}\right)^2$$

Substitution of these expressions in equation (4.50) together with those for S and w gives the result:

$$\frac{E_S}{E_S - E_C} = \left(\frac{2h}{d} - 1\right)^2\left(4\frac{h}{d} + 1 + \frac{3\rho_C}{\rho_S - \rho_C}\right) \tag{4.51}$$

Thus three values of h/d are produced and the solutions to equation (4.51) must be examined to determine if a valid optimum is obtained. Approximations for $h/d \ll 1$ are possible but care must be exercised in their application. For example, if equation (4.51) is expanded we have:

$$\frac{E_S}{E_S - E_C} = 16\left(\frac{h}{d}\right)^3 - 12\left(\frac{\rho_S - 2\rho_C}{\rho_S - \rho_C}\right)\left(\frac{h}{d}\right)^2 - 12\left(\frac{\rho_C}{\rho_S - \rho_C}\right)\left(\frac{h}{d}\right) + \left(\frac{\rho_S + 2\rho_C}{\rho_S - \rho_C}\right)$$

and if we consider steel skins and a polymer core then we can say, approximately that by weight, $\rho_S = 8$, $\rho_C = 1$, and $E_C \ll E_S$

$$\therefore \quad 1 = 16\left(\frac{h}{d}\right)^3 - \frac{72}{7}\left(\frac{h}{d}\right)^2 - \frac{12}{7}\left(\frac{h}{d}\right) + \frac{10}{7}$$

An examination of the solution of this equation shows that the cubic term may be ignored and since these figures are quite typical the general solution for $E_C \ll E_S$ is:

$$\frac{h}{d} = \frac{\rho_C}{2(\rho_S - 2\rho_C)}\left[-1 + \sqrt{\frac{\rho_S}{\rho_C} - 1}\right] \tag{4.52}$$

i.e. for this example $h/d = 0{\cdot}137$. The nature of the solution is shown graphically in Figure 4.8 for the parameters:

$$\frac{12S}{bd^3E_S} = 1-\left(1-\frac{2h}{d}\right)^3 = S_0$$

and

$$\frac{w}{bd} = \rho_C+\frac{2h}{d}(\rho_S-\rho_C) = w_0$$

At the optimum condition $S = 0{\cdot}61$ and $w_0 = 2{\cdot}9$ and a beam of the same stiffness as one of entirely skin material when made in the sandwich form is only 18% deeper but is lighter by a factor of 2·3; which represents a considerable gain.

The calculation process is sometimes more complex when costs are used since the prices of sheet are often a function of sheet thickness. The optimisation procedure is the same but a graphical or numerical solution is necessary. In addition other factors limiting the value of h may have to be included such as skin wrinkling.

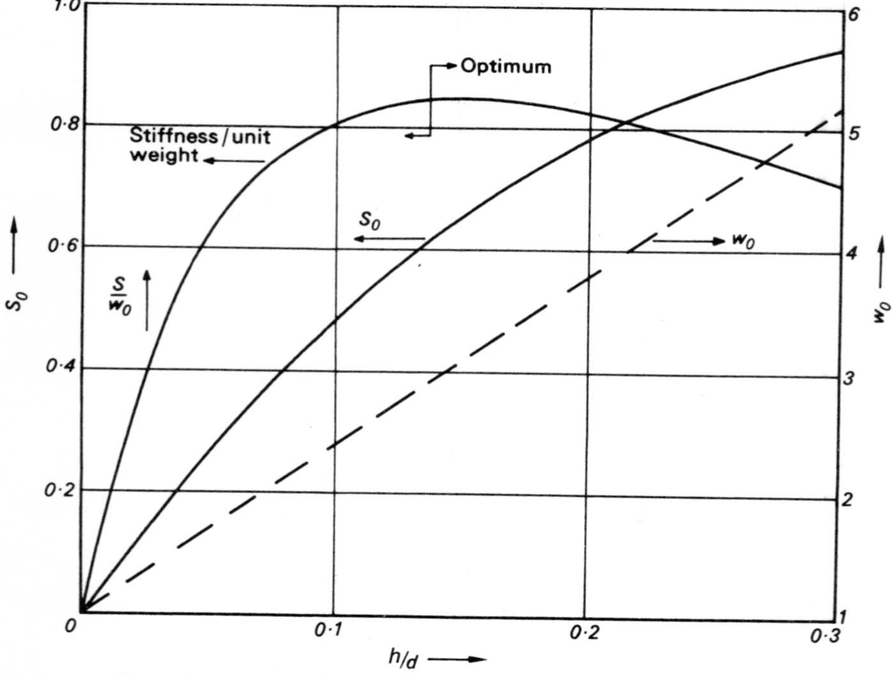

Figure 4.8 Sandwich beam optimisation: $E_c \ll E_s$, $\rho_s = 8$, $\rho_c = 1$.

4.6.4. *Shear deflections*

The problem here is to determine the shear stress on the neutral axis and we will consider the general case of different skin thicknesses. The total shear force on any section is given by:

$$Q = b\int_0^{h_1} p_{xy}{}^S \,dy + b\int_{h_1}^{d-h_2} p_{xy}{}^C \,dy + b\int_{d-h_2}^{d} p_{xy}{}^S \,dy \tag{4.53}$$

Now the direct stresses are given by:

$$p_{xx}{}^S = E_S \frac{y-y_0}{R} \text{ and } p_{xx}{}^C = E_C \frac{y-y_0}{R}$$

and thus by using the equilibrium condition:

$$p_{xy}{}^S = -\frac{Q}{S} E_S \left(\frac{y^2}{2} - y_0 y\right) + A$$

and

$$p_{xy}{}^C = -\frac{Q}{S} E_C \left(\frac{y^2}{2} - y_0 y\right) + B$$

where S is given by equation (4.46) and A and B are constants. Now the shear stress in the skin $p_{xy}{}^S = 0$ at $y = 0$ and hence $A = 0$ but B must be determined by substituting in equation (4.53) and solving using the definition of y_0 from equation (4.44). Considering the simple case of $h_1 = h_2 = h$ the shear stress p_{xy} at the neutral axis $y_0 = d/2$ is given by:

$$p_{xy0}^{C} = \frac{Q}{bd\left(1-\frac{2h}{d}\right)} + \frac{Q}{bd}\frac{E_S}{E_S-(E_S-E_C)\left(1-\frac{2h}{d}\right)^3} \times \left[\frac{3}{2}\frac{E_C}{E_S} - \frac{E_S-(E_S-E_C)\left(1-4\left(\frac{h}{d}\right)^3\right)}{E_S\left(1-\frac{2h}{d}\right)}\right] \tag{4.54}$$

The shear deflection may therefore be calculated from the relationship:

$$\frac{dv_S}{dx} = -\left(\frac{2+\nu_C}{E_C}\right)\frac{Q}{bd}\phi \tag{4.55}$$

where ϕ is the function:

$$\phi = \left[\frac{1}{\left(1-\frac{2h}{d}\right)} + \frac{E_S}{E_S-(E_S-E_C)\left(1-\frac{2h}{d}\right)^3}\left\{\frac{3}{2}\frac{E_C}{E_S} - \frac{E_S-(E_S-E_C)\left(1-4\left(\frac{h}{d}\right)^3\right)}{E_S\left(1-\frac{2h}{d}\right)}\right\}\right]$$

For the case of three point loading, we have:

$$\delta = \frac{WL^3}{48E_S I\psi}\left(1+3\left(1+\frac{\nu_C}{2}\right)\phi\frac{d^2}{L^2}\psi\frac{E_S}{E_C}\right) \tag{4.56}$$

where
$$\psi = \frac{E_S-(E_S-E_C)\left(1-\frac{2h}{d}\right)^3}{E_S}$$

For very thin skins the deflection may be approximated to:

$$\delta = \frac{WL^3}{24E_Sbd^2h}+\frac{3}{8}\left(\frac{2+\nu_C}{E_C}\right)\frac{WL}{bd} \qquad (4.57)$$

The importance of the shear term can be seen to be strongly dependent on the span. If the span is large then the bending analysis will be quite adequate but for short deep beams the shear must be included. Optimisation can be carried out as before and for thin skins the optimum skin thickness is given by:

$$h = \frac{L}{3\sqrt{2}}\sqrt{\left\{\frac{E_C}{E_S(2+\nu_C)}\frac{\rho_C}{\rho_S-\rho_C}\right\}} \qquad (4.58)$$

This is independent of the beam dimensions and for typical polymer cores with metal skins will be:

$$h \simeq 0{\cdot}005L$$

For the shear to be significant, the span to depth ratio must be less than about 20:1 giving values of h/d of around 0·1. The inclusion of shear in optimisation calculations thus requires careful consideration of the particular situation involved.

4.6.5. *Multilayered beams*

The analytical solutions presented for sandwich beams represent a reasonable limit to this method. For beams composed of more layers it is preferable to write computer programs and determine the various parameters numerically. The process is exactly the same as in the analytical solutions involving the determination of y_0 by the zero axial load condition, hence the bending moment and finally the shear forces.

4.7. Wide beams and plates

A basic assumption of beam theory is that the stress system is one of plane stress, i.e. in the coordinate system used the stresses in the z direction are all zero. The consequences of this assumption are of interest when the resulting strain distributions are examined. Let us consider only the simple beam theory (i.e. ignoring p_{yy} and p_{xy}) and thus from equations (4.7) and (4.11), we have:

$$e_{xx} = \frac{y-y_0}{R} \quad \text{and} \quad p_{xx} = \frac{E}{R}(y-y_0)$$

Now since p_{zz} is assumed to be zero we may write:

$$e_{zz} = -\frac{v}{E} p_{xx} = -\frac{v}{R}(y - y_0) \tag{4.59}$$

This means that when there is a local radius of curvature R in the xy plane giving the e_{xx} strain distribution there is an induced radius $-R/v$ in the yz plane. This is illustrated in Figure 4.9 and is an effect familiar to anyone who has bent

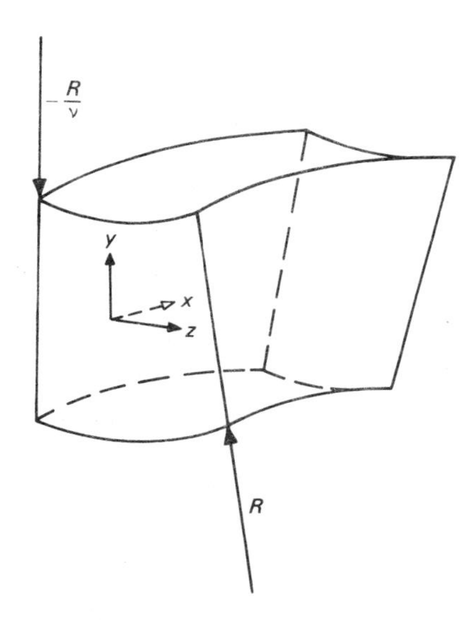

Figure 4.9 Anticlastic curvature in a bent beam.

a rubber block. Since $v < 1$, $R/v > R$ and for most beams the curvature is not easily seen except for large deflections, as with rubber. The effect is known as anticlastic curvature and for the plane stress conditions to be maintained it must be present. However, if the beam is very wide, it is often clear that it is not present, in which case, $e_{zz} = 0$ and to maintain this p_{zz} must have a value given by:

$$0 = \frac{1}{E}(p_{zz} - vp_{xx})$$

i.e.

$$p_{zz} = vp_{xx}$$

Substituting in the expression for e_{xx} we now have:

$$e_{xx} = \frac{1}{E}(p_{xx} - vp_{zz}) = \frac{p_{xx}}{E}(1 - v^2) \tag{4.60}$$

Thus for wide sections the beam formulae are altered by a factor $(1 - v^2)$. The exact width to depth ratio where this occurs is difficult to determine but for rectangular sections it is probably when $b > 6d$. The design of panels, etc., usually involves the inclusion of this term and in testing in bending if observation suggests the plane strain condition ($e_{zz} = 0$) then the results should be corrected accordingly.

Since the stress p_{zz} is induced there is also a bending moment given by:

$$M_z = \int_0^d p_{zz} y\, by\, dy$$

and since $p_{zz} = \nu p_{xx} = E\nu/(y - y_0)R$ we have:

$$M_z = \frac{\nu}{R} IE \tag{4.61}$$

The applied bending moment is now given by:

$$M = (1 - \nu^2) \frac{EI}{R}$$

and hence

$$M_z = \frac{\nu}{1 - \nu^2} M \tag{4.62}$$

i.e. the condition of plane strain induces a bending moment of this value in a plane at right angles to the applied moment. In fact this suggests a general analysis for a plate in two-dimensional bending as shown in Figure 4.10 in

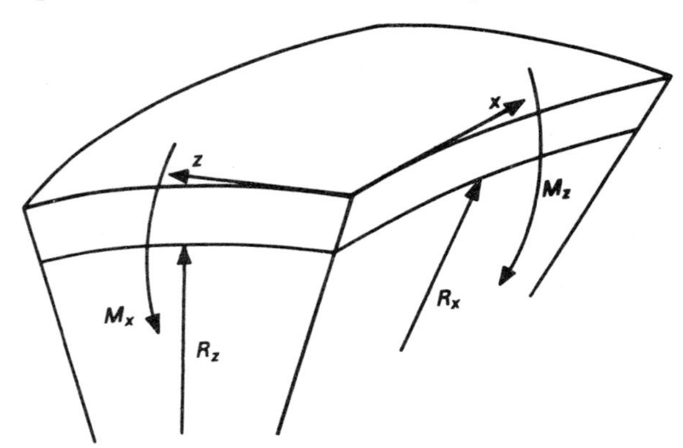

Figure 4.10 An element of a plate in bending.

which the radii of curvature of the neutral axis in the x and z direction are R_x and R_z respectively. The bending moments per unit length are given by:

$$M_x = \int_0^d p_{xx}\, y\, dy \text{ and } M_z = \int_0^d p_{zz}\, y\, dy$$

The strains become:

$$e_{xx} = \frac{y - y_0}{R_x} \text{ and } e_{zz} = \frac{y - y_0}{R_z}$$

and from Hooke's law:

$$e_{xx} = \frac{1}{E}(p_{xx} - \nu p_{zz}) \quad \text{and} \quad e_{zz} = \frac{1}{E}(p_{zz} - \nu p_{xx})$$

Thus for p_{xx}, for example, we have:

$$p_{xx} = \frac{E(e_{xx} + \nu e_{zz})}{1 - \nu^2}$$

and hence:

$$M_x = \int_0^d \frac{E}{1-\nu^2}\left(\frac{y-y_0}{R_x} + \frac{\nu(y-y_0)}{R_z}\right) y \, dy$$

$$M_x = \frac{E\,d^3}{(1-\nu^2)\,12}\left(\frac{1}{R_x} + \frac{\nu}{R_z}\right)$$

and

$$M_z = \frac{Ed^3}{(1-\nu^2)\,12}\left(\frac{1}{R_z} + \frac{\nu}{R_x}\right)$$

It is usual to write:

$$D = \frac{Ed^3}{12(1-\nu^2)}$$

which is termed the flexural rigidity of the plate and since:

$$\frac{1}{R_x} = \frac{\partial^2 v}{\partial x^2} \quad \text{and} \quad \frac{1}{R_z} = \frac{\partial^2 v}{\partial z^2}$$

we have:

$$\left.\begin{aligned} M_x &= D\left(\frac{\partial^2 v}{\partial x^2} + \nu \frac{\partial^2 v}{\partial z^2}\right) \\ \text{and} \quad M_z &= D\left(\frac{\partial^2 v}{\partial z^2} + \nu \frac{\partial^2 v}{\partial x^2}\right) \end{aligned}\right\} \tag{4.63}$$

As a simple example consider a plate in which two constant moments are applied:

$$M_x = M_1 \text{ and } M_z = M_2$$

which gives:

$$\frac{\partial^2 v}{\partial x^2} = \frac{1}{D(1-\nu^2)}(M_1 - \nu M_2)$$

Integrating gives:

$$v = \left[\frac{M_1 - \nu M_2}{D(1-\nu^2)}\right]\frac{x^2}{2} + C_1(z)\,x + C_2(z)$$

and similarly:

$$v = \left[\frac{M_2 - \nu M_1}{D(1-\nu^2)}\right]\frac{z^2}{2} + C_3(x)\,z + C_4(x)$$

If we take the conditions that the deflection and slopes are all zero at the point $x = 0$ and $z = 0$ then the solution becomes:

$$v = \left[\frac{M_1 - \nu M_2}{D(1-\nu^2)}\right]\frac{x^2}{2} + \left[\frac{M_2 - \nu M_1}{D(1-\nu^2)}\right]\frac{z^2}{2}$$

i.e.

$$v = \frac{(M_1 - \nu M_2)\,x^2 + (M_2 - \nu M_1)\,z^2}{2D(1-\nu^2)}$$

Special cases of interest are:

(a) $M_2 = 0$, simple bending of beams:

$$v = \frac{M_1}{2D(1-\nu^2)}(x^2 - \nu z^2)$$

which is the anticlastic curvature effect.

(b) $M_2 = M_1$, equal bending which gives:

$$v = \frac{M_1(1-\nu)}{2D(1-\nu^2)}(x^2 + z^2)$$

the cap of a sphere.

The solution given here took no account of the twisting of the plate which arises when M_x and M_z vary as functions of x and z. The previously derived equations are still correct but in addition the equilibrium of the system must be examined. The reader is referred to specialised texts for this analysis since there are few simple solutions possible. The exception to this is when the plates are circular where there is no twisting and we will consider this as a special case.

4.8. Circular plates

If we consider a circular plate with axisymmetric loading we may analyse the problem in precisely the same way as the beam. Consider a small element at a radius r of length δr projected by an angle $\delta\psi$ from the central axis as shown in Figure 4.11(a) in which the shear force per unit length Q varies from Q at a radius r to $Q + \delta Q$ at $r + \delta r$. The side faces carry no shear since this would violate the condition of axial symmetry. The bending moments are M_r and M_ψ per unit length. M_ψ does not vary, again to retain symmetry and M_r varies from M_r at r to $M_r + \delta M_r$ at $r + \delta r$. Equilibrium of moments may be taken in two directions; radially we have:

$$r\,\delta\psi M_r - (M_r + \delta Mr)(r + \delta r)\,\delta\psi + (Q + \delta Q)(r + \delta r)\,\delta\psi\,\delta r$$
$$+ 2M_\psi\,\delta r \sin\frac{\delta\psi}{2} + wr\,\delta\psi\,\delta r\,\frac{\delta r}{2} = 0$$

where w = force per unit area.

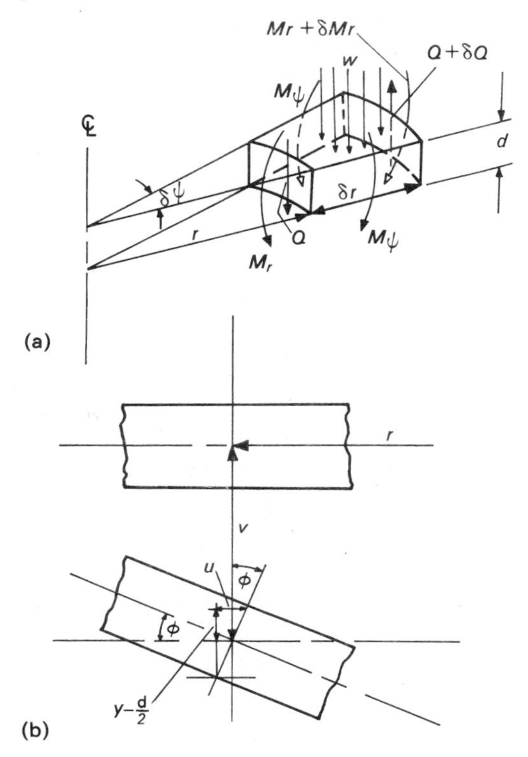

Figure 4.11 Moments and deflections in circular plates.

Ignoring terms of higher order than the second we have:

$$M_\psi - \frac{rdM_r}{\mathrm{d}r} - M_r + rQ = 0$$

i.e.
$$\frac{\mathrm{d}(rM_r)}{\mathrm{d}r} - M_\psi = rQ \tag{4.64}$$

Equilibrium in the tangential direction gives:

$$Qr\,\delta\psi - (Q + \delta Q)(r + \delta r)\,\delta\psi + w\left[(r + \delta r)^2\,\frac{\delta\psi}{2} - \frac{r^2\,\delta\psi}{2}\right] = 0$$

i.e.
$$\frac{\mathrm{d}}{\mathrm{d}r}(rQ) = rw \tag{4.65}$$

and thus with equation (4.64) we have:

$$\frac{d^2(rM_r)}{dr^2} - \frac{dM_\psi}{dr} = \frac{d}{dr}(rQ) = rw \tag{4.66}$$

which may be compared with equation (4.3) for simple beams.

Again, as for beams, the moments may be written as:

$$\left.\begin{aligned} M_r &= \int_0^d p_{rr}\, y\, dy \\ \text{and} \quad M_\psi &= \int_0^d p_{\psi\psi}\, y\, dy \end{aligned}\right\} \tag{4.67}$$

and from elasticity we have:

$$\left.\begin{aligned} p_{rr} &= \frac{E}{1-\nu^2}(e_{rr} + \nu e_{\psi\psi}) \\ \text{and} \quad p_{\psi\psi} &= \frac{E}{1-\nu^2}(e_{\psi\psi} + \nu e_{rr}) \end{aligned}\right\} \tag{4.68}$$

Figure 4.11(b) shows a section of the plate deflected an amount v and with a slope $\phi = dv/dr$ in the deflected state. Assuming plane sections remain plane, as before, the displacement in the r direction, u, at a distance y from the bottom of the plate is given by:

$$u = \phi(y - d/2)$$

Now from the analysis of strains in axial symmetry, equations (1.59), we have:

$$e_{\psi\psi} = \frac{u}{r} \quad \text{and} \quad e_{rr} = \frac{\partial u}{\partial r}$$

and hence:

$$e_{\psi\psi} = \frac{1}{r}\frac{dv}{dr}(y - d/2) \quad \text{and} \quad e_{rr} = \frac{d^2v}{dr^2}(y - d/2)$$

Substitution in equations (4.67) and (4.68) gives:

$$M_\psi = \frac{E}{1-\nu^2}\int_0^d \left[\frac{1}{r}\frac{dv}{dr}(y - d/2) + \frac{\nu\, d^2v}{dr^2}(y - d/2)\right] y\, dy$$

i.e.

$$M_\psi = \frac{E}{1-\nu^2}\frac{d^3}{12}\left[\frac{1}{r}\frac{dv}{dr} + \frac{\nu\, d^2v}{dr^2}\right]$$

or

$$M_\psi = D\left[\frac{1}{r}\phi + \nu\frac{d\phi}{dr}\right] \tag{4.69}$$

and similarly:

$$M_r = D\left[\frac{d\phi}{dr} + \frac{\nu}{r}\phi\right] \tag{4.70}$$

These relationships may be compared with equation (4.15) which may be written as:

$$M = EI \frac{d\phi}{dx}$$

where ϕ can be expressed in terms of shear force as:

$$\frac{d^2\phi}{dx^2} = \frac{Q}{EI}$$

Similarly by differentiation of equation (4.70) and substitution in equation (4.66) we have:

$$\frac{d^2\phi}{dr^2} + \frac{1}{r}\frac{d\phi}{dr} - \frac{\phi}{r^2} = \frac{Q}{D}$$

which reduces to:

$$\frac{d}{dr}\left[\frac{1}{r}\frac{d}{dr}(r\phi)\right] = \frac{Q}{D} \tag{4.71}$$

The method of solution is simply to write the appropriate expression for Q and then deduce the constants in terms of the boundary conditions. The solution of the deflection equation is easily obtained once ϕ is known and the boundary conditions are used to define the constants in the expression for v.

As an example consider a plate of diameter L with a central load W and with a simply supported perimeter such that:

$$M_r = 0 \text{ at } r = L/2$$

By symmetry the slope at the centre must be zero:

$$\phi = 0 \text{ at } r = 0$$

and the shear force per unit length Q is given by:

$$Q = -\frac{W}{2\pi r}$$

Substituting in equation (4.71) and integrating twice gives:

$$\phi = -\frac{W}{2\pi D}\frac{r}{2}(\ln r - \tfrac{1}{2}) + C_1\frac{r}{2} + \frac{C_2}{r}$$

In this case, since $\phi = 0$ at $r = 0$, $C_2 = 0$ and we are left to find C_1 from the expression for M_r which becomes:

$$M_r = D\left[\frac{C_1}{2}(1+\nu) - \frac{W}{2\pi D}\frac{1}{2}\left\{(1+\nu)\ln r + \frac{1-\nu}{2}\right\}\right]$$

and using the zero moment condition:

$$C_1 = \frac{W}{2\pi D}[(1-\nu)+\ln(L/2)]$$

The central deflection can be found by integrating ϕ and is:

$$\delta = \frac{WL^2}{4Ed^3}\left[\frac{(2-\nu)(1-\nu^2)3}{\pi}\right] \tag{4.72}$$

which may be compared with the narrow beam solution of equation (4.19):

$$\delta = \frac{WL^2}{4Ed^3}\frac{L}{b}$$

and that for wide panels (section 4.7):

$$\delta = \frac{WL^2}{4Ed^3}\left[\frac{L(1-\nu^2)}{b}\right]$$

Thus the circular section is equivalent to a panel supported down two edges of width:

$$b = \left(\frac{\pi}{2-\nu}\right)L/3$$

4.9. Large deflections and stability

4.9.1. *A simply supported beam with a central load*

It is a basic assumption of the simple bending theory that the distortions are so small that no changes in the geometry of the beams need be included in the analysis. This assumption is valid for most practical problems but some exceptions are important. The low modulus of polymers often leads to beam tests with large deflections and it is of interest to examine the nature of their effects.

Consider a simply supported beam of length L and a central load W as shown in Figure 4.12. Let the deflections be large such that the contact angle at the supports is α and the reaction X is normal to the beam. It should be noted that friction at the supports will produce a reaction parallel to the beam but this will be ignored here.

Equilibrium gives the condition for X:

$$2X\cos\alpha = W \tag{4.73}$$

The bending moment at A is given by:

$$M = -(X\cos\alpha . x + X\sin\alpha . v) \tag{4.74}$$

and if we assume that conventional bending theory is applicable we have:

$$M = \frac{EI}{R}$$

where R is the local radius of curvature given by:

$$R = \frac{\left[1+\left(\frac{\mathrm{d}v}{\mathrm{d}x}\right)^2\right]^{3/2}}{\frac{\mathrm{d}^2v}{\mathrm{d}x^2}}$$

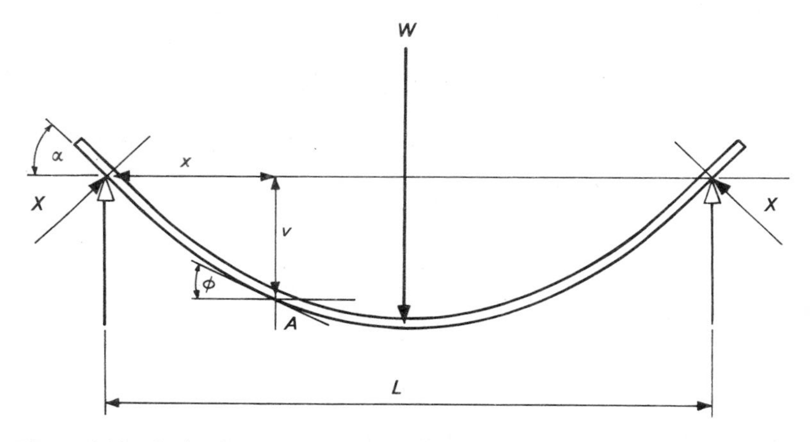

Figure 4.12 A simply supported, centrally loaded beam with large deflections.

The equations are most easily managed in terms of angles and since:

$$\tan\phi = \frac{\mathrm{d}v}{\mathrm{d}x} \qquad \text{and} \qquad \frac{\mathrm{d}s}{\mathrm{d}x} = \left[1+\left(\frac{\mathrm{d}v}{\mathrm{d}x}\right)^2\right]^{1/2}$$

where s is the length of the beam, we may write:

$$R = \frac{\mathrm{d}s}{\mathrm{d}\phi} \tag{4.75}$$

Now:

$$\frac{EI}{R} = -(X\cos\alpha . x + X\sin\alpha . v)$$

and differentiating and noting that:

$$\frac{\mathrm{d}x}{\mathrm{d}s} = \cos\phi \qquad \text{and} \qquad \frac{\mathrm{d}v}{\mathrm{d}s} = \sin\phi \tag{4.76}$$

we have:

$$-\frac{EI}{R^2}\frac{\mathrm{d}R}{\mathrm{d}s} = -X(\cos\alpha.\cos\phi+\sin\alpha.\sin\phi)$$

i.e.

$$\frac{EI}{R^3}\frac{\mathrm{d}R}{\mathrm{d}\phi} = X\cos(\alpha-\phi)$$

Integrating this gives:

$$-\frac{EI}{2R^2} = X\sin(\alpha-\phi)+C_1$$

and since $M = 0$ when $\phi = \alpha$ and $1/R = 0$, $C_1 = 0$ and we have:

$$\frac{1}{R} = \left[\frac{2X}{EI}\sin(\alpha-\phi)\right]^{1/2} \tag{4.77}$$

Combining equations (4.75), (4.76) and (4.77) yields the following results:

$$x = \left(\frac{EI}{2X}\right)^{1/2}\int_{\alpha}^{\phi}\cos\phi.\sin^{-1/2}(\alpha-\phi)\,\mathrm{d}\phi$$

and

$$v = \left(\frac{EI}{2X}\right)^{1/2}\int_{\alpha}^{\phi}\sin\phi.\sin^{-1/2}(\alpha-\phi)\,\mathrm{d}\phi \tag{4.78}$$

The central deflection may be found from:

$$v = \delta \qquad \text{when} \qquad \phi = 0$$

and since $x = L/2$ when $\phi = 0$ and $X = W/(2\cos\alpha)$ we have the following conditions:

$$\frac{1}{2}\left(\frac{WL^2}{EI}\frac{1}{\cos\alpha}\right)^{1/2} = -\int_{0}^{\alpha}\cos\phi.\sin^{-1/2}(\alpha-\phi)\,\mathrm{d}\phi$$

and

$$\frac{\delta}{L}\left(\frac{WL^2}{EI}\frac{1}{\cos\alpha}\right)^{1/2} = -\int_{0}^{\alpha}\sin\phi.\sin^{-1/2}(\alpha-\phi)\,\mathrm{d}\phi \tag{4.79}$$

Thus a solution is obtained by choosing any value of α from which a value of WL^2/EI is found from the first equation. Substituting this together with α gives δ/L from the second equation and hence δ/L may be plotted versus WL^2/EI.

The integrals in equation (4.79) are not soluble analytically and it is probably simpler to solve them by numerical methods if a computer is available. Care is necessary in their evaluation since as $\phi \to \alpha$ the function being integrated tends to infinity. However standard numerical integration methods will cope. An alternative method is to rearrange the equations into the forms of elliptic integrals which are tabulated. For example we may write the integrals as:

$$I_1 = -\int_{0}^{\alpha}\cos\phi.\sin^{1/2}(\phi-\alpha)\,\mathrm{d}\phi$$

$$= -2\cos\alpha.\sqrt{\sin\alpha} - 2\sin\alpha\left[\int_0^{\pi/2}\sqrt{(1-\tfrac{1}{2}\sin^2\gamma)}\,d\gamma\right.$$

$$\left.-\int_0^{\cos^{-1}(\sqrt{\sin\alpha})}\sqrt{(1-\tfrac{1}{2}\sin^2\gamma)}\,d\gamma\right]$$

and

$$I_2 = -\int_0^{\alpha}\sin\phi.\sin^{1/2}(\phi-\alpha)\,d\phi$$

$$= -2\sin\alpha\sqrt{\sin\alpha} + 2\cos\alpha\left[\int_0^{\pi/2}\sqrt{(1-\tfrac{1}{2}\sin^2\gamma)}\,d\gamma\right.$$

$$\left.-\int_0^{\cos^{-1}(\sqrt{\sin\alpha})}\sqrt{(1-\tfrac{1}{2}\sin^2\gamma)}\,d\gamma\right]$$

The function:

$$E = \int_0^{\pi/2}\sqrt{(1-k^2\sin^2\gamma)}\,d\gamma \tag{4.80}$$

is a complete elliptic integral of the second kind and is tabulated in several standard texts usually in terms of θ when:

$$k = \sin\theta$$

From such tables we have:

$$\int_0^{\pi/2}\sqrt{\left(1-\frac{1}{2}\sin^2\gamma\right)}\,d\gamma = 1{\cdot}467 \quad (\theta = 30°)$$

and for any value of α the incomplete elliptic integral must be found from tables with $\cos^{-1}(\sqrt{\sin\alpha})$ replacing $\pi/2$ in equation (4.80). Figure 4.13 shows the result of the complete computation with WL^2/EI plotted as a function of δ/L. For small values of δ/L the solution tends to the linear case:

i.e.
$$\frac{WL^2}{EI} = 48\left(\frac{\delta}{L}\right)$$

as shown but for δ/L values above 0·05 the errors become significant reaching 10% at $\delta/L = 0{\cdot}1$. The line is, of course, curved, indicating a non-linear effect which in this case is entirely geometrical in origin. This non-linearity results in the load reaching a maximum and falling thereafter for:

$$\delta/L > 0{\cdot}24$$

Thus the deflection will increase for a decreasing load which is an unstable system, since, if the load is not reduced the system will collapse at the maximum load condition:

i.e.
$$\frac{\hat{W}L^2}{EI} = 6{\cdot}72$$

since no load greater than $\hat{W}$ can be carried. Because we included the changes of geometry an instability condition has been predicted which would not have resulted from the simpler theory.

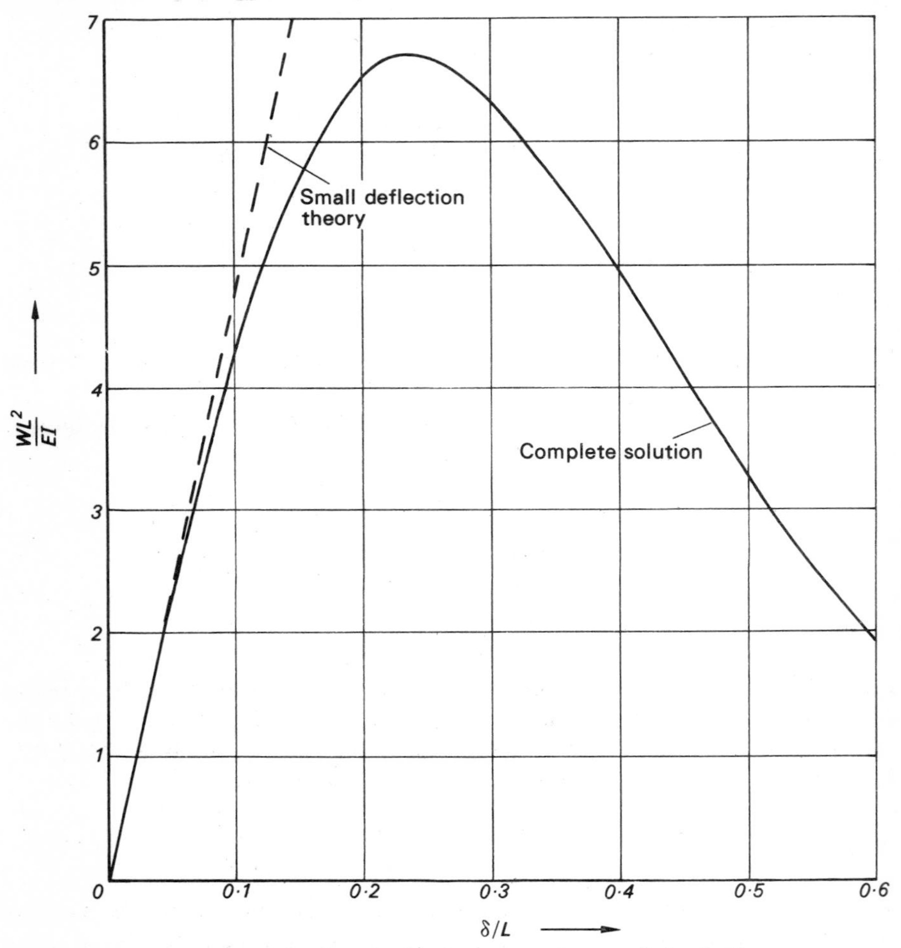

Figure 4.13 Deflection of a simply supported, centrally loaded beam.

4.9.2. *Strut stability*

A simple extension of the large displacement analysis is illustrated in Figure 4.14 where a simply supported beam or strut with a distance L' between the supports is subjected to an end thrust of W. The moment at the point A is given by:

$$M = -Wv \tag{4.81}$$

By using this result in the general bending theory as before we have the results:

$$\left.\begin{aligned}
\left(\frac{2W}{EI}\right)^{1/2}.x &= \int_{\alpha}^{\phi} (\cos\phi-\cos\alpha)^{-1/2}\cos\phi\, d\phi \qquad &\text{(a)}\\
\left(\frac{2W}{EI}\right)^{1/2}.v &= \int_{\alpha}^{\phi} (\cos\phi-\cos\alpha)^{-1/2}\sin\phi\, d\phi \qquad &\text{(b)}\\
\left(\frac{2W}{EI}\right)^{1/2}.s &= \int_{\alpha}^{\phi} (\cos\phi-\cos\alpha)^{-1/2}\, d\phi \qquad &\text{(c)}
\end{aligned}\right\} \quad (4.82)$$

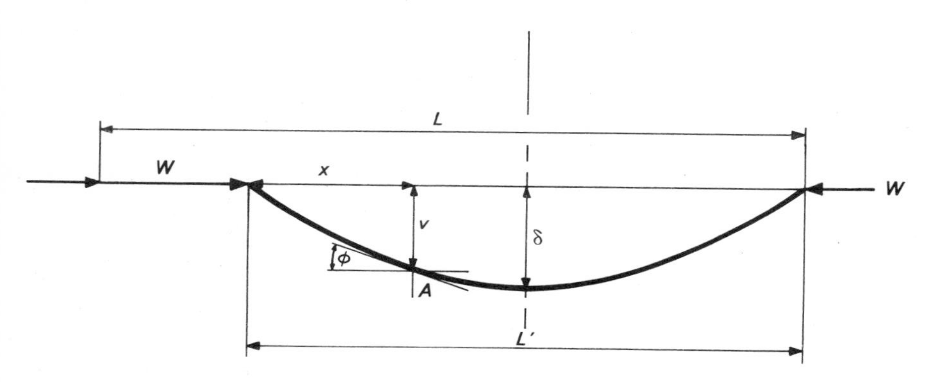

Figure 4.14 A strut with an end load and large deflections.

This third condition is necessary since in this case the span decreases from the initial L to L' while the curved length remains at L. Thus we may substitute the appropriate boundary conditions in equation (4.82) to give the results:

$$\left(\frac{2W}{EI}\right)^{1/2}.\frac{L'}{2} = -\int_{0}^{\alpha} (\cos\phi-\cos\alpha)^{-1/2}\cos\phi\, d\phi = J_1$$

$$\left(\frac{2W}{EI}\right)^{1/2}.\delta = -\int_{0}^{\alpha} (\cos\phi-\cos\alpha)^{-1/2}\sin\phi\, d\phi = J_2$$

$$\left(\frac{2W}{EI}\right)^{1/2}.\frac{L}{2} = -\int_{0}^{\alpha} (\cos\phi-\cos\alpha)^{-1/2}\, d\phi = J_3$$

Only J_2 yields an analytical solution:

i.e.
$$J_2 = -\int_0^{\alpha} \frac{\sin\phi\, d\phi}{(\cos\phi-\cos\alpha)^{1/2}} = \left[2(\cos\phi-\cos\alpha)^{1/2}\right]_0^{\alpha}$$

and
$$J_2 = -2\sqrt{2}\sin\frac{\alpha}{2} \qquad (4.83(a))$$

Both J_1 and J_3 can be written in terms of complete elliptic integrals such that:

$$J_1 = -\sqrt{2}\,(2E-K) \qquad (4.83(b))$$

and

$$J_3 = -\sqrt{2}\,K \qquad (4.83(c))$$

where

$$E = \int_0^{\pi/2} \frac{d\gamma}{\sqrt{(1-\sin^2(\alpha/2)\sin^2\gamma)}}$$

an elliptic integral of the second kind and:

$$K = \int_0^{\pi/2} \sqrt{(1-\sin^2(\alpha/2)\sin^2\gamma)}\,d\gamma$$

an elliptic integral of the first kind. These two functions may be obtained from tables and hence J_1, J_2 and J_3 evaluated for any value of α. The parameters of interest may be written as:

$$\frac{\delta}{L} = \frac{1}{2}\left(\frac{J_2}{J_3}\right) = -\frac{\sin\frac{\alpha}{2}}{K}, \quad \frac{WL^2}{EI} = 2(J_3)^2 = 4K^2$$

and

$$\frac{L-L'}{L} = \frac{\Delta}{L} = \frac{J_3-J_1}{J_3} = -\frac{2E}{K} \qquad (4.84)$$

Figure 4.15 shows these results plotted as WL^2/EI versus δ/L and Δ/L. The pattern of the results is the same for both deflections and shows that for this initially straight beam the deflection remains at zero until

$$\frac{WL^2}{EI} = \pi^2$$

at which point there is a very rapid increase in deflection. The system is not truly unstable since an increasing load is required to increase the deflection but the marked change of behaviour is of obvious practical significance and is termed elastic buckling. In practice there is usually some deflection due to initial curvature. The effect is not generally large and the behaviour described here can be verified easily with a thin steel strip.

If it is the onset of elastic buckling which is of interest then the large deflection theory is no longer necessary and we may write:

$$\frac{1}{R} = \frac{d^2v}{dx^2} = \frac{M}{EI} = -\frac{Wv}{EI} \qquad (4.85)$$

It should be noted, however, that a change of geometry must be included in order to give this solution. The Euler theory, equation (4.85), gives the solution:

$$v = C_1 \sin\sqrt{\frac{W}{EI}}\,x + C_2 \cos\sqrt{\frac{W}{EI}}\,x \qquad (4.86(a))$$

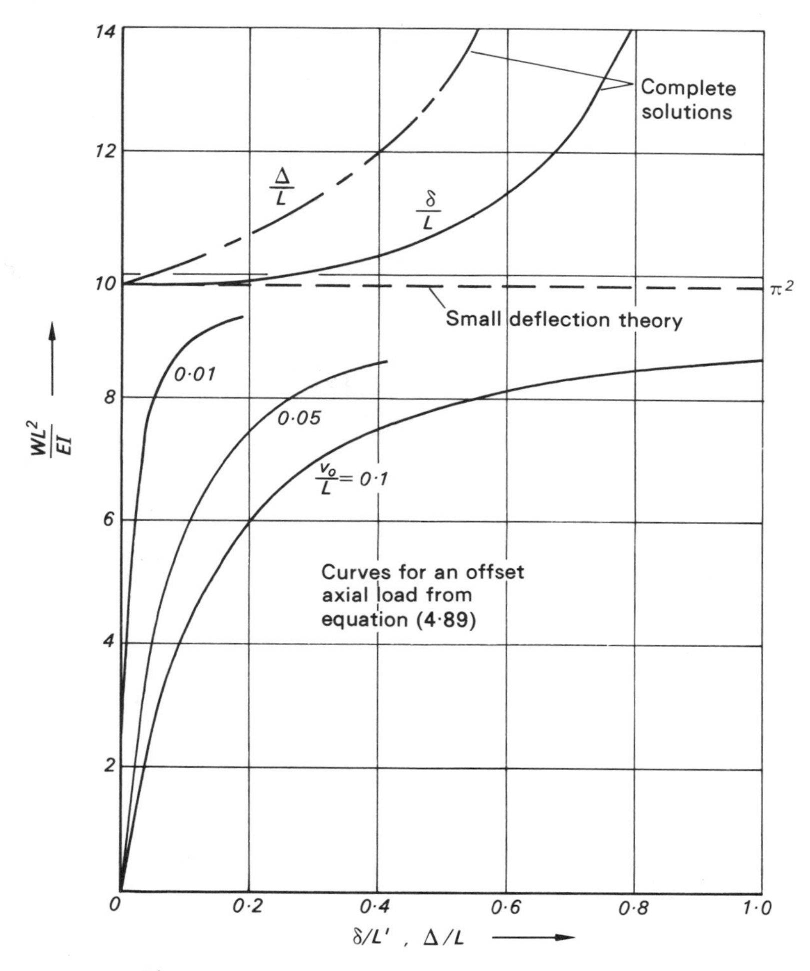

Figure 4.15 The deflection of an axially loaded beam.

and
$$\frac{dv}{dx} = C_1\sqrt{\frac{W}{EI}}\cos\sqrt{\frac{W}{EI}}\,x - C_2\sqrt{\frac{W}{EI}}\sin\sqrt{\frac{W}{EI}}\,x \qquad (4.86(b))$$

The boundary conditions for the problem are:

$$v = 0 \text{ at } x = 0$$

and
$$v = 0 \text{ at } x = L$$

Substitution of these two conditions into (4.86(a)) gives the results that:

$$C_2 = 0 \quad \text{and } 0 = C_1 \sin\sqrt{\frac{W}{EI}}\,L$$

The second of these conditions can be met for any value of C_1 if:

$$\sqrt{\frac{W}{EI}}\,L = n\pi$$

where $n = 0, 1, 2, 3, \ldots$
The deflected shape is a curve of the form:

$$v = C_1 \sin\sqrt{\frac{W}{EI}}\,x = C_1 \sin n\pi\frac{x}{L}$$

and Figure 4.16 shows the shapes indicated by the various values of n. That for $n = 0$ is trivial and $n = 1$ is a single sine wave. The nature of the solution is such that C_1 cannot be determined but the load is given by:

$$\frac{WL^2}{EI} = \pi^2$$

and is valid for all deflections. This results in the horizontal line in Figure 4.15 indicating the indeterminancy of the solution but showing that it is the limit of the large deflection analysis. It is possible to achieve equilibrium with different

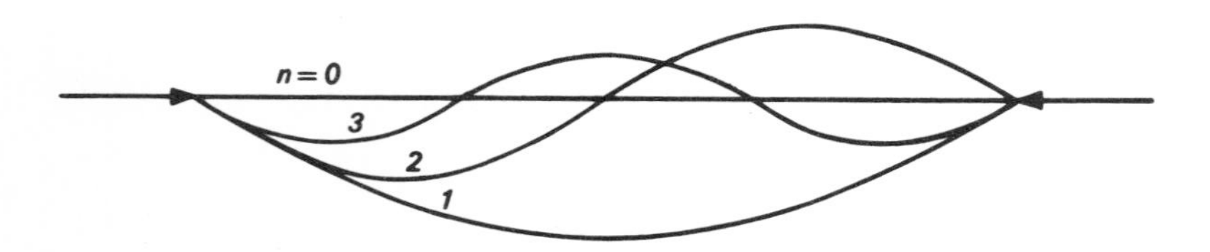

Figure 4.16 Buckling modes from the Euler theory.

configurations of sine waves for $n = 2, 3$, etc., as previously discussed and these give higher values of WL^2/EI, i.e. $4\pi^2$, $9\pi^2$, etc. The lowest value, i.e. π^2, is of greatest practical interest. Solutions for Euler's theory may be obtained for a whole range of boundary conditions and loadings and these may be found in any standard text.

An interesting extension of the buckling concept is to consider a linear visco-elastic material in which the modulus is time dependent. The case of interest is where the modulus decreases with time and thus, for a fixed W, the parameter WL^2/EI will increase until it reaches the critical value, say π^2. Thus the time at which this happens, the buckling time t_B may be calculated. For a power law material with a creep modulus of the form:

$$E(t) = E_0\, t^{-n}$$

we have,
$$E_0\, t_B{}^{-n} = \frac{WL^2}{\pi^2 I}$$

i.e.
$$t_B = \left(\frac{WL^2}{\pi^2 E_0 I}\right)^{-1/n} \tag{4.87}$$

In practice there is always a continuous increase in deflection from zero load and this may be represented by including an initial offset of v_0 for the applied load as shown in Figure 4.17. The moment now becomes:

$$M = -W(v+v_0)$$

and the solution to equation (4.85), is now:

$$v = C_1 \sin\sqrt{\frac{W}{EI}}\,n + C_2 \cos\sqrt{\frac{W}{EI}}\,x - v_0$$

If we now assume the first mode of deformation we may write the boundary conditions:

$$v = 0 \text{ at } x = 0 \text{ and } v = 0 \text{ at } x = L$$

which give the result:

$$v = v_0\left(\tan\sqrt{\frac{W}{EI}}\frac{L}{2}\,.\,\sin\sqrt{\frac{W}{EI}}\,x + \cos\sqrt{\frac{W}{EI}}\,x - 1\right) \tag{4.88}$$

If we take $v = \delta$ at $x = L/2$ we have:

$$\frac{\delta}{L} = \frac{v_0}{L}\left(\sec\sqrt{\frac{W}{EI}}\frac{L}{2} - 1\right) \tag{4.89}$$

Clearly as $WL^2/EI \rightarrow \pi^2$, $\delta/L \rightarrow \infty$ as in the limiting solution but in this case there is a curve of δ/L versus WL^2/EI. Examples of these are shown in Figure 4.15 for v_0/L values of 0·01, 0·05 and 0·10.

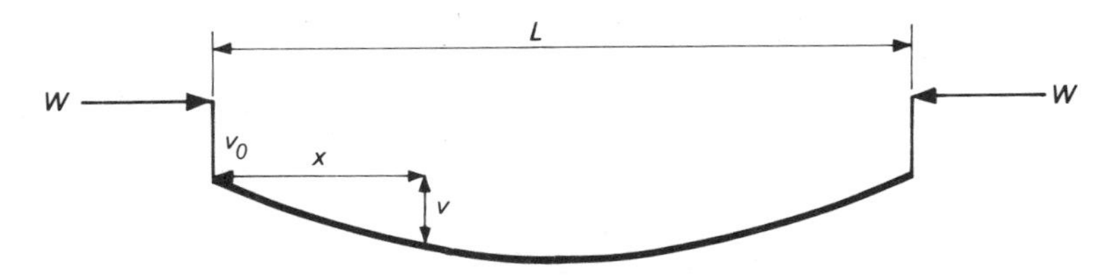

Figure 4.17 A beam with an offset axial load.

In the visco-elastic case discussed previously the creep modulus is used which assumes that the stress remains constant. This is clearly not so since the moment at any section and hence the stress at any point must increase as v increases. Thus equation (4.87) must be regarded as only a crude first approximation to the real behaviour. In practice it has been shown to work reasonably well.

4.10. Redundancy, curved beams and rings

4.10.1. *Simple beam problems with redundancy*

Redundant structures are those in which it is not possible to define the bending moments and shear forces without recourse to a deflection analysis. The simply supported beam with the central load discussed previously is statically determinate since all the bending moments and shear forces may be derived without any reference to the deflections.

Figure 4.18 shows a simple redundant structure (i.e. one which is not statically determinate) in which a centrally loaded beam has built in supports such that the slopes are zero at each end. The reactions at the supports must be $W/2$ as

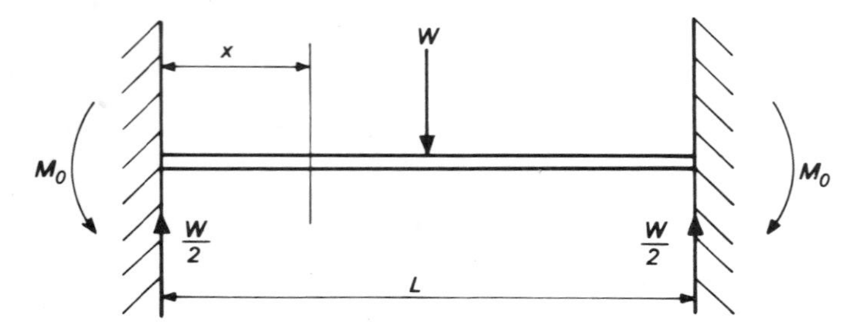

Figure 4.18 A centrally loaded beam with built in supports.

before but in addition there is an unknown bending moment M_0 at each support. Clearly any value of M_0 can give a system in equilibrium and the true value must be determined by using the properties of the particular beam. The standard method of solution is to write down the expression for the bending moment:

$$M = -\frac{W}{2}x + M_0 \tag{4.90}$$

and incorporate this in equation (4.15):

$$\frac{d^2v}{dx^2} = \frac{M}{EI}$$

giving:

$$\frac{d^2v}{dx^2} = \frac{1}{EI}\left(-\frac{W}{2}x + M_0\right) \tag{4.91}$$

Integration gives the expressions for deflections and slopes:

$$\frac{dv}{dx} = \phi = \frac{1}{EI}\left(-\frac{W}{2}\frac{x^2}{2} + M_0 x\right) + C_1$$

and

$$v = \frac{1}{EI}\left(-\frac{W}{2}\frac{x^3}{6} + M_0\frac{x^2}{2}\right) + C_1 x + C_2$$

where C_1 and C_2 are constants. The boundary conditions for the beam are: $\phi = 0$ at $x = 0$ and $L/2$ and $v = 0$ at $x = 0$ which gives both $C_1 = 0$ and $C_2 = 0$ and in addition:

$$0 = \frac{1}{EI}\left(-\frac{W}{2}\frac{L^2}{8} + M_0\frac{L}{2}\right)$$

i.e.
$$M_0 = \frac{WL}{8} \tag{4.92}$$

The central deflection is therefore given by:

$$\delta = \frac{1}{EI}\left(-\frac{W}{2}\frac{L^3}{48} + \frac{L^2}{8}\frac{WL}{8}\right)$$

i.e.
$$\delta = \frac{WL^3}{192EI} \tag{4.93}$$

A slightly different form of representation which can be used is to conduct the analysis in terms of ϕ. Thus the radius of curvature can be written

$$R = \frac{dx}{d\phi}$$

and equation (4.15) becomes:

$$\frac{d\phi}{dx} = \frac{M}{EI} \tag{4.94}$$

Now the change in deflection at any point, say A in Figure 4.19(a), can be deduced by considering the rotation of the element at O, a distance x from A:

i.e.
$$\delta v = x\,\delta\phi$$

and integrating this equation and substituting for $d\phi$ from equation (4.94) we have:

$$v = \int x\,d\phi = \int \frac{Mx\,dx}{EI} \tag{4.95}$$

The centrally loaded, simply supported beam may now be solved by taking x from the support as shown in Figure 4.19(b) and the moment becomes:

$$M = -\frac{Wx}{2}$$

giving the central deflection:

$$\delta = \int_{L/2}^{0} -\frac{W}{2EI}x^2\,dx = \frac{WL^3}{48EI}$$

For the built-in beam we must first integrate equation (4.94) to define M_0:

i.e.
$$\phi = \frac{1}{EI}\int\left(-\frac{W}{2}x+M_0\right)dx+C_1$$

$$\phi = -\frac{Wx^2}{4EI}+\frac{M_0}{EI}x+C_1$$

Now $\phi = 0$ at $x = 0$ and $L/2$

$\therefore$
$$M_0 = \frac{WL}{8} \quad \text{as before.}$$

Similarly
$$\delta = \int_{L/2}^{0}\left(-\frac{Wx}{2}+M_0\right)\frac{x}{EI}dx = \frac{WL^3}{192EI} \quad \text{as before.}$$

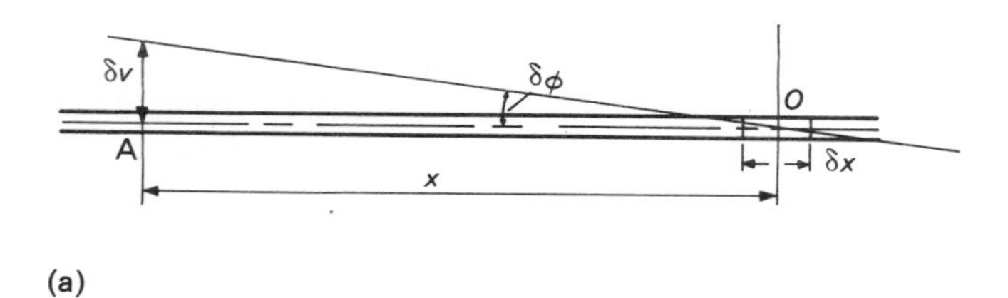

(a)

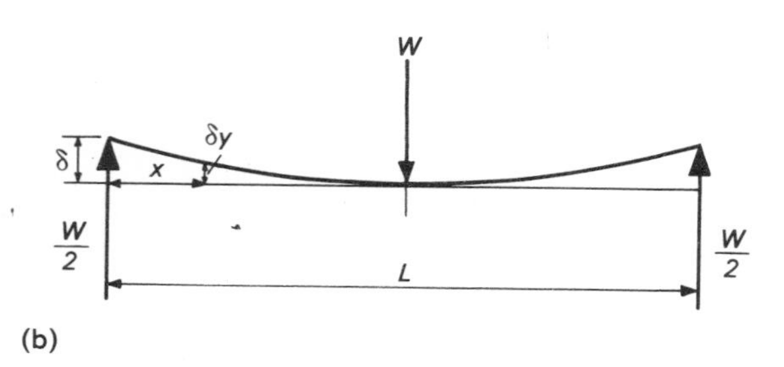

(b)

Figure 4.19 Deflection analysis in terms of slope.

4.10.2. *Curved thin beams*

The method of using slopes described in the previous section is useful for solving problems in which the beams are not initially straight. Figure 4.20 shows a section of such a beam and the slope at O is changed by $\delta\phi$ as before. The deflection at A now has two components in directions x and y as shown. As before we have:

$$\phi = \int\frac{M\,ds}{EI}$$

(noting that δx now becomes δs)

and in this case:

$$v = \int \frac{Mx \, \mathrm{d}s}{EI} \quad \text{and} \quad u = \int \frac{My \, \mathrm{d}s}{EI} \tag{4.96}$$

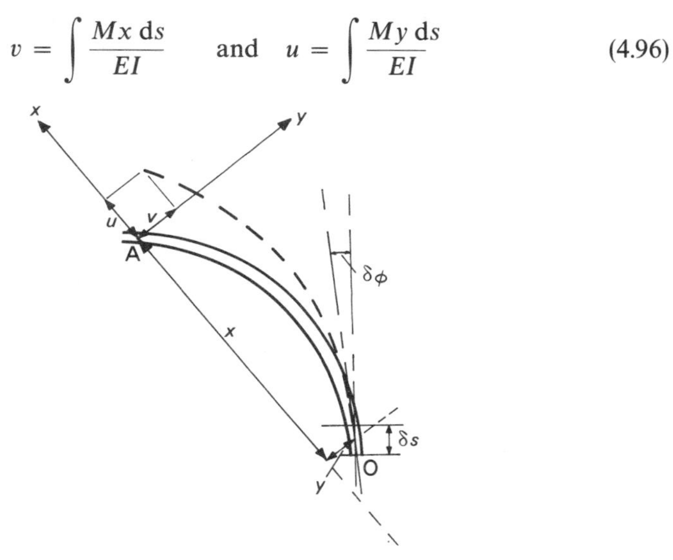

Figure 4.20 A section of a curved beam.

As a simple example consider a cantilever in the form of a quarter of a circle of radius a as shown in Figure 4.21. If we define x and y as shown we have:

$$x = a(1-\cos\theta), \qquad y = a\sin\theta, \qquad \mathrm{d}s = a\,\mathrm{d}\theta$$

and

$$M = -Wx = -Wa(1-\cos\theta)$$

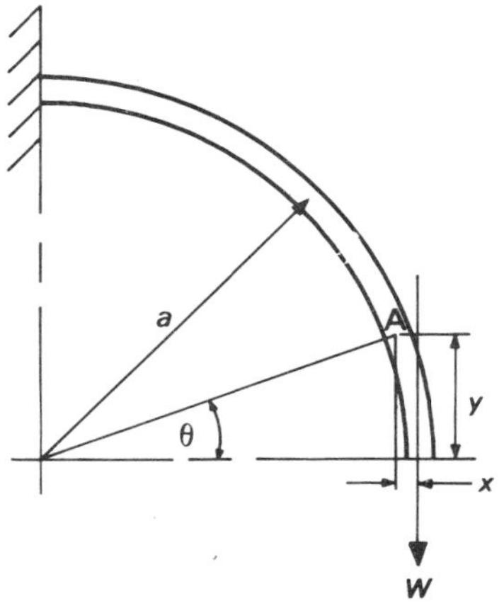

Figure 4.21 A quarter circle cantilever.

Hence the deflections at O become:

$$u = \int_{\pi/2}^{0} -\frac{Wa^3}{EI}(1-\cos\theta)\sin\theta\,\mathrm{d}\theta$$

and

$$v = \int_{\pi/2}^{0} -\frac{Wa^3}{EI}(1-\cos\theta)^2 \, d\theta$$

i.e.

$$u = \frac{Wa^3}{2EI} \quad \text{and} \quad v = \left(\frac{3\pi}{4} - 2\right)\frac{Wa^3}{EI}$$

4.10.3 *Circular rings*

When designing pipes there is often a need to analyse the case of a circular ring subjected to a diametral load as shown in Figure 4.22(a). In practice this may

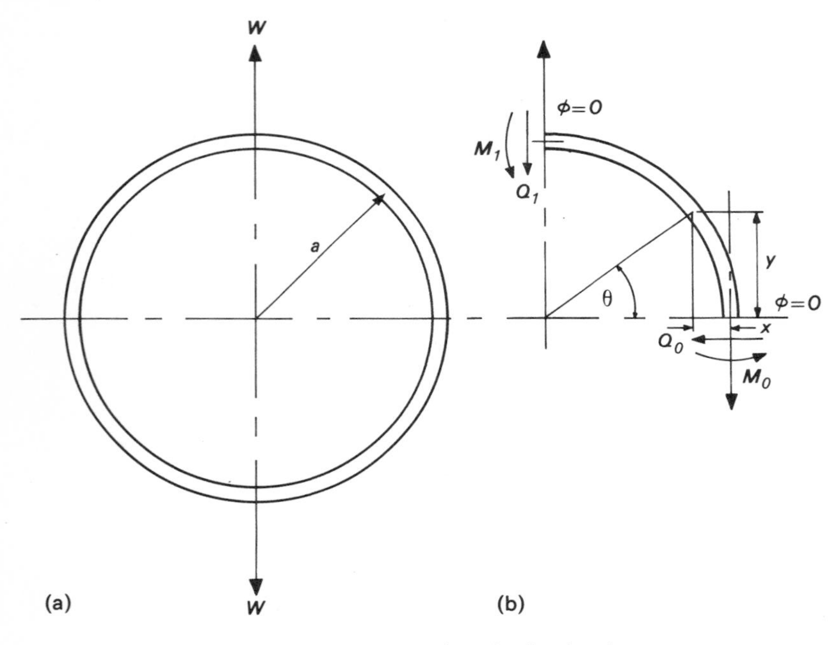

Figure 4.22 The deflection of a circular ring.

be a load on a section of pipe or it may be a load applied during a test of a ring cut from a section of pipe. Because of symmetry only one quadrant need be considered and we must introduce unknown shear forces and bending moments to maintain equilibrium as shown in Figure 4.22(b). Clearly, because of symmetry the slope $d\phi/ds = 0$ at each end of the quadrant which means that no shear forces are possible at these points:

i.e.

$$Q_0 = Q_1 = 0$$

The structure is redundant however since M_0 is still unknown and we must first examine the expression for ϕ. Now:

$$M = -\frac{Wx}{2} + M_0 = -\frac{Wa}{2}(1-\cos\theta) + M_0$$

and hence
$$\phi = \int \frac{1}{EI}\left(-\frac{Wa}{2}(1-\cos\theta) + M_0\right) a\,\mathrm{d}\theta$$

i.e.
$$\phi = -\frac{Wa}{2EI}(\theta - \sin\theta) + \frac{M_0\theta}{EI} + C_1$$

Since $\phi = 0$ at $\theta = 0$ and $\pi/2$ we have:

$$C_1 = 0 \qquad \text{and} \quad M_0 = Wa.\frac{1}{\pi}\left(\frac{\pi}{2} - 1\right)$$

i.e.
$$M_0 = Wa\left(\frac{1}{2} - \frac{1}{\pi}\right) \tag{4.97}$$

If we regard the top of the ring as fixed then the deflections at the other end become:

$$v = \int_{\pi/2}^{0} \left[-\frac{Wa}{2}(1-\cos\theta)^2 + M_0(1-\cos\theta)\right]\frac{a^2}{EI}\,\mathrm{d}\theta$$

and
$$u = \int_{\pi/2}^{0} \left[-\frac{Wa}{2}\sin\theta(1-\cos\theta) + M_0\sin\theta\right]\frac{a^2}{EI}\,\mathrm{d}\theta$$

i.e.
$$v = \left(\frac{\pi}{8} - \frac{1}{\pi}\right)\frac{Wa^3}{EI} \qquad \text{and} \quad u = \left(\frac{1}{\pi} - \frac{1}{4}\right)\frac{Wa^3}{EI} \tag{4.98}$$

The change in diameter in the direction of the load is thus given by:

$$\Delta = 2v = \left(\frac{\pi}{4} - \frac{2}{\pi}\right)\frac{Wa^3}{EI}$$

and the modulus of the material is:

$$E = \left(\frac{\pi}{4} - \frac{2}{\pi}\right)\frac{Wa^3}{\Delta I} = 0{\cdot}148\,\frac{Wa^3}{\Delta I}$$

A variant on this form of test is to split the ring on one side thus making one side ineffective. In this case, $M_1 = 0$ and $M_0 = Wa$ giving a deflection:

$$v = \frac{\pi}{4}\frac{Wa^3}{EI}$$

i.e.
$$E = \frac{\pi}{2}\frac{Wa^3}{\Delta I} = 1{\cdot}571\,\frac{Wa^3}{\Delta I} \tag{4.99}$$

This modification is frequently carried out in tests since it gives larger deflections and thus easier measurements. In addition the maximum bending moment is away from the support, i.e. it is M_0, which is advantageous if collapse loads are to be determined. For the closed ring we have:

$$M_1 = -\frac{Wa}{2} + M_0 = -\frac{1}{\pi} Wa = -0{\cdot}318 Wa$$

and

$$M_0 = \left(\frac{1}{2} - \frac{1}{\pi}\right) Wa = 0{\cdot}182 Wa$$

Thus M_1 is the greater and failure comes under the support.

If the rings used are wide then it is advisable to use the plane strain correction factor, i.e. multiply the deflection by the factor $(1-\nu^2)$, as discussed in section 4.7.

If the thickness of the ring is significant with respect to the radius then the thin ring theory must be modified. Cases where this is necessary in problems involving polymers are not common but if required the solution will be found in elasticity texts under the heading of the Winkler theory of curved beams.

4.11. Plastic collapse in bending

4.11.1. *The basic concepts*

Most polymers below their glass transition temperature can be described approximately as elastic perfectly plastic materials as described in chapter 2. The concept of the plastic collapse of a beam is illustrated in Figure 4.23, where (a) is the linear elastic stress distribution and (b) the distribution at first yield in the section. The bending moment on a rectangular section at first yield is given by:

$$M_E = \frac{bd^3}{12} \cdot p_Y \cdot \frac{2}{d} = \frac{bd^2}{6} p_Y \qquad (4.100)$$

where p_Y = the tensile yield stress.

As the moment is increased above M_E the stress cannot exceed p_Y and hence the distribution takes the form shown in (c) where the shaded sections are plastically deformed. The limit is when the plastic condition has reached the centre as shown in (d) in which case the fully plastic moment is given by:

$$M_P = \int_0^d p_Y b y \, dy \qquad (4.101)$$

which, for a rectangular section, gives:

$$M_P = -bp_Y \int_0^{d/2} y \, dy + bp_Y \int_{d/2}^{d} y \, dy$$

$$= -\frac{bd^2}{8} p_Y + \frac{3bd^2}{8} p_Y$$

i.e.
$$M_P = \frac{bd^2}{4} p_Y \tag{4.102}$$

No section of a beam may take a greater moment than this and thus the total collapse system may be computed from a bending moment distribution.

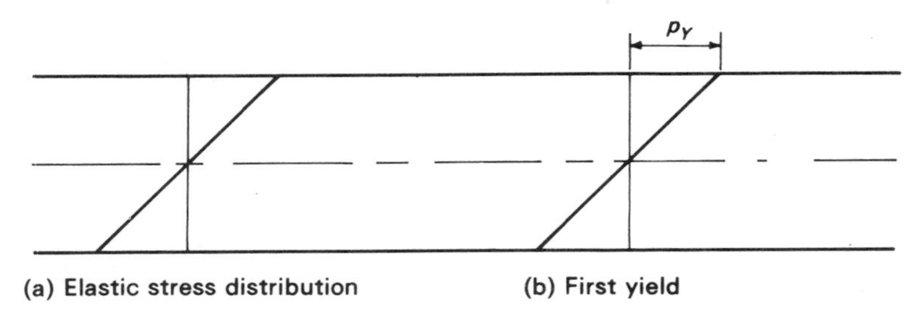

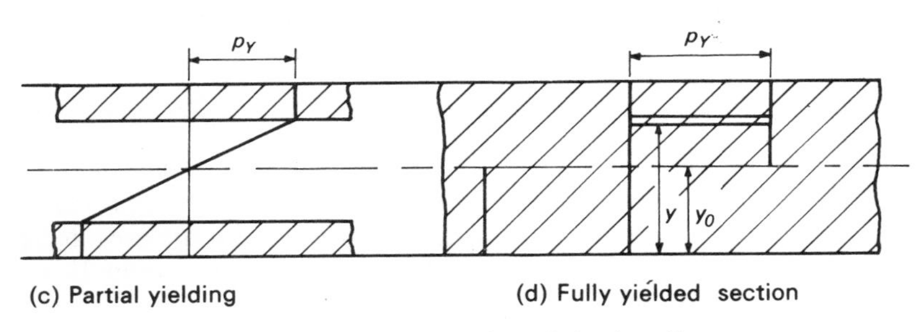

Figure 4.23 The mechanism of plastic collapse.

A simple example is the centrally loaded, simply supported beam in which the maximum moment at the centre is:

$$M = \frac{WL}{4}$$

Clearly when this reaches M_P the system can resist no increase in load since a 'plastic hinge' forms at the centre of the beam and the structure collapses. The maximum load is given by:

$$W_P = \frac{4}{L}\frac{bd^2}{4} p_Y = \frac{bd^2}{L} p_Y \tag{4.103}$$

for a rectangular section. The load at first yield is given by:

$$W_E = \frac{2}{3}\frac{bd^2}{L} p_Y = \frac{2}{3} W_P \tag{4.104}$$

A typical graph of load versus central deflection is shown in Figure 4.24. The initial slope is obtained from the elastic solution and for loads greater than W_E the curve becomes non-linear (the shape is undefined by this analysis). When the load reaches W_P the deflection increases indefinitely as shown.

In some polymers yielding in tension produces whitening and it is often

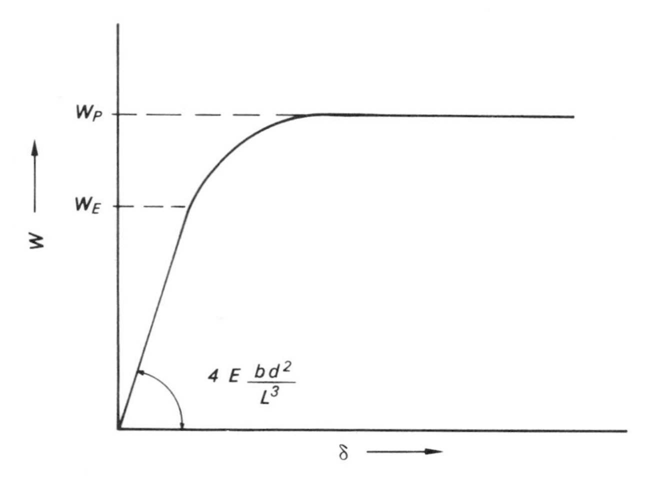

Figure 4.24 Plastic collapse in a centrally loaded simply supported beam.

possible to see the extent of the plastic region in areas of tensile stress. In the beam discussed above the bending moment decreases linearly from the centre so that for a fully plastic moment at the centre the fully plastic area decreases until the section becomes fully elastic at some point along the beam. If we let s be the distance from the centre of the beam then:

$$M = \frac{W_P}{2}\left(\frac{L}{2} - s\right)$$

Now $M = M_P$ at $s = 0$ and if $M = M_E$ at $s = \hat{s}$ we have:

$$\hat{s} = \frac{2}{W_P}(M_P - M_E)$$

which, for a rectangular section, becomes:

$$\hat{s} = \frac{2}{W_P}\left(\frac{bd^2 p_Y}{4} - \frac{bd^2 p_Y}{6}\right) = \frac{bd^2 p_Y}{6W_P} = \frac{L}{6} \tag{4.105}$$

Thus there will be two wedges of plastic deformation as shown in Figure 4.25 with the lower one, being in tension, in general well defined by whitening.

4.11.2. *Redundant structures*

Plastic collapse in redundant structures may be illustrated by reference to a built-in beam with a uniformly distributed load as shown in Figure 4.26. The elastic solution (section 4.10.1) gives the bending moments at the ends as:

$$M_0 = \frac{wL^2}{12}$$

and at the centre:

$$M_1 = -\frac{wL^2}{24}$$

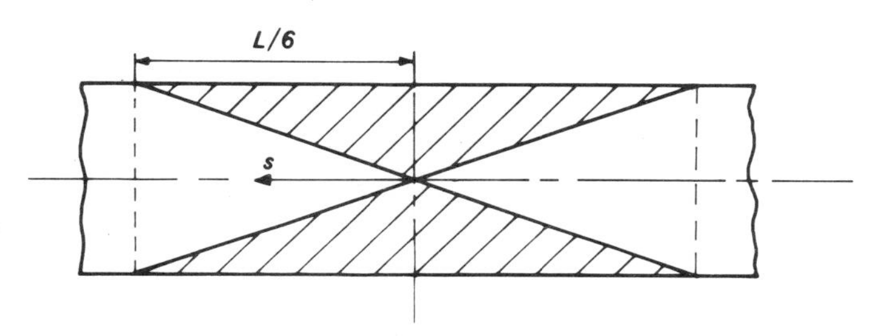

Figure 4.25 Extent of plastic flow in a centrally loaded simply supported beam at collapse.

The distribution of bending moments is also shown in Figure 4.26 and clearly the fully plastic moment will be reached at the ends first. However, this will not constitute collapse and the load can increase until it reaches M_P at the centre when the necessary three hinges for collapse occur. The load for this condition is given by:

$$-M_P = -\frac{wL}{2}\frac{L}{2} + M_P$$

i.e.

$$M_P = \frac{wL^2}{8}$$

and the collapse load is given by:

$$w = \frac{8M_P}{L^2} \qquad (4.106)$$

where for a rectangular section, M_P is given by equation (4.102).

4.11.3. *Circular rings*

The circular ring in section 4.10.3 is a further example of interest. The split ring gives the simple condition:

$$M_P = Wa \qquad (4.107)$$

while for the complete ring the collapse condition is with M_P at each end of the quadrant giving the result:

$$M_P = \frac{Wa}{4} \tag{4.108}$$

It may be noted that at 90° to the applied load there is an axial load, $W/2$, as well as a bending moment which will influence the yielding behaviour. It is easily

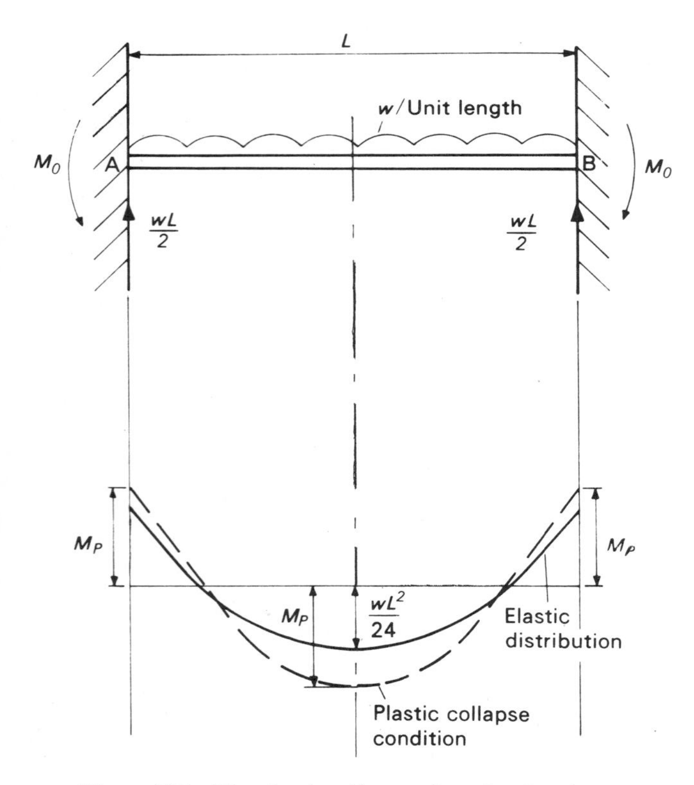

Figure 4.26 The plastic collapse of a redundant beam.

shown that, for a rectangular section, with a moment M and an axial load F, the fully plastic condition is:

$$\frac{M}{M_P} = 1 - \left(\frac{F}{F_P}\right)^2 \tag{4.109}$$

where $F_P = p_Y bd$, the full plastic load. Thus a more exact solution is to say that collapse takes place when the moment:

$$M = \frac{Wa}{2} - M_P$$

reaches a critical value with $F = W/2$. Solving this condition gives the series:

$$\frac{W}{W_P} = \frac{d}{a} - \frac{1}{8}\left(\frac{d}{a}\right)^3 + \cdots$$

The simple solution ignoring the axial load is,

$$\frac{W}{W_P} = \frac{d}{a}$$

and since $d/a \ll 1$ for thin rings, this is a good approximation.

4.12. Vibrating visco-elastic beams

Vibrations of beams frequently occur in practice and the use of the inherent damping capacity of polymers in controlling them is well established. In addition, the in-phase and out-of-phase moduli of polymers are often determined by vibrating beams. We shall confine our attention here to beam systems in which the spatial arrangements remain unaltered with time and only the load and deflections undergo harmonic variations. As discussed in chapter 3 this type of system may be analysed by simply modifying the elastic solution with an appropriate time function and in this case we shall consider that the vibrations are harmonic in form.

4.12.1. *Beams with input cyclic stress*

The method of solution is best explained by the example of the simply supported, centrally loaded beam used in previous sections. The elastic solution gives the following expression for the central deflection:

$$\delta = \frac{WL^3}{48EI} \tag{4.110}$$

The expression may be written in terms of the maximum stress in the beam given by:

$$\hat{p} = \frac{d}{2}\frac{WL}{4I} \tag{4.111}$$

and the maximum strain (equation (4.28)):

$$\hat{e} = 6\frac{\delta d}{L^2} \tag{4.112}$$

Clearly the substitution of equations (4.111) and (4.112) in equation (4.110) gives the elastic relationship:

$$\hat{e} = \frac{\hat{p}}{E} \tag{4.113}$$

If the applied stress has a sinusoidal form $p = p_0 \sin \omega t$ it has been shown in sections 3.4.5 and 3.5.3 that after the transients have died away the strain response becomes:

$$e(t) = \frac{p_0}{E'(\omega)} \sin \omega t + \frac{p_0}{E''(\omega)} \cos \omega t$$

i.e. the first term is in-phase and the second exactly out-of-phase. Direct substitution of this result in equation (4.110) gives:

$$\delta(t) = \frac{WL^3}{48I}\left(\frac{1}{E'(\omega)} \sin \omega t + \frac{1}{E''(\omega)} \cos \omega t\right) \tag{4.114}$$

Thus if a loading system can be designed with this form of input the in-phase deflection $\delta'(t)$ given by the first term and the out-of-phase $\delta''(t)$ given by the second may be found.

An example of this is the use of rotating circular beams under a dead load. In the section shown in Figure 4.27 the stress at a distance y from the central axis X–X is given by:

$$p = y \cdot \frac{M}{I}$$

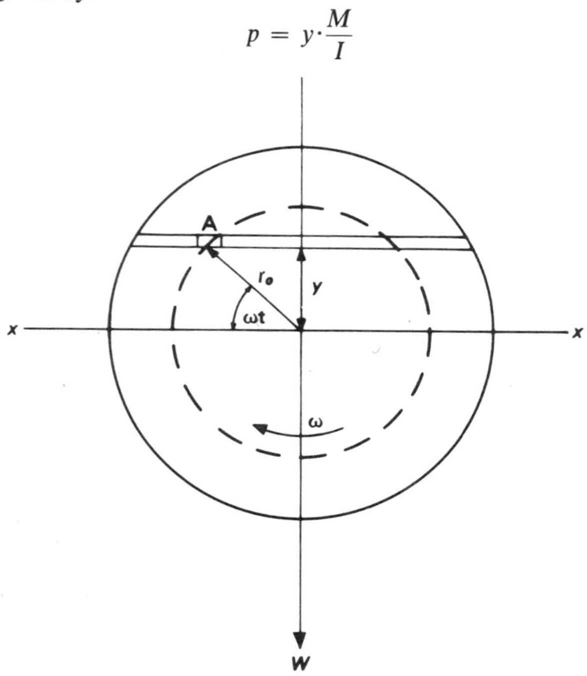

Figure 4.27 The cross-section of a rotating circular beam.

Now at any point in the section, say the small element at A, the radius from the centre is r_0 and as the beam rotates at an angular velocity ω, y varies in the form:

$$y = r_0 \sin \omega t$$

and hence:

$$p = \frac{r_0 M}{I} \sin \omega t = p_0 \sin \omega t$$

This means that at every point in the section the stress is undergoing a forced sinusoidal variation as required by the analysis. Observation of this system shows that the deflection is as shown in Figure 4.28, with the in-phase deflection:

$$\delta' = \frac{WL^3}{48I} \frac{1}{E'(\omega)} \tag{4.115}$$

in the direction of W and the out-of-phase component

$$\delta'' = \frac{WL^3}{48I} \frac{1}{E''(\omega)} \tag{4.116}$$

lagging behind. The loss factor is given by the tangent of the angle θ:

i.e.
$$\tan \theta = \frac{\delta''}{\delta'} = \frac{E'(\omega)}{E''(\omega)} \tag{4.117}$$

Note that as this is stress cycling the loss factor is the ratio $E'(\omega)/E''(\omega)$ and not $E''(\omega)/E'(\omega)$ as it is usually defined.

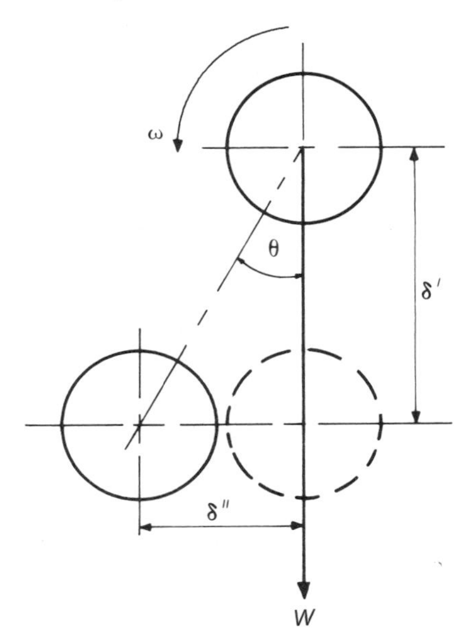

Figure 4.28 Deflection of a rotating visco-elastic beam.

Such a test is sometimes used to measure $E'(\omega)$ and $E''(\omega)$ although it is quite difficult to perform for small values of $\tan \theta$ because of instability effects.

4.12.2. *Beams with input cyclic deflections*

A more common type of test is based on some form of beam with an additional mass. For example there may be a mass M at the centre of the simply supported beam. In this case it is assumed that the variations with time will be sinusoidal and hence the material may be regarded as having an elastic component given by E' and a viscous component E''. By direct analogy with the spring and dash-pot models the force due to a central deflection δ is given by:

$$W' = \frac{\delta\, E' I 48}{L^3} \tag{4.118}$$

and the force due to the viscous effect is:

$$W'' = \frac{\dot{\delta}}{\omega} \frac{E'' I 48}{L^3} \tag{4.119}$$

By the equilibrium of forces acting on the mass M we have:

$$M\ddot{\delta} + \frac{\dot{\delta}}{\omega} \frac{E'' I 48}{L^3} + \frac{\delta E' I 48}{L^3} = F_0 \cos \omega t$$

where $F_0 \cos \omega t$ is the resulting forcing function on M.

i.e.
$$m\,\ddot{\delta} + \frac{\dot{\delta} E''}{\omega} + \delta E' = f \cos \omega t \tag{4.120}$$

where
$$m = \frac{ML^3}{48I} \quad \text{and} \quad f = \frac{F_0 L^3}{48I}$$

This is a general form and different geometries simply produce different forms for m and f. For example for a cantilever of length L we have:

$$m = \frac{ML^3}{3I} \quad \text{and} \quad f = \frac{F_0 L^3}{3I}$$

It should be noted that in this case the deflection is assumed to be prescribed and the loss factor is given by:

$$\tan \theta = \frac{E''}{E'} \tag{4.121}$$

The full solution of equation (4.120) can be expressed as:

$$\delta = \exp\left(-\frac{E''}{2m\omega} t\right)(C_1 \cos \omega_1 t + C_2 \sin \omega_1 t) + \frac{f}{E'} \frac{1}{\left(1 - \dfrac{\omega^2 m}{E'}\right)^2 + E''^2/E'^2} \left[\left(1 - \frac{\omega^2 m}{E'}\right) \cos \omega t + \frac{E''}{E'} \sin \omega t\right] \tag{4.122}$$

where
$$\omega_1^2 = \frac{E'}{m} - \left(\frac{E''}{2\omega m}\right)^2 \tag{4.123}$$

The nature of the solution may be illustrated by considering the free vibration case where $f = 0$. If we take $\delta = \delta_0$ at $t = 0$ a solution is:

$$\delta = \delta_0 \exp\left(-\frac{E''}{2m\omega}t\right)\cos \omega_1 t \tag{4.124}$$

which is shown in Figure 4.29, as a decaying harmonic function. Thus ω_1 is the natural frequency of the system which is sinusoidal as assumed in the initial derivation and hence:

$$\omega = \omega_1$$

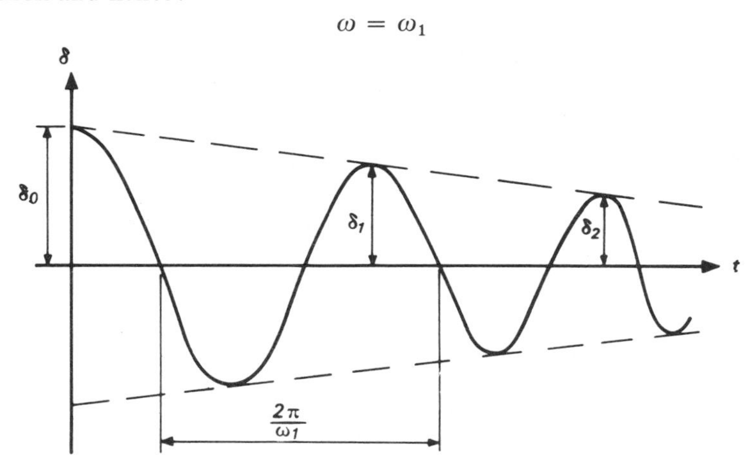

Figure 4.29 The variation of amplitude with time for the free vibration of a beam.

In fact, the wave form is not truly sinusoidal because of the damping effect but providing this is not too large it does not affect the conclusion greatly.

From equation (4.123) we have:

$$\omega_1^2 = \frac{1}{2}\frac{E'}{m}\left[1 + \sqrt{1 - \left(\frac{E''}{E'}\right)^2}\;\right] \tag{4.125}$$

and for a system with small damping:

$$E'' \ll E'$$

and
$$\omega_1^2 \simeq E'/m \tag{4.126}$$

This is, of course, the standard solution for the natural frequency of a beam of modulus E' and says that the effect of E'' on the frequency as given in equation (4.125) is negligible. Hence E' may be determined from the measured value of ω_1 and a range of values may be obtained by varying m to obtain E' as a function of ω_1.

Additional measurements which may be taken are the values of two adjacent peaks, shown as δ_1 and δ_2 in Figure 4.29, so that:

$$\delta_1 = \delta_0 \exp\left(-\frac{E''}{2m\omega_1}t\right)\cos\omega_1 t$$

and

$$\delta_2 = \delta_0 \exp\left(-\frac{E''}{2m\omega_1}\left(t+\frac{2\pi}{\omega_1}\right)\right)\cos\omega_1\left(t+\frac{2\pi}{\omega_1}\right)$$

As these are peaks $\cos\omega_1 t$ and $\cos\omega_1(t+2\pi/\omega_1)$ are both unity giving:

$$\frac{\delta_1}{\delta_2} = \exp\left(+\frac{E''\pi}{m\omega_1^{\,2}}\right)$$

The parameter logarithmic decrement is now usually introduced as:

$$\Delta = \ln\frac{\delta_1}{\delta_2} = \frac{E''\pi}{m\omega_1^{\,2}}$$

i.e.

$$\frac{\Delta\omega_1^{\,2}}{\pi} = \frac{E''}{m} \tag{4.127}$$

and hence if both Δ and ω_1 are determined from a free vibration E'' may be found. The inclusion of Δ in equation (4.125) gives the result for general damping:

$$\frac{E'}{m} = \omega_1^{\,2}\left[1+\frac{\Delta^2}{4\pi^2}\right] \tag{4.128}$$

When a forcing function is present we may consider the form of the solution for large values of t when the exponential term of equation (4.122) has died out and we are left with the expression:

$$\delta = \frac{f}{E'}\frac{1}{\left(1-\dfrac{\omega^2 m}{E'}\right)^2 + E''^2/E'^2}\left[\left(1-\frac{\omega^2 m}{E'}\right)\cos\omega t+\frac{E''}{E'}\sin\omega t\right]$$

Two special cases are of practical significance. Firstly there is the resonance test where a beam is vibrated through its natural frequency ω_1 and the amplitude measured and plotted as shown in Figure 4.30. Now for the case of small damping we may write:

$$\omega_1^{\,2} \simeq \frac{E'}{m}$$

and hence:

$$\delta = \frac{f}{E'}\frac{1}{\left\{1-\left(\dfrac{\omega}{\omega_1}\right)^2\right\}^2+\left(\dfrac{E''}{E'}\right)^2}\left[\left(1-\left(\frac{\omega}{\omega_1}\right)^2\right)\cos\omega t+\frac{E''}{E'}\sin\omega t\right]$$

At the resonance condition $\omega = \omega_1$ and hence we have

$$\hat{\delta}_0 = \frac{f}{E'}\frac{E'^2}{E''^2}\frac{E''}{E'} = \frac{f}{E''} \tag{4.129}$$

which provides a measure of E''. Secondly E' may be determined from measuring the frequency spread $2\Delta\omega$ for an amplitude $\hat{\delta}_0/2$. The frequencies for this amplitude are given by:

$$\frac{\hat{\delta}_0}{2} = \frac{f}{2E''} = \frac{f}{E'}\frac{1}{\left(1-\left(\frac{\omega}{\omega_1}\right)^2\right)^2+\left(\frac{E''}{E'}\right)^2}\left[\left(1-\left(\frac{\omega}{\omega_1}\right)^2\right)^2+\left(\frac{E''}{E'}\right)^2\right]^{1/2}$$

i.e.

$$\omega^2 = \omega_1{}^2\left[1\pm\sqrt{3}\,\frac{E''}{E'}\right]$$

and we have:

$$\frac{\Delta\omega}{\omega_1} = \frac{\sqrt{3}}{2}\frac{E''}{E'} \tag{4.130}$$

and hence E' may be found.

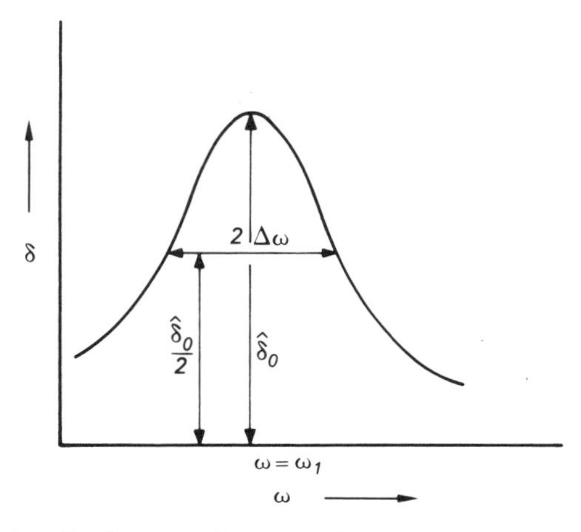

Figure 4.30 Amplitude versus frequency for the forced vibration of a beam.

A further experimental method is to test at frequencies very much less than ω_1 in which case we have:

$$\delta_0 = \frac{f}{\sqrt{(E'^2+E''^2)}} \tag{4.131}$$

The general phase angle between force and deformation is given by:

$$\tan\phi = \frac{E''/E'}{(1-\omega^2/\omega_1{}^2)} \tag{4.132}$$

and hence if $\omega \ll \omega_1$ for this case:

$$\tan \phi = E''/E'$$

It is of interest to note that otherwise $\tan \phi$ is not generally equal to the loss factor E''/E' and indeed for resonance

$$\tan \phi \to \infty$$

i.e.

$$\phi \to \pi/2$$

for all values of loss factor.

The vibration of beams without additional mass will not be considered here although they are used in some tests. In these tests one end is vibrated and various nodal patterns may be produced. The loads arise from the inertia of the self weight of the beam and the analysis is quite complex. The reader is referred to specialised texts for this.

4.12.3. *Sandwich beams*

In many practical design problems the core is employed to provide a built-in damping in a sandwich beam structure. This may be illustrated by the case of a simply supported, centrally loaded sandwich beam with thin skins in which the central deflection is given by equation (4.57) as:

$$\delta = \frac{WL^3}{24bd^2h}\left[\frac{1}{E_S} + \frac{9(2+\nu_S)}{E_C}\left(\frac{dh}{L^2}\right)\right]$$

Now if the skin is of metal then the out-of-phase component is small so that we may write:

$$E' = E_S$$

and

$$E'' = \left(\frac{L^2}{dh}\right)\frac{E_C''}{9(2+\nu_S)}$$

and hence the loss factor is:

$$\tan \theta = \frac{E''}{E'} = \frac{E_C''}{E_S}\frac{1}{9(2+\nu_S)}\left(\frac{L^2}{dh}\right)$$

i.e. the loss factor for the beam contains a geometrical factor and can be designed to a particular value.

Bibliography

1. Den Hartog, J. P. *Strength of Materials*, Dover Publications Inc., New York (1949).
2. Ford, H. *Advanced Mechanics of Materials*, Longmans, London (1963).

3. Allen, Howard G. *Analysis and Design of Structural Sandwich Panels*, Pergamon Press (1969).
4. Freeman, J. G. 'Mathematical Theory of Deflection of Beams'. *Phil. Mag.* (1947) **37**.
5. Staverman, A. J. and Schwarzl, F. *Die Physik der Hochpolymeren*, Stuart, H. A. (ed.), Vol. 4, Chapters 1 and 2, Springer-Verlag (1956).

5 Problems with axial symmetry

5.1. Introduction

This chapter will be concerned with the solution of problems in which there is an axis of geometric symmetry in the bodies and loading considered. Such a condition produces considerable simplifications in the analysis of two- and three-dimensional problems and is also of wide practical interest because many engineering components have circular sections. The largest class of problems is concerned with right circular cylinders which will be considered under various loading systems and for a range of material behaviour. A general description of axisymmetric membrane shells will also be given together with illustrations of their applications in design and thermoforming.

5.2. Thin-walled circular cylinders

Thin-walled circular cylinders are practically important systems and are defined by an initial internal diameter D_0 and initial wall thickness h_0 as shown in Figure 5.1(a), where:

$$h_0 \ll D_0$$

but a consideration of stress is required to define a limit on this condition. Consider a cylinder, shown in Figure 5.1(b) of diameter D, wall thickness h and length L with an internal pressure P. The average hoop stress p_θ is derived from consideration of equilibrium across a diameter:

$$p_\theta\, 2Lh = PDL$$

i.e.

$$p_\theta = \frac{PD}{2h} \tag{5.1}$$

Clearly the radial stress in the wall, p_r, is zero at the outer surface and equal to $-P$ at the inner so that the mean value is of order $P/2$. Thus p_θ is greater than p_r by the factor D/h, which for thin cylinders is a large number. Thus the condition used in solutions is that $p_r \ll p_\theta$ and that p_r may be assumed zero. This assumption is examined later under the heading of thick-walled cylinders where p_r is not taken as zero.

5.2.1. *Small strain solutions*

Small strain solutions are extremely simple since, as the strains are assumed to be small, the changes of dimensions in the cylinder may be ignored and the stresses are determined directly from the initial dimensions. Since the average stress is used in the thin-walled theory the displacements are taken at the mid-section of the thickness such that:

$$e_\theta = \frac{2u_r}{D_0 + h_0} \tag{5.2}$$

where u_r is the radial displacement.

For the special case of Hookean materials we may write:

$$e_\theta = \frac{1}{E}[p_\theta - \nu p_z] \tag{5.3}$$

since p_r is assumed small.

(i) *Internal pressure loading*
For the case of internal pressure we have:

$$p_\theta = \frac{PD_0}{2h_0}$$

and we may take three end conditions:

(a) Open ends: $p_z = 0$ and hence there is simple tension in the wall:

i.e.
$$e_\theta = \frac{PD_0}{2h_0 E}$$

and from equation (5.2):

$$u_r = \frac{P}{4E}\frac{D_0}{h_0}(D_0 + h_0) \tag{5.4}$$

(b) Closed ends. In this case we consider axial equilibrium giving:

$$p_z\,\pi(D_0 + h_0)h_0 = P\frac{\pi D_0^2}{4}$$

i.e.
$$p_z = \frac{P\,D_0^2}{4h_0(D_0 + h_0)}$$

and hence:

$$u_r = \frac{P}{4E}\frac{D_0}{h_0}(D_0 + h_0)\left(1 - \frac{\nu}{2}\frac{D_0}{D_0 + h_0}\right) \tag{5.5}$$

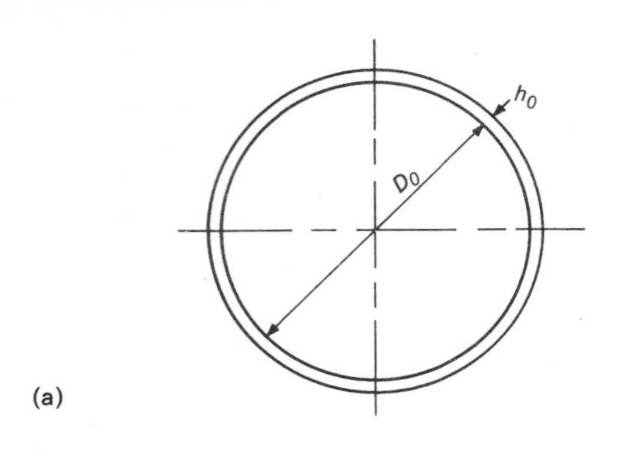

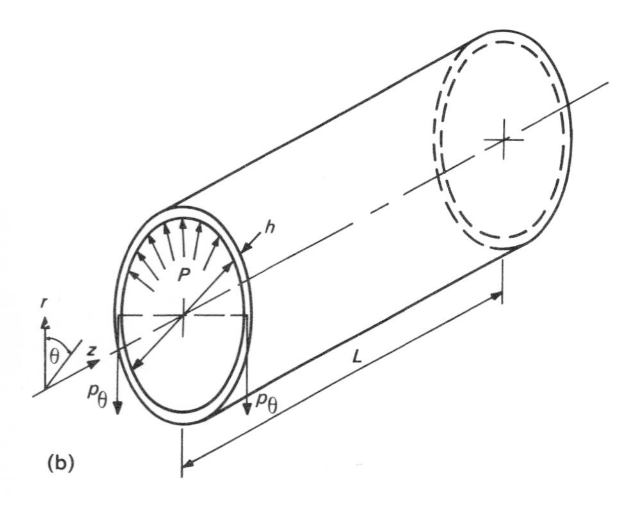

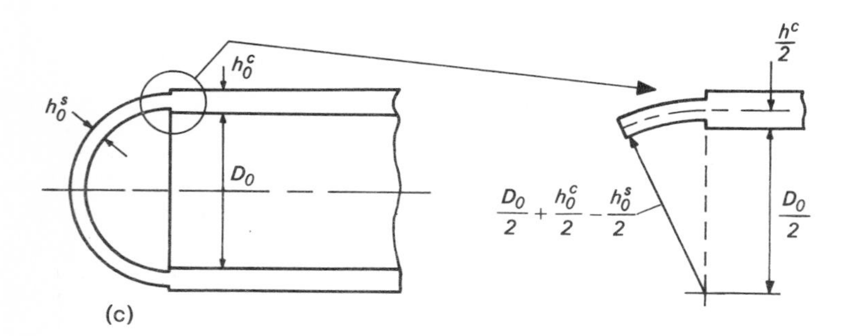

Figure 5.1 Thin-walled circular cylinders.

(c) Plane strain. Here we assume axial restraint giving $e_z = 0$ and hence:

$$0 = e_z = \frac{1}{E}[p_z - \nu p_\theta]$$

i.e.

$$p_z = \nu p_\theta$$

and

$$u_r = \frac{P}{4E}\frac{D_0}{h_0}(D_0 + h_0)(1 - \nu^2) \tag{5.6}$$

Simple calculations may be performed with this sort of analysis such as the design of a spherical end cap for a pressure vessel (e.g. a fire extinguisher body) as shown in Figure 5.1(c). If local bending stresses are to be avoided at section A then the displacements must be made equal. If the centre lines are made coincident and the cylinder and sphere thicknesses are $h_0{}^c$ and $h_0{}^s$ respectively then the sphere internal diameter is:

$$D_0 + h_0{}^c - h_0{}^s$$

By equilibrium the stresses in the sphere are:

$$p_\psi = p_\theta = \frac{P}{4h_0{}^s}\frac{(D_0 + h_0{}^c - h_0{}^s)^2}{(D_0 + h_0{}^c)}$$

and the displacement is:

$$u_r^s = \frac{P}{4E}\frac{(D_0 + h_0{}^c - h_0{}^s)^2}{h_0{}^s}\frac{(1-\nu)}{2}$$

The cylinder now has the closed end condition giving, from equation (5.5):

$$u_r^c = \frac{P}{4E}\frac{D_0}{h_0{}^c}(D_0 + h_0{}^c)\left(1 - \frac{\nu}{2}\frac{D_0}{D_0 + h_0{}^c}\right)$$

Equating the displacements we have:

$$\frac{(D_0 + h_0{}^c - h_0{}^s)^2}{h_0{}^s}(1-\nu) = \frac{D_0}{h_0{}^c}\left(2 - \nu\frac{D_0}{D_0 + h_0{}^c}\right)(D_0 + h_0{}^c)$$

which may be solved for $h_0{}^s$ for any value of $h_0{}^c$ as a quadratic. For very small wall thicknesses ($h_0{}^s, h_0{}^c \ll D_0$) this reduces to:

$$h_0{}^s = h_0{}^c\left(\frac{1-\nu}{2-\nu}\right) \tag{5.7}$$

Thus the thickness ratio depends only on Poisson's ratio and since the time dependence of this is not large the ratio will give the required condition even when the modulus varies with time.

It should be noted that the expressions derived here contain terms in $(D_0 + h_0)$ and since h_0 is generally very much less than D_0 this can often be taken as D_0. However in many calculations the extra precision is justified since the stress error is decreased which is particularly important in correlating calculated and experimental results.

(ii) *Yielding with internal pressure*

The yield condition for the thin-walled case is obtained from the Von Mises criterion:

$$p_Y^{\,2} = p_\theta^{\,2} + p_z^{\,2} - p_\theta p_z$$

where p_Y is the tensile yield stress.
For open ends this gives:

$$p_\theta = p_Y$$

and for closed ends:

$$p_\theta = \frac{2}{\sqrt{3}} p_Y \qquad (5.8)$$

and for plane strain:

$$p_\theta = \frac{p_Y}{1 - v + v^2}$$

for $h_0 \ll D_0$.

It should be noted that when $v = 1/2$, equations (5.8) give the same p_θ for plane strain and closed end conditions indicating a zero axial strain for closed ends when $v = 1/2$.

The yielding of the vessel with hemispherical ends is an interesting example in that yield pressure for the cylindrical portion is:

$$P^c = \frac{4}{3} p_Y \frac{h_0^{\,c}}{D_0}$$

and for the spherical region: $p_\theta = p_Y$ and hence:

$$P^S = \frac{4 p_Y h_0^{\,s}}{D_0}$$

for small wall thicknesses. If the equal displacement criterion is used then from equation (5.8):

$$P^S = \frac{4 p_Y h_0^{\,c}}{D_0}\left(\frac{1-v}{2-v}\right)$$

i.e.

$$P^S = \sqrt{3}\left(\frac{1-v}{2-v}\right) P^C \qquad (5.9)$$

and for the practical range of v from 1/3 to 1/2, P^S is less than P^C and yielding occurs in the spherical dome first.

(iii) *Torsion*

If a torque T is applied to a thin-walled cylinder then we again assume that there is an average stress over the wall section, in this case $p_{\theta z}$, as shown in Figure 5.2. By equilibrium we have:

$$T = \frac{\pi(D_0+h_0)^2}{2} h_0 \, p_{\theta z}$$

i.e.

$$p_{\theta z} = \frac{2T}{\pi(D_0+h_0)^2 \, h_0} \tag{5.10}$$

The shear strain is given by:

$$e_{\theta z} = \frac{u_\theta}{L_0} = \left(\frac{D_0+h_0}{2L_0}\right)\theta$$

where θ is the angle of twist over a length L_0. For the special case of a linearly elastic material we have:

$$e_{\theta z} = \frac{p_{\theta z}}{G}$$

and hence

$$\theta = \frac{4L_0 T}{\pi(D_0+h_0)^3 \, h_0}\frac{1}{G} \tag{5.11}$$

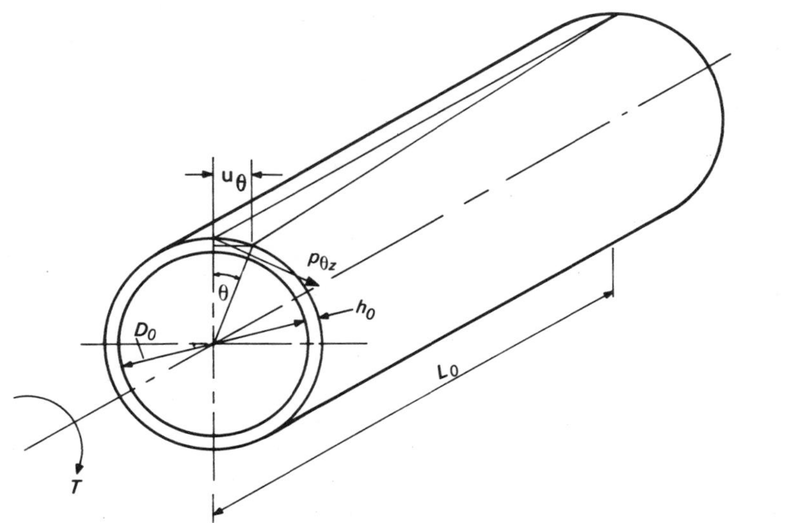

Figure 5.2 Torsion of a thin-walled circular cylinder.

The yielding condition is again given by Von Mises and since all other stress components are zero:

$$2p_Y^2 = 6{p_{\theta z}}^2$$

i.e.

$$p_{\theta z} = \sqrt{3}\, p_Y$$

and the yield torque is:

$$T^C = \frac{\pi}{2\sqrt{3}} p_Y (D_0 + h_0)^2 h_0 \tag{5.12}$$

(iv) *Combined loading*

The stresses produced by a torque, a pressure acting on closed ends and an axial load F may be easily computed from equilibrium:

$$p_\theta = \frac{PD_0}{2h_0}, \quad p_{\theta z} = \frac{2T}{\pi(D_0 + h_0)^2 h_0}$$

and

$$p_z = \frac{D_0^2}{(D_0 + h_0)} \frac{P}{4h_0} + \frac{F}{(D_0 + h_0)h_0} \tag{5.13}$$

The strains may be expressed in the usual way in terms of u_z, the axial deflection, θ the angle of twist and u_r the radial displacement, giving the results:

$$e_\theta = \frac{2u_r}{(D_0 + h_0)}, \quad e_{\theta z} = \left(\frac{D_0 + h_0}{2L_0}\right)\theta \quad \text{and} \quad e_z = \frac{u_z}{L_0} \tag{5.14}$$

In experimental work the results of these tests are frequently compared on the basis of equivalent values (see section 2.3.) and here these may be calculated from:

$$\bar{p}^2 = p_\theta{}^2 + p_z{}^2 - p_\theta p_z + 3p_{\theta z}{}^2 \tag{5.15}$$

for the stresses. The strains present an extra problem in that e_r is required and is not generally measured. If a linear elastic material is assumed then it may be expressed in terms of Poisson's ratio as:

$$e_r = -\frac{\nu}{1 - \nu}(e_\theta + e_z)$$

and

$$\bar{e}^2 = \frac{1}{(2 - \nu)(1 - \nu)}\left[e_\theta{}^2 + e_z{}^2 + \frac{3\nu - 1}{\nu} e_\theta e_z + \frac{1 - \nu}{\nu} e_{\theta z}^2\right] \tag{5.16}$$

The special case of constant volume is frequently used in which $\nu = 1/2$ and hence:

$$\bar{e}^2 = \tfrac{4}{3}\left[e_\theta{}^2 + e_z{}^2 + e_\theta e_z + e_{\theta z}{}^2\right] \tag{5.17}$$

5.2.2. *Large strain solutions*

In this section we shall consider finite strains where the changes in the dimensions during straining are significant. Attention will be restricted to internal pressure loading with closed ends although the methods used have a general application. The hoop stress is given by:

$$p_\theta = \frac{PD}{2h} \tag{5.18}$$

as described in section 5.2.1.(i) where D and h are the current values and for the purposes of the large strain analysis no thickness correction will be made to the expression for p_z so that we have:

$$p_z = \frac{PD}{4h} \tag{5.19}$$

and hence $p_z = p_\theta/2$.

The strains are most conveniently expressed in terms of extension ratios

$$\lambda_\theta = \frac{D}{D_0}, \lambda_z = \frac{L}{L_0} \quad \text{and} \quad \lambda_h = \frac{h}{h_0} \tag{5.20}$$

A constant volume material will be assumed in all cases such that:

$$\lambda_\theta \lambda_z \lambda_h = 1 \tag{5.21}$$

and combining equations (5.18), (5.20) and (5.21) we have:

$$p_\theta = \left(\frac{PD_0}{2t_0}\right) \lambda_\theta^2 \lambda_z \tag{5.22}$$

The stress-strain relationships appropriate to polymers at large strains depend very much on the material and conditions of use and two typical forms of constitutive relationship will be used here. The first involves the use of equivalent stress and strain and the total strain relationship as derived from plasticity theory, (section 2.3), which works well with polymers in the glassy state. These will be called equivalence solutions. The second will be rubber elasticity to describe polymers in their rubbery state above the glass transition temperature (see introduction to chapter 2).

(i) *Equivalence solution*

This method of solution was discussed under the headings of plasticity and rubber elasticity theory but no distinction need be made here since we will deal only with stress systems with constant stress ratios with no unloading. The analysis assumes a unique relationship between the equivalent stress (equation (2.56)):

$$\bar{p} = \frac{1}{\sqrt{2}} [(p_1 - p_2)^2 + (p_2 - p_3)^2 + (p_3 - p_1)^2]^{\frac{1}{2}}$$

and the equivalent strain (equation (2.56)):

$$\bar{\varepsilon} = \frac{\sqrt{2}}{3} [(\varepsilon_1 - \varepsilon_2)^2 + (\varepsilon_2 - \varepsilon_3)^2 + (\varepsilon_3 - \varepsilon_1)^2]^{\frac{1}{2}}$$

for constant stress ratios i.e. $p^{**} = 0$ and $p^* = p$ in the elastic analysis.

For the plane stress system considered here the stress-strain relationships (equation (2.57)) are:

$$\left.\begin{aligned} \varepsilon_\theta &= \frac{\bar{\varepsilon}}{\bar{p}}(p_\theta - \tfrac{1}{2}p_z) \qquad &\text{(a)} \\ \text{and} \qquad \varepsilon_z &= \frac{\bar{\varepsilon}}{\bar{p}}(p_z - \tfrac{1}{2}p_\theta) \qquad &\text{(b)} \end{aligned}\right\} \tag{5.23}$$

Clearly from equation (5.23b), since $p_z = \frac{1}{2}p_\theta$, then $\varepsilon_z = 0$ and there is no axial strain, i.e. $\lambda_z = 1$. From equation (5.23a) we have:

$$\varepsilon_\theta = \frac{3}{4}\frac{\bar{\varepsilon}}{\bar{p}}p_\theta$$

and from the definition of $\bar{p}$:

$$\bar{p} = \frac{\sqrt{3}}{2}p_\theta \tag{5.24}$$

since p_z and p_θ are principal values, and hence:

$$\varepsilon_\theta \doteq \frac{\sqrt{3}}{2}\bar{\varepsilon} \tag{5.25}$$

Substituting from equation (5.24) in (5.22) and noting that $\lambda_z = 1$ gives the result:

$$\left.\begin{aligned} \frac{PD_0}{2h_0} &= \frac{2}{\sqrt{3}}\frac{\bar{p}}{\lambda_\theta^2} \\ \text{and from equation (5.25)} \qquad \bar{\varepsilon} &= \frac{2}{\sqrt{3}}\ln\lambda_\theta \end{aligned}\right\} \tag{5.26}$$

If there is an experimental curve connecting $\bar{p}$ and $\bar{\varepsilon}$, obtained in a simple tension test say, then for each value of λ_θ, $\bar{\varepsilon}$ may be found. Reference to the curve will give $\bar{p}$ and substitution will give the parameter $PD_0/2h_0$. Thus it is possible to construct a general solution for any stress-strain relationship.

A simple analytical form often used is:

$$\bar{p} = p_0\,\bar{\varepsilon}^n$$

and substitution in equation (5.26) gives the result:

$$\frac{PD_0}{2h_0} = p_0\left(\frac{2}{\sqrt{3}}\right)^{1+n}\frac{(\ln\lambda_\theta)^n}{\lambda_\theta^2}$$

or in a dimensionless form:

$$\frac{1}{p_0}\left(\frac{PD_0}{2h_0}\right) = \left(\frac{2}{\sqrt{3}}\right)^{1+n} \frac{(\ln \lambda_\theta)^n}{\lambda_\theta^{\,2}} \tag{5.27}$$

Figure 5.3 shows this equation plotted for various values of n together with the corresponding p/p_0 versus λ curves for simple tension. The shapes of the stress-strain curves are typical of polymers in the glassy state which yield and have convex stress-strain curves. $n = 0$ is a rigid-plastic material and the

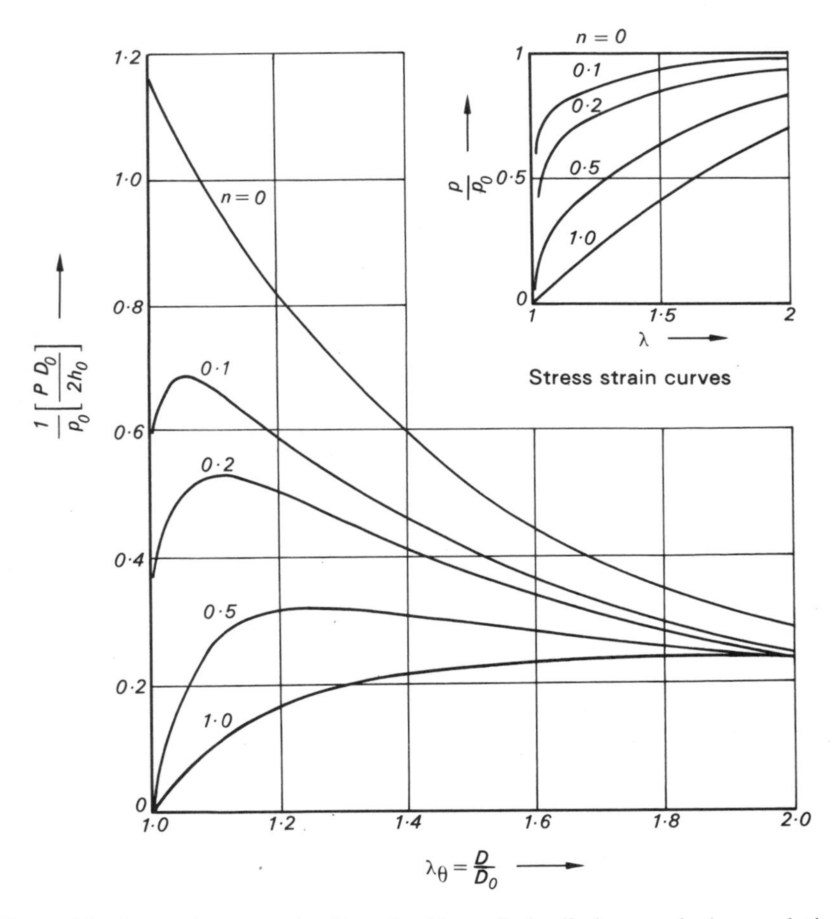

Figure 5.3 Internal pressure loading of a thin-walled cylinder—equivalence solution.

cylinder curve shows, as expected, that there is no diametral change until the yield condition, i.e. $1/p_0\,(PD_0/2h_0) = 2/\sqrt{3}$, is reached. As the diameter increases the pressure drops since there is a decrease in wall thickness but a constant stress. For other values of n the initial increase in stress more than compensates for the changes in dimensions and the pressure increases for increasing diameter. As the stress-strain curve becomes flatter at larger strains, however, the dimensional changes become predominant and the pressure falls with increasing diameter. Thus there is a peak in the curve defining the onset of unstable behaviour. Differentiation of equation (5.27) and equating to zero gives the condition:

$$\hat{\lambda}_\theta = \exp(n/2) \tag{5.28}$$

This instability condition is analogous to the onset of necking in simple tension and is often the point at which bulging and thinning take place in cylinders.

(ii) *Rubber elasticity*

The basic stress-strain relationships for rubber elasticity were described in section 2.2.2 and for the case of pure deformation, i.e. without shear, appropriate to this axisymmetric case, equations (2.34), we have:

$$p_\theta = \lambda_\theta^2\phi - \frac{\psi}{\lambda_\theta^2} + p$$

$$p_z = \lambda_z^2\phi - \frac{\psi}{\lambda_z^2} + p$$

$$p_r = \lambda_h^2\phi - \frac{\psi}{\lambda_h^2} + p.$$

where p is the arbitrary hydrostatic pressure for the assumed constant volume deformation and ϕ and ψ are constants. For the thin-walled assumption we have:

$$p_r = 0$$

and hence:

$$p = -\lambda_h^2\phi + \frac{\psi}{\lambda_h^2} = -\frac{\phi}{\lambda_\theta^2\lambda_z^2} + \psi\lambda_\theta^2\lambda_z^2$$

and substitution gives:

$$\left.\begin{aligned} p_\theta &= (\phi + \psi\lambda_z^2)\left(\lambda_\theta^2 - \frac{1}{\lambda_\theta^2\lambda_z^2}\right) \\ \text{and} \qquad p_z &= (\phi + \psi\lambda_\theta^2)\left(\lambda_z^2 - \frac{1}{\lambda_\theta^2\lambda_z^2}\right) \end{aligned}\right\} \tag{5.29}$$

The basic condition for the solution is:

$$p_z = \tfrac{1}{2}\,p_\theta$$

and hence:

$$(\phi+\psi\lambda_\theta^2)\left(\lambda_z^2-\frac{1}{\lambda_\theta^2\lambda_z^2}\right) = \tfrac{1}{2}(\phi+\psi\lambda_z^2)\left(\lambda_\theta^2-\frac{1}{\lambda_\theta^2\lambda_z^2}\right)$$

Rearranging this gives a quadratic for λ_z^2 in terms of λ_θ:

$$\lambda_z^4\left(\phi+\psi\frac{\lambda_\theta^2}{2}\right)-\lambda_z^2\left(\phi\frac{\lambda_\theta^2}{2}-\psi\frac{1}{2\lambda_\theta^2}\right)-\left(\phi\frac{1}{2\lambda_\theta^2}+\psi\right) = 0 \qquad (5.30)$$

and substitution from equation (5.22) into equation (5.29) gives the result:

$$\frac{PD_0}{2h_0} = (\phi+\psi\lambda_z^2)\left(\frac{1}{\lambda_z}-\frac{1}{\lambda_\theta^4\lambda_z^3}\right) \qquad (5.31)$$

Thus a complete solution is obtained since for any value of λ_θ a value of λ_z may be found from equation (5.30) and substituting in equation (5.31) will give the appropriate value of $PD_0/2h_0$.

The general solution is rather cumbersome and does not yield a simple result but the nature of the solution may be illustrated with the assumption that $\psi = 0$. This is not unreasonable since ψ is considerably less than ϕ for most polymers. Equation (5.30) now reduces to:

$$\lambda_z^4-\frac{1}{2}\lambda_\theta^2\lambda_z^2-\frac{1}{2\lambda_\theta^2} = 0$$

and hence:

$$\lambda_z^2 = \frac{\lambda_\theta^2}{4}\left(1+\sqrt{1+\frac{8}{\lambda_\theta^6}}\right) \qquad (5.32)$$

and on substituting into equation (5.31) we have:

$$\frac{1}{\phi}\left(\frac{PD_0}{2h_0}\right) = \frac{2}{\lambda_\theta[1+(1+8/\lambda_\theta^6)^{1/2}]^{1/2}}\left[1-\frac{4}{\lambda_\theta^6[1+(1+8/\lambda_\theta^6)^{1/2}]}\right] \qquad (5.33)$$

This equation is shown plotted in Figure 5.4 and the maximum in the pressure is evident as in the equivalence solution (Figure 5.3). The maximum is at a similar value of λ_θ to the $n = 1$ case of Figure 5.3 and the similarity of the simple tension stress-strain curves, as shown in Figures 5.3 and 5.4, would explain this. In fact the rubber elasticity curves are predominantly concave but this is only a slight effect for $\psi = 0$. For simple tension, the stress-strain curve has the form (equation (2.36)):

$$p_T = \phi\left(\lambda^2-\frac{1}{\lambda}\right)$$

The predominant difference in the solutions is in the behaviour of λ_z as shown in Figure 5.5 and derived from equation (5.32). λ_z has an appreciable variation with λ_θ as opposed to the equivalence solution in which $\lambda_z = 1$.

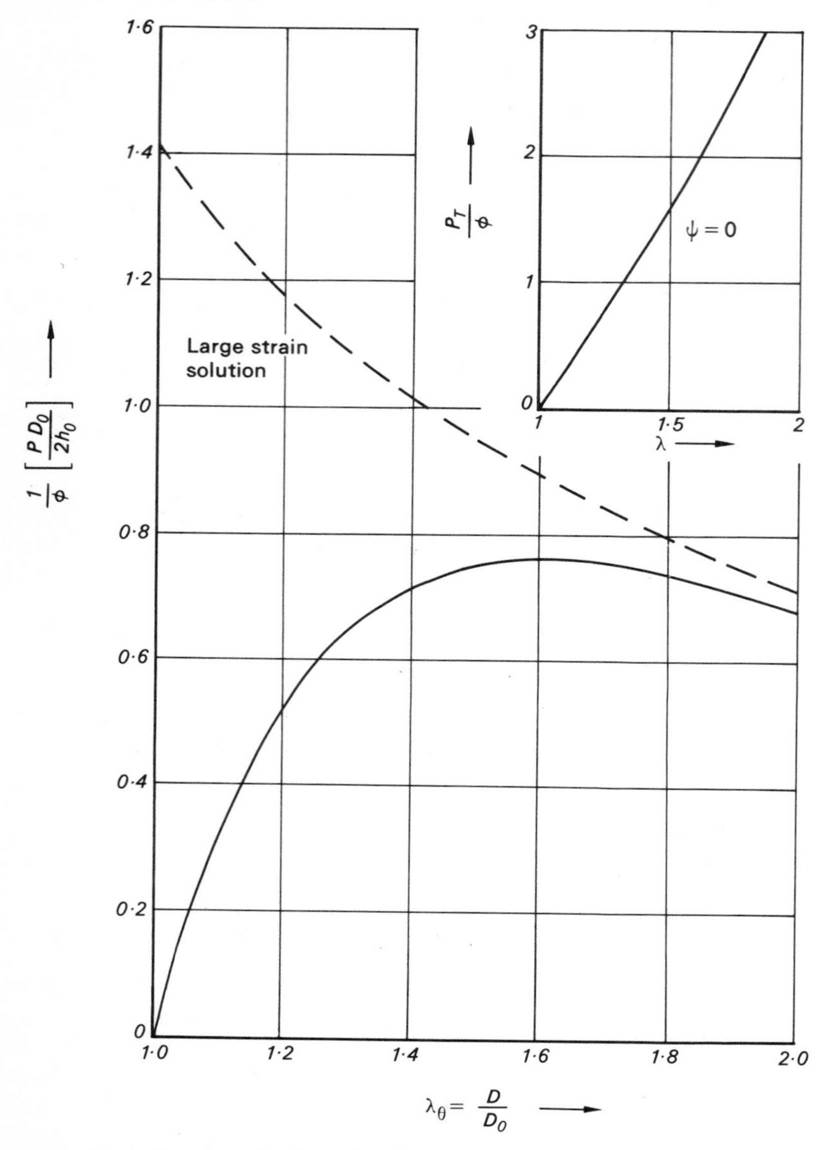

Figure 5.4 Internal pressure loading of a thin-walled cylinder—rubber elasticity solution.

A further result of some interest is a large strain simplification in which, for large values of λ_θ, equation (5.32), reduces to:

$$\lambda_z = \frac{\lambda_\theta}{\sqrt{2}}$$

giving the result:

$$\frac{1}{\phi}\left(\frac{PD_0}{2h_0}\right) = \frac{\sqrt{2}}{\lambda_\theta} \tag{5.34}$$

Figures 5.4 and 5.5 include this simplification and it can be seen that at $\lambda_\theta = 2$ the agreement is quite good. The maximum is not predicted but if interest is confined to very large strain behaviour it can be useful.

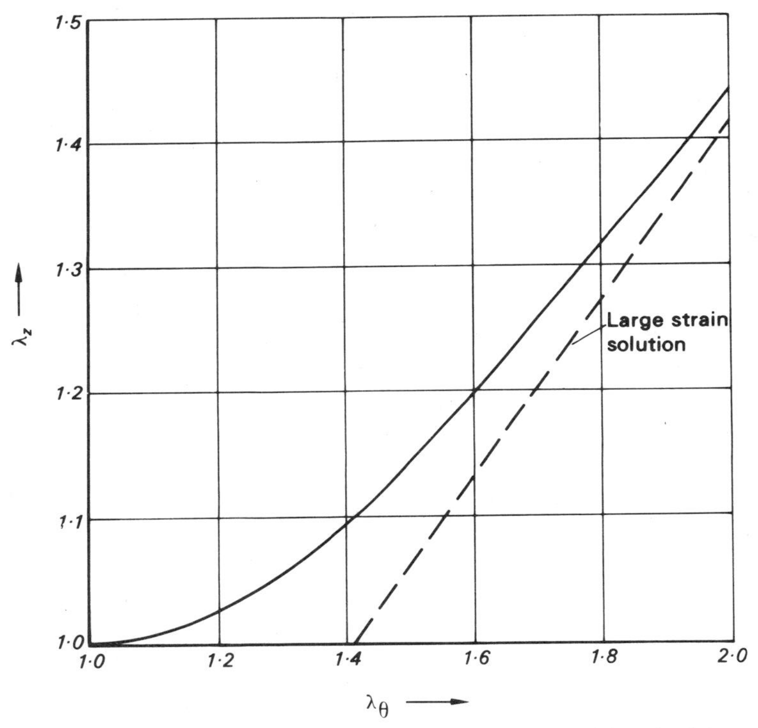

Figure 5.5 Increase in axial length in a thin-walled cylinder—rubber elasticity solution.

A similar, but simpler, problem is that of a thin-walled sphere in which the two stresses, and hence the extension ratios, are equal. Thus from equations (5.29) the stress-strain relationship becomes:

$$p_\theta = (\phi + \psi\lambda_\theta^2)\left(\lambda_\theta^2 - \frac{1}{\lambda_\theta^4}\right) \tag{5.35}$$

From equilibrium as in equation (5.22), we have:

$$p_\theta = \frac{PD_0}{2h_0}\lambda_\theta^3$$

and hence:

$$\frac{PD_0}{2h_0} = \frac{\phi + \psi\lambda_\theta^2}{\lambda_\theta}\left(1 - \frac{1}{\lambda_\theta^6}\right) \tag{5.36}$$

This again has a maximum which is an effect known to anyone who has blown a balloon. This is usually difficult to start with but then becomes easier. The maximum value of pressure is given at a value of λ_θ determined by differentiating equation (5.36) and equating to zero which gives:

$$\psi\hat{\lambda}_\theta^8 - \phi\hat{\lambda}_\theta^6 + 5\psi\hat{\lambda}_\theta^2 + 7\phi = 0$$

For the special case of $\psi = 0$ the solution is:

$$\hat{\lambda}_\theta = 7^{1/6} = 1{\cdot}384$$

5.3. Thick-walled circular cylinders

When the cylinder walls are thick the assumption that p_r may be ignored is no longer valid and the stress system within the wall of the cylinder has the three components p_θ, p_r and p_z. No axial variations will be included in the analysis and hence from equilibrium (equation 1.29) we have the condition that:

$$\frac{\mathrm{d}p_r}{\mathrm{d}r} + \frac{p_r - p_\theta}{r} + F_r = 0 \tag{5.37}$$

where F_r is the radial body force.

5.3.1. *Small strain solutions*

Unlike the thin-walled case the stresses in the thick cylinder cannot be determined directly from equlibrium and hence a strain analysis must be used. If u_r is the radial displacement and u_z the axial displacement then, from equations (1.59):

$$e_\theta = \frac{u_r}{r}, \qquad e_r = \frac{\partial u_r}{\partial r} \qquad \text{and} \qquad e_z = \frac{\partial u_z}{\partial z} \tag{5.38}$$

We shall consider the linear elastic solution here which gives:

$$\left.\begin{aligned} e_\theta &= \frac{1}{E}[p_\theta - \nu(p_r + p_z)] \\ e_r &= \frac{1}{E}[p_r - \nu(p_\theta + p_z)] \\ e_z &= \frac{1}{E}[p_z - \nu(p_\theta + p_r)] \end{aligned}\right\} \tag{5.39}$$

Now from equation (5.38) we may write:

$$\frac{\partial e_\theta}{\partial r} = \frac{1}{r}\frac{\partial u_r}{\partial r} - \frac{u_r}{r^2} = \frac{e_r - e_\theta}{r}$$

and substitution from equations (5.39) gives:

$$\frac{\partial}{\partial r}[p_\theta - \nu(p_r + p_z)] = \frac{(1+\nu)}{r}(p_r - p_\theta)$$

The equilibrium condition, equation (5.37), gives a substitution for the right-hand side and thus:

$$\frac{\partial}{\partial r}[p_\theta - \nu(p_r + p_z)] = -(1+\nu)\frac{\partial p_r}{\partial r} - F_r(1+\nu)$$

and integrating we have:

$$\nu p_z = (p_r + p_\theta) + (1+\nu)\int F_r \, \mathrm{d}r - C(z) \qquad (5.40)$$

where $C(z)$ is determined by the boundary conditions of the problem. Any further progress requires an additional condition and the most common is the Lamé assumption. This is that the cylinder is very long in comparison with other dimensions, and, since shear and axial variations are absent, there will be no axial strain variations across the wall thickness. Thus e_z is not a function of r and, since all variations are with r in this axisymmetric case, it may be regarded as a constant. This is equivalent to saying that plane sections will remain plane. Thus from equations (5.39) we may write:

$$p_z = Ee_z + \nu(p_r + p_\theta)$$

and substitution into equation (5.40) gives p_θ as a function of p_r and r.

i.e.
$$p_\theta = -p_r + \frac{1}{(1-\nu^2)}[\nu Ee_z - (1+\nu)\int F_r \, \mathrm{d}r + C]$$

It is usual to ignore body forces in this case and to group the constants together:

i.e.
$$2A = \frac{\nu Ee_z + C}{1-\nu^2}$$

giving
$$p_\theta = -p_r + 2A \qquad (5.41)$$

Substitution of this into the equilibrium equation (5.37) gives:

$$\frac{\mathrm{d}p_r}{\mathrm{d}r} = \frac{1}{r}(-2p_r + 2A)$$

and integrating we have:

$$-\tfrac{1}{2}\ln(2A - 2p_r) = \ln r - \tfrac{1}{2}\ln(2B)$$

where B is a constant of integration.

Rearranging we have:

$$p_r = A - B/r^2 \tag{5.42}$$

and from equation (5.41):

$$p_\theta = A + B/r^2 \tag{5.43}$$

These are the well-known Lamé equations and the constants A and B may be determined for any particular problem.

(i) *Internal pressure loading*

Consider the long cylinder shown in Figure 5.6 with an internal radius of a_0 and

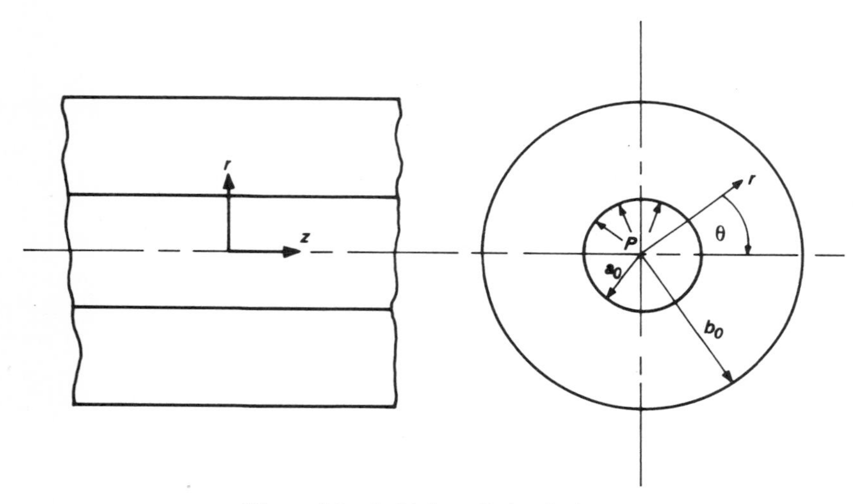

Figure 5.6 A thick-walled cylinder.

an external radius b_0. Since the strains are assumed small these dimensions will not vary significantly. The boundary conditions for internal pressure are:

$$p_r = -P \quad \text{at} \quad r = a_0$$

and

$$p_r = 0 \quad \text{at} \quad r = b_0$$

and by substitution into equation (5.42) we have:

$$A = \frac{Pa_0{}^2}{b_0{}^2 - a_0{}^2} \quad \text{and} \quad B = \frac{Pa_0{}^2 b_0{}^2}{b_0{}^2 - a_0{}^2}$$

The expressions for the stresses are thus:

$$p_r = \frac{Pa_0{}^2}{b_0{}^2 - a_0{}^2}\left(1 - \frac{b_0{}^2}{r^2}\right) \quad \text{and} \quad p_\theta = \frac{Pa_0{}^2}{b_0{}^2 - a_0{}^2}\left(1 + \frac{b_0{}^2}{r^2}\right) \tag{5.44}$$

The forms of the distributions are shown in Figure 5.7 for a cylinder where:

$$b_0/a_0 = K = 2$$

and it is clear that the stress system at the inside wall is of interest where:

$$p_r = -P \qquad \text{and} \quad p_\theta = P\left(\frac{K^2+1}{K^2-1}\right) \tag{5.45}$$

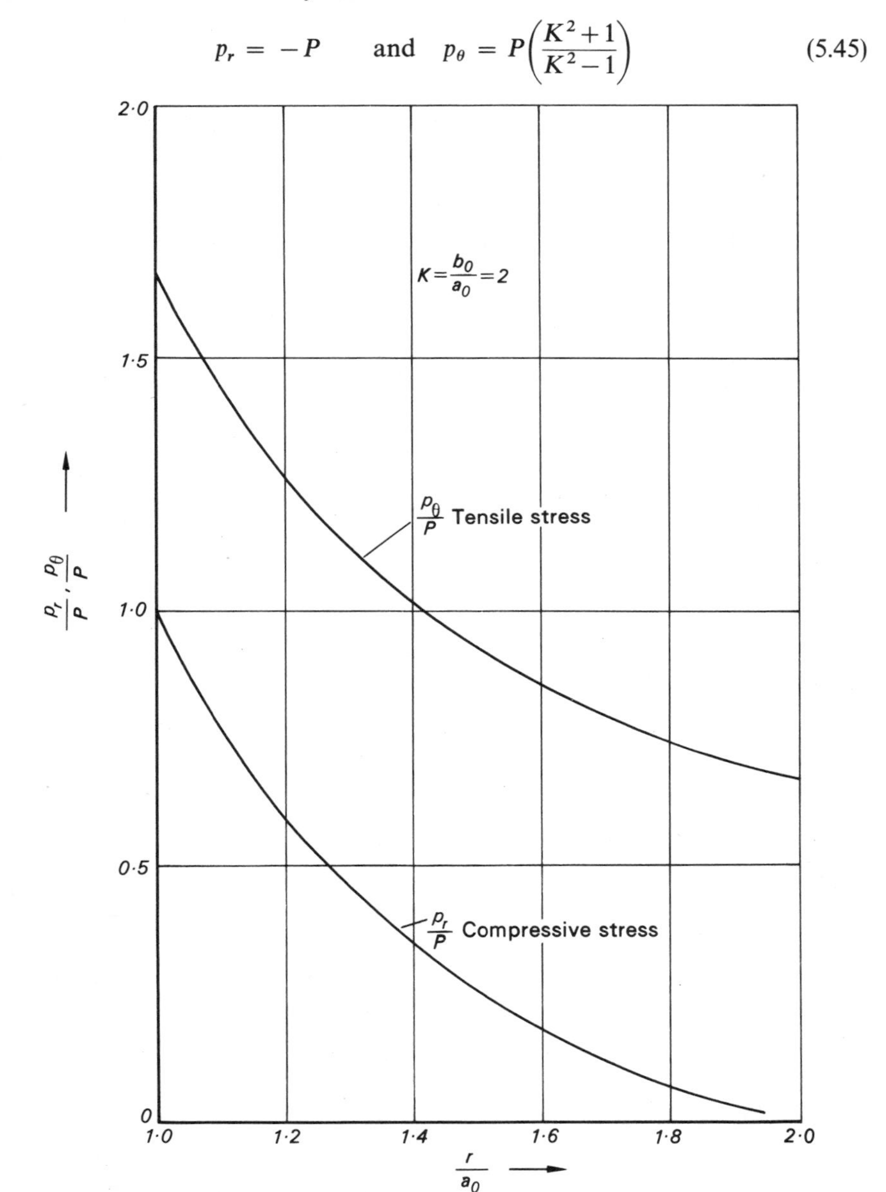

Figure 5.7 Stresses in a thick-walled cylinder.

The axial stress p_z is determined by the particular end conditions of the problem. For example, the closed end case gives the condition:

$$\pi a_0{}^2 p = \int_{a_0}^{b_0} p_z 2\pi r \, dr$$

i.e.

$$a_0{}^2 p = 2 \int_{a_0}^{b_0} p_z r \, dr$$

Now from equation (5.40) we have:

$$\nu p_z = p_r + p_\theta - C \quad (\text{for } F_r = 0)$$

i.e. $p_z = 1/\nu(2A - C)$, a constant for this solution and hence:

$$p_z = \frac{P}{K^2 - 1} \tag{5.46}$$

Comparison with the thin-walled solutions is of interest since $a_0 = D_0/2$ and $b_0 - a_0 = h_0$ giving:

$$p_z = \frac{PD_0}{4h_0} \frac{1}{[1 + h_0/D_0]}$$

which is the accurate form used in section 5.2.1 and reduces to the familiar form for $h_0/D_0 \ll 1$. The hoop stress used in the thin-walled theory is obviously the mean value which is obtained by integrating equation (5.44):

i.e.

$$(p_\theta)_{\text{mean}} = \frac{1}{b_0 - a_0} \int_{a_0}^{b_0} p_\theta \, dr = \frac{Pa_0}{b_0 - a_0} = \frac{P}{K - 1}$$

Since $b_0 = a_0 + t_0$ then $K = 1 + 2h_0/D_0$ and:

$$(p_\theta)_{\text{mean}} = \frac{PD_0}{2h_0}$$

as used in the thin-walled theory. For the calculation of deflections the thin-walled theory will be quite a good approximation but for maximum stresses it is in error by the factor:

$$\frac{p_{\theta\,\max}}{p_{\theta\,\text{mean}}} = \frac{K^2 + 1}{K + 1} = \frac{1 + 2h_0/D_0 + 2(h_0/D_0)^2}{1 + h_0/D_0} \tag{5.47}$$

which approximates, for small h_0/D_0 ratios, to:

$$\frac{p_{\theta\,\max}}{p_{\theta\,\text{mean}}} = 1 + h_0/D_0$$

(ii) *Hose end coupling*

A method of fixing end attachments to thick-walled rubber and plastic hoses is as shown in Figure 5.8. A metal tube fits inside the hose and an outer metal sleeve is swaged over the outer diameter. If we take the radial displacement at zero at $r = a_0$ and u_0 at $r = b_0$ then using the Lamé theory it is possible to calculate the interfacial pressures at both the inner and outer radii. Thus the strains at boundaries are:

$$\left.\begin{aligned} e_\theta &= 0 \quad \text{at} \quad r = a_0 \\ e_\theta &= \frac{u_0}{b_0} \quad \text{at} \quad r = b_0 \end{aligned}\right\}$$

and

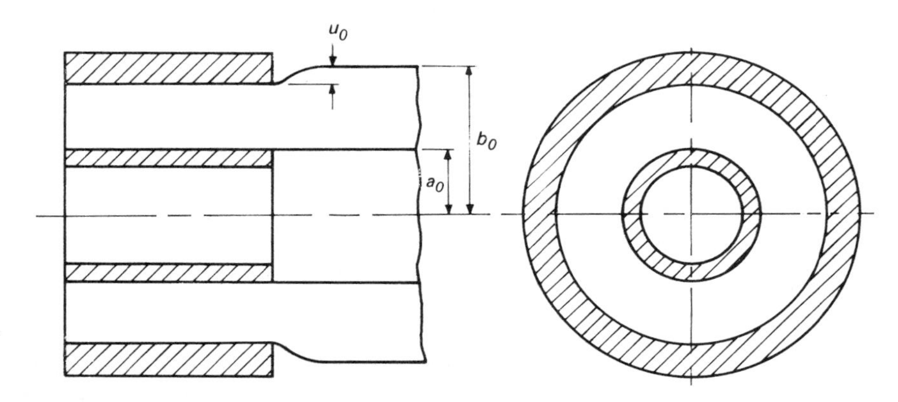

Figure 5.8 Hose end coupling.

The ends of the cylinder are free and thus there is zero axial load and since p_z is a constant in this solution it must also be zero. The hoop strain is given by:

$$e_\theta = \frac{1}{E}(p_\theta - \nu p_r) = \frac{1}{E}\left[(1-\nu)\,A + (1+\nu)\,\frac{B}{r^2}\right] \tag{5.48}$$

and using the strain boundary conditions we have:

$$A = \frac{E}{1-\nu}\,\frac{K^2}{K^2-1}\,\frac{u_0}{b_0} \quad \text{and} \quad B = -\frac{E}{1+\nu}\,\frac{u_0 b_0}{K^2-1}$$

The interfacial pressures are thus given by:

$$\left.\begin{aligned} (p_r)_{a_0} &= \frac{2E}{1-\nu^2}\,\frac{K^2}{K^2-1}\left(\frac{u_0}{b_0}\right) \\ (p_r)_{b_0} &= \frac{2E}{1-\nu^2}\,\frac{K^2}{K^2-1}\left(\frac{u_0}{b_0}\right)\left[\left(\frac{1+\nu}{2}\right)+\left(\frac{1-\nu}{2K^2}\right)\right] \end{aligned}\right\} \tag{5.49}$$

Since $\nu \leqslant 1/2$ and $K > 1$, $(p_r)_{a_0} > (p_r)_{b_0}$. If we take a criterion of leaking when the internal pressure is equal to the interfacial pressure then leaking is governed by $(p_r)_{a_0}$ and thus the critical condition is:

$$P = \frac{2E}{1-\nu^2}\frac{K^2}{K^2-1}\left(\frac{u_0}{b_0}\right)$$

For a visco-elastic material the decrease in $E(t)$ will give a time at which this condition is reached.

A slight variant of this case is a solid cork in a bottle for which $a_0 = 0$. For all solid rods the Lamé solution is:

$$p_r = p_\theta = A$$

and $B = 0$ since a singularity at $r = 0$ is inadmissable. Using this with the same outer radius boundary condition we have:

$$(p_r)_{b_0} = \frac{E}{1-\nu}\frac{u_0}{b_0} \tag{5.50}$$

and the leakage criterion is now $(p_r)_{b_0} = P$. A more complete solution to this problem, where the pressure, end load and frictional effects at the interface are included, is a complex problem requiring a detailed analysis.

(iii) *Solid fuel rocket motor*

A solid fuel rocket motor as shown in Figure 5.9 consists of a thick-walled polymer cylinder in a metal casing. The burning of the fuel produces an internal

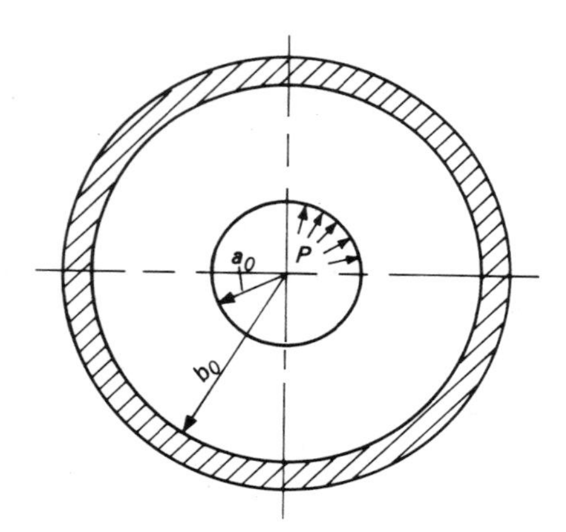

Figure 5.9 Solid fuel rocket motor.

pressure P and increases the inner radius of the propellant cylinder a_0 until it reaches b_0. Again it is assumed that the containing shell is rigid so that the boundary conditions are:

$$e_\theta = 0 \text{ at } r = b_0 \quad \text{and} \quad p_r = -P \text{ at } r = a_0$$

and the resulting stresses become:

$$\left.\begin{aligned} p_r &= \frac{-P}{1+K^2\left(\dfrac{1-\nu}{1+\nu}\right)}\left[1+\left(\frac{1-\nu}{1+\nu}\right)\frac{b_0{}^2}{r^2}\right] \\ p_\theta &= \frac{-P}{1+K^2\left(\dfrac{1-\nu}{1+\nu}\right)}\left[1-\left(\frac{1-\nu}{1+\nu}\right)\frac{b_0{}^2}{r^2}\right] \end{aligned}\right\} \quad (5.51)$$

assuming that $p_z = 0$. It is of interest to note that $-P$ is the largest stress encountered since the containing effect of the shell decreases p_θ compared with the free cylinder. The pressure transmitted to the shell is given by p_r at $r = b_0$:

i.e.
$$(p_r)_{b_0} = \frac{-2P}{(1-\nu)K^2+(1+\nu)}$$

and increases to $-P$ when $K = 1$.

(iv) *Rotating disc*

Rotating discs are a special case involving a body force which requires returning to equation (5.40), i.e.

$$\nu p_z = (p_r+p_\theta)+(1+\nu)\int F_r\,\mathrm{d}r - C$$

The disc is assumed to have a very short axial length such that $p_z = 0$ and the loading is provided by the body force due to rotation at an angular speed ω:

$$F_r = \rho\omega^2 r$$

where ρ is the density of the material.

$$\therefore \qquad p_r+p_\theta = C-(1+\nu)\rho\omega^2\frac{r^2}{2} \quad (5.52)$$

Substituting in the equilibrium equation (5.37) gives:

$$\frac{\mathrm{d}p_r}{\mathrm{d}r}+\frac{2p_r-C+(1+\nu)\rho\omega^2 r^2/2}{r}+\rho\omega^2 r = 0$$

Integrating produces the solutions:

$$\left.\begin{aligned} p_r &= C/2-B/r^2-\left(\frac{3+\nu}{8}\right)\rho\omega^2 r^2 \\ \text{and} \qquad p_\theta &= C/2+B/r^2-\left(\frac{1+3\nu}{8}\right)\rho\omega^2 r^2 \end{aligned}\right\} \quad (5.53)$$

The similarity with the Lamé equations is evident. The most usual boundary conditions are $p_r = 0$ at $r = a_0$ and b_0 giving the results:

$$p_r = \left(\frac{3+v}{8}\right)\rho\omega^2\left[b_0{}^2 + a_0{}^2 - \frac{a_0{}^2 b_0{}^2}{r^2} - r^2\right]$$

and

$$p_\theta = \left(\frac{3+v}{8}\right)\rho\omega^2\left[b_0{}^2 + a_0{}^2 + \frac{a_0{}^2 b_0{}^2}{r^2} - \left(\frac{1+3v}{3+v}\right)r^2\right]$$

The maximum value of p_r is not at the centre or inner value as in most solutions but at a radius given by:

$$\hat{r} = \sqrt{a_0 b_0}$$

5.3.2. *Predicting first yield*

In the thin-walled cylinder case, when the yield condition was reached the whole system was at yield since the stress system was uniform throughout the vessel. In the case of thick-walled vessels, the stress system at some radius reaches the yield criterion and yielding takes place while the rest of the vessel remains elastic. If the load continues to increase the amount of plastically deformed material increases until eventually the whole wall is plastic, i.e. the fully plastic state has been reached.

The prediction of first yield requires that the stresses at the appropriate radius be chosen to give the onset of yield. For the thick-walled cylinder under internal pressure an examination of equations (5.44) and (5.46) shows that the most severe stress system is at the inner radius giving:

$$p_r = -P, \qquad p_\theta = P\left(\frac{K^2+1}{K^2-1}\right)$$

For closed ends we have:

$$p_z = \frac{P}{K^2-1}$$

and if we use the Von Mises criterion:

$$2p_Y{}^2 = (p_r - p_\theta)^2 + (p_\theta - p_z)^2 + (p_z - p_r)^2$$

the critical pressure becomes:

$$\hat{P} = \frac{p_Y}{\sqrt{3}}\left(1 - \frac{1}{K^2}\right) \tag{5.54}$$

It is of interest to note that if the Tresca criterion is used the largest shear stress is given by the difference between the hoop and radial stresses giving a result independent of end conditions:

$$\hat{P} = \frac{p_Y}{2}\left(1 - \frac{1}{K^2}\right) \tag{5.55}$$

This solution differs from the more accurate Von Mises case by 15·5%.

The other cases may be solved by direct substitution. The solid fuel rocket motor is of interest since P remains constant and K decreases. The yielding is again at the inner surface and is given by the roots of the equation:

$$K^4+2\left(\frac{1+v}{1-v}\right)\frac{K^2}{(1-3(P/p_Y)^2)}+\left(\frac{1+v}{1-v}\right)^2\left[\frac{1-(P/p_Y)^2}{1-3(P/p_Y)^2}\right] = 0 \tag{5.56}$$

The problem is somewhat more complex when the position of the yield radius is not known. The expressions for the stress are substituted in the yield criterion which is then differentiated with respect to radius to find the critical value. For example with the rotating disc we may use the Tresca criterion for simplicity:

$$p_Y = p_\theta - p_r = \left(\frac{3+v}{8}\right)\rho\omega^2\left[\frac{2a_0{}^2b_0{}^2}{r^2}+\frac{2(1-v)}{3+v}r^2\right]$$

The yield radius is given by:

$$\frac{\mathrm{d}p_Y}{\mathrm{d}r} = \left(\frac{3+v}{8}\right)\rho\omega^2\left[-\frac{4a_0^2b_0{}^2}{r^3}+\frac{4(1-v)}{3+v}r\right] = 0$$

i.e.

$$\hat{r}^4 = \left(\frac{3+v}{1-v}\right)a_0{}^2b_0{}^2$$

and the critical speed is given by:

$$\hat{\omega}^2 = \frac{2p_Y}{[(3+v)(1-v)]^{1/2}}\frac{1}{\rho a_0 b_0} \tag{5.57}$$

5.3.3. *Fully plastic, non-work hardening solutions*

Fully plastic solutions assume that the entire section of the cylinder has yielded and that the yield criterion is obeyed everywhere. For ease of manipulation it is usual to consider the Tresca criterion in which the largest shear stress from the three principal stresses is used. For most cylinders under internal pressure p_θ and p_r are of opposite sign and thus the appropriate condition is:

$$p_\theta - p_r = p_Y \tag{5.58}$$

and in addition the equilibrium equation must be satisfied:

$$\frac{\mathrm{d}p_r}{\mathrm{d}r}+\frac{p_r-p_\theta}{r}+F_r = 0 \tag{5.59}$$

If we consider the case of the thick-walled cylinder under internal pressure with $F_r = 0$ and substituting from equation (5.58) we have:

$$\frac{\mathrm{d}p_r}{\mathrm{d}r}-\frac{p_Y}{r} = 0$$

Integrating gives:

$$p_r = p_Y \ln r + A$$

where A is a constant. The boundary conditions are $p_r = -P$ at $r = a_0$ and $p_r = 0$ at $r = b_0$ as before giving:

$$P = p_Y \ln K \tag{5.60}$$

Such a solution contains no strain analysis and is called a limit solution. It may be regarded as a method giving a lower bound solution (see section 5.3.6 for a discussion of upper and lower bounds) since equilibrium is satisfied but not necessarily compatibility. It should be noted that problems containing displacement boundary conditions cannot be considered since the displacements are not defined.

An interesting example of the choice of the appropriate Tresca condition is provided by the rotating disc with the equilibrium condition:

$$\frac{dp_r}{dr} + \frac{p_r - p_\theta}{r} + \rho\omega^2 r = 0$$

The stresses are p_r and p_θ with $p_z = 0$ and we have a choice of three yield criteria:

$$(1)\quad p_r - p_\theta = \pm p_Y, \qquad (2)\quad p_\theta = \pm p_Y, \qquad (3)\quad p_r = \pm p_Y$$

Since the boundary conditions are $p_r = 0$ at $r = a_0$ and b_0 the third criterion is obviously invalid but the other two both give solutions. The results are:

$$\left.\begin{aligned} &(1) \quad \frac{\omega^2 \rho a_0{}^2}{p_Y} = \frac{\ln K^2}{K^2 - 1} \\ &\text{and } (2) \quad \frac{\omega^2 \rho a_0{}^2}{p_Y} = \frac{3}{K^2 + K + 1} \end{aligned}\right\} \tag{5.61}$$

An examination of these results shows that the stresses predicted in (2) are always greater, for a given ω, than (1) and hence (2) is the better lower bound solution, i.e. since it is a lower bound the highest stress values must be nearer the true solution. It is also useful to note that for a solid disc only the second condition gives a sensible result, it being:

$$\omega^2 = \frac{3p_Y}{\rho b_0{}^2} \tag{5.62}$$

5.3.4. *Elastic-plastic interfaces*

The result of the last two sections may be combined for the case when the plasticity has spread part-way through the thickness, to a radius c as shown in

Figure 5.10. Since the inner section is non-work hardening we may consider it as in the previous section with the boundary condition:

$$\left.\begin{aligned} p_r &= -P \quad \text{at} \quad r = a_0 \\ \text{and} \qquad p_r &= -p_c \quad \text{at} \quad r = c \end{aligned}\right\}$$

Substitution in the solution (section 5.3.3) gives:

$$p_c - P = -p_Y \ln (c/a_0) \tag{5.63}$$

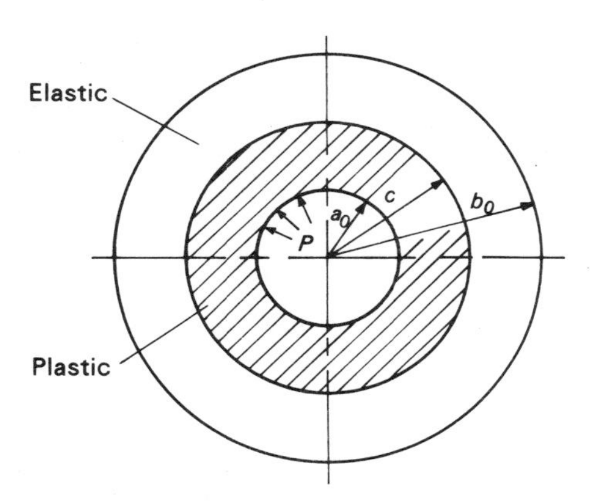

Figure 5.10 Elastic-plastic thick-walled cylinder.

The outer section is elastic and hence is described by Lamé's equations and since the inner surface must be at the yield condition then from equation (5.55) we have:

$$p_c = \frac{p_Y}{2}\left(1 - \frac{c^2}{{b_0}^2}\right) \tag{5.64}$$

Eliminating p_c from equations (5.63) and (5.64) gives the result:

$$P/p_Y = \ln c/a_0 + \tfrac{1}{2}\left(1 - \frac{c^2}{{b_0}^2}\right) \tag{5.65}$$

Hence c may be found for any value of P.

5.3.5. *Large strain solutions*

Figure 5.11(a) shows a cross-section of a thick-walled cylinder of initial internal and external radii of a_0 and b_0. Figure 5.11(b) shows the same section strained so that the dimensions become a and b and, for the ring shown, the initial radius r_0 and initial width δr_0 become r and δr. The expressions for the extension ratios are thus:

$$\lambda_\theta = \frac{r}{r_0} \quad \text{and} \quad \lambda_r = \frac{\delta r}{\delta r_0} \tag{5.66}$$

In all the solutions described here it will be assumed that the cylinder is long compared with its other dimensions so that plane sections will remain plane on

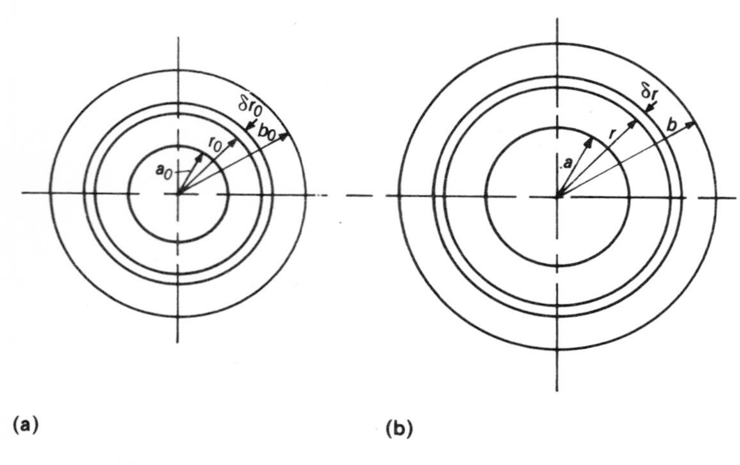

Figure 5.11 Large strains in a thick-walled cylinder.

straining. In effect this means that the axial extension ratio λ_z is independent of r and r_0. It is further assumed that the material deforms at constant volume so that:

$$\lambda_\theta \lambda_r \lambda_z = 1$$

and hence

$$\frac{r}{r_0}\frac{\delta r}{\delta r_0}\lambda_z = 1 \tag{5.67}$$

Since λ_z is a constant this may be integrated to give:

$$r^2 \lambda_z = r_0{}^2 + A$$

where A is a constant. Hence we may write:

$$\left.\begin{aligned} \lambda_\theta{}^2 &= \frac{r^2}{r_0{}^2} = \frac{r^2}{\lambda_z r^2 - A} \\ \text{and} \qquad \lambda_r &= \frac{1}{\lambda_\theta \lambda_z} \end{aligned}\right\} \tag{5.68}$$

In addition the usual equilibrium equations must be satisfied at every point in the strained state so that:

$$\frac{\mathrm{d}p_r}{\mathrm{d}r} + \frac{p_r - p_\theta}{r} = 0 \tag{5.69}$$

assuming that there are no body forces. The end conditions in the cylinder may be typified by considerations of the end load given by:

$$F = \int_a^b p_z \, 2\pi r \, dr \tag{5.70}$$

For example we may have:
(a) Free ends: $F = 0$
(b) Closed ends with internal pressure:

$$F = \pi a^2 p$$

A strain end condition is also possible such as fully restrained axial movement giving $\lambda_z = 1$ resulting in some value of F.

As in the case of thin-walled cylinders there are two forms of constitutive relationship which are of interest and the solutions for each type will be considered.

(i) *Equivalence solution*

As discussed in previous sections this method supposes a unique relationship between the equivalent stress $\bar{p}$ and strain $\bar{\varepsilon}$ and is confined to cases of constant stress ratios throughout the deformation. The form of the constitutive equations is:

$$\left.\begin{aligned} \varepsilon_\theta &= \frac{\bar{\varepsilon}}{\bar{p}}\left[p_\theta - \frac{1}{2}(p_r + p_z)\right] \\ \varepsilon_r &= \frac{\bar{\varepsilon}}{\bar{p}}\left[p_r - \frac{1}{2}(p_\theta + p_z)\right] \\ \text{and} \quad \varepsilon_z &= \frac{\bar{\varepsilon}}{\bar{p}}\left[p_z - \frac{1}{2}(p_\theta + p_r)\right] \end{aligned}\right\} \tag{5.71}$$

If we take the example of the cylinder under internal pressure it is of interest to look at the condition:

$$\varepsilon_z = 0$$

i.e. a zero axial strain.

From equations (5.71):

$$p_z = \tfrac{1}{2}(p_\theta + p_r)$$

and by substituting from the equilibrium condition (equation (5.69)) we have:

$$p_z = p_r + \frac{1}{2} r \frac{dp_r}{dr} = \frac{1}{2r}\frac{d}{dr}(r^2 p_r)$$

Substitution in equation (5.70) gives the expression for the end load:

$$F = \int_a^b 2\pi p_z r \, dr = \pi \int_a^b \frac{d(r^2 p_r)}{dr} \, dr$$

i.e.
$$F = \left[r^2\, p_r\right]_{r=a,\ p_r=-P}^{r=b,\ p_r=0}$$

$\therefore\ F = \pi a^2 P$, the closed end condition. This result is not surprising since the thick cylinder may be regarded as a series of thin-walled cylinders and since each has the zero axial strain condition the summation would be expected to have this condition also.

Thus the closed end case may be analysed by taking $\varepsilon_z = 0$, i.e. $\lambda_z = 1$ in equation (5.68) giving the expression for λ_θ of the form:

$$\lambda_\theta^2 = \frac{r^2}{r^2 - A}$$

Thus if we consider a particular bore extension ratio:

$$\left.\begin{aligned} \lambda_0 &= a/a_0 \\ \text{then}\quad A &= a_0{}^2(\lambda_0{}^2 - 1) \\ \text{and}\quad b &= \sqrt{(b_0{}^2 - a_0{}^2) + a^2} \end{aligned}\right\} \tag{5.72}$$

For any value of r between a and b, λ_θ may be calculated and hence ε_θ since $\varepsilon_\theta = \ln \lambda_\theta$.

i.e.
$$\varepsilon_\theta = \ln\left(\frac{r^2}{r^2 - A}\right) \tag{5.73}$$

If we substitute for p_z in the expression for $\bar{p}$:

$$2\bar{p}^2 = (p_r - p_\theta)^2 + (p_\theta - p_z)^2 + (p_z - p_r)^2$$

then:

$$\bar{p} = \frac{\sqrt{3}}{2}(p_\theta - p_r)$$

and substituting in the equilibrium equation (5.69) gives:

$$\frac{\mathrm{d}p_r}{\mathrm{d}r} = -\frac{2}{\sqrt{3}}\frac{\bar{p}}{r}$$

i.e.
$$\int_{-P}^{0} \mathrm{d}p_r = -\frac{2}{\sqrt{3}}\int_a^b \frac{\bar{p}}{r}\,\mathrm{d}r$$

and
$$P = -\frac{2}{\sqrt{3}}\int_a^b \bar{p}\,\frac{\mathrm{d}r}{r} \tag{5.74}$$

Similarly since $\varepsilon_z = 0$ and $\varepsilon_r = -\varepsilon_\theta$ we have from equations (5.71):

$$\bar{\varepsilon} = \frac{2}{\sqrt{3}}\varepsilon_\theta \tag{5.75}$$

Thus a numerical solution may be effected in the following way:

(1) Choose a value of λ_0 (a_0 and b_0 given).

(2) Calculate a, A and b from equations (5.72).
(3) At each of a range of values of r between a and b, ε_θ may be calculated from equation (5.73) and hence $\bar{\varepsilon}$ from equation (5.75).
(4) From these values of $\bar{\varepsilon}$, $\bar{p}$ may be found from the stress-strain curve and hence the function $\bar{p}/r$ may be deduced.
(5) This function may then be integrated between a and b to give P from equation (5.74).
(6) The whole process may be repeated for a range of values of λ_0 so that a graph of P versus λ_0 may be constructed.

This method is ideal for a computer solution and most practical problems are best solved in this way. In general there is a maximum in the P versus λ_0 curve which is a useful design criterion.

The solution is somewhat more difficult to obtain for other end conditions since ε_z cannot be predetermined. The equations become modified to the form:

$$\bar{\varepsilon} = \frac{2}{\sqrt{3}}\left[\varepsilon_\theta^2 + \varepsilon_\theta\varepsilon_z + \varepsilon_z^2\right]^{1/2} \tag{5.76}$$

and

$$P = -\frac{2}{\sqrt{3}}\int_a^b \bar{p}\left(1 - \frac{\varepsilon_z^2}{\bar{\varepsilon}^2}\right)^{1/2}\frac{\mathrm{d}r}{r} \tag{5.77}$$

The solution process is the same as before except that a value of λ_z must be guessed with λ_0 and hence a, A and b are found from the equations including λ_z, i.e. (5.68). The whole process is carried out as before but with ε_z included until a value of P is computed. Reference must now be made to the end condition which may be rearranged to give:

$$F = \varepsilon_z\left[2\pi\int_a^b \frac{\bar{p}}{\bar{\varepsilon}}\, r\, \mathrm{d}r\right] + P\pi a^2 \tag{5.78}$$

and by a similar process this integral may be evaluated and in conjunction with P give a value for F. If the end condition is $F = 0$, for example, then the whole process must be repeated for different values of ε_z until the correct condition is reached. Convergence can usually be achieved by a halving process.

(ii) *Rubber elasticity*

Rubber elasticity solutions are of practical importance because of the use of rubber in seals and flexible hoses. The stress-strain relationship (see section 2.2.2) are:

$$\left.\begin{aligned} p_\theta &= \lambda_\theta^2\phi - \frac{1}{\lambda_\theta^2}\psi + p \\ p_r &= \lambda_r^2\phi - \frac{1}{\lambda_r^2}\psi + p \\ p_z &= \lambda_z^2\phi - \frac{1}{\lambda_z^2}\psi + p \end{aligned}\right\} \tag{5.79}$$

where p is an arbitrary hydrostatic pressure for constant volume deformation. Using the constant volume condition we may take the first two equations to give:

$$p_\theta - p_r = \lambda_\theta^2\phi - \frac{\psi}{\lambda_\theta^2} - \frac{\phi}{\lambda_\theta^2\lambda_z^2} + \psi\,\lambda_\theta^2\lambda_z^2$$

i.e.

$$p_\theta - p_r = (\phi + \lambda_z^2\,\psi)\left(\lambda_\theta^2 - \frac{1}{\lambda_\theta^2\lambda_z^2}\right) \tag{5.80}$$

Substituting this in the equilibrium equation (5.69) and replacing λ_θ by equation (5.68) we have:

$$\frac{\mathrm{d}p_r}{\mathrm{d}r} = (\phi + \lambda_z^2\,\psi)\left(\frac{r^2}{\lambda_z r^2 - A} - \frac{1}{\lambda_z^2}\frac{\lambda_z r^2 - A}{r^2}\right)\frac{1}{r}$$

This may be integrated to give:

$$p_r = B - \frac{1}{2}\left(\frac{\phi}{\lambda_z^2} + \psi\right)\left[\frac{A}{r^2} + \lambda_z \ln\left(\lambda_z - \frac{A}{r^2}\right)\right] \tag{5.81}$$

where B is a constant. There are effectively three unknown constants A, B and λ_z which depend on the boundary conditions and the similarity of the Lamé equations with the first two terms is apparent. The additional term means that p_r is not fixed by two pressure boundary conditions however and a further condition must be sought.

The most simple solution is that for assumed plane strain, i.e. $\lambda_z = 1$ for which equation (5.81) is a complete solution as in the Lamé case. For example the cylinder with internal pressure gives the two equations:

$$O = B - \frac{A}{2}(\phi + \psi)\frac{1}{b^2} + \frac{1}{2}(\phi + \psi)\ln\left(\frac{b^2 - A}{b^2}\right)$$

and

$$-P = B - \frac{A}{2}(\phi + \psi)\frac{1}{a^2} + \frac{1}{2}(\phi + \psi)\ln\left(\frac{a^2 - A}{a^2}\right)$$

i.e.

$$P = \frac{A}{2}(\phi + \psi)\left(\frac{1}{a^2} - \frac{1}{b^2}\right) + \frac{1}{2}(\phi + \psi)\ln\left[\left(\frac{b^2 - A}{a^2 - A}\right)\left(\frac{a^2}{b^2}\right)\right]$$

Now since $\lambda_z = 1$, a and b are related and from equations (5.72) in the previous section we have:

$$A = a_0^2(\lambda_0^2 - 1)$$

and

$$b^2 = (b_0^2 - a_0^2) + a^2$$

giving the final result:

$$P = \left[\frac{\phi + \psi}{2}\right]\left[\frac{(\lambda_0^2 - 1)(K^2 - 1)}{\lambda_0^2((K^2 - 1) + \lambda_0^2)} + \ln\frac{\lambda_0^2 K^2}{(K^2 - 1) + \lambda_0^2}\right] \tag{5.82}$$

This expression does not maximise but continues to rise tending to a value of:

$$p = (\phi+\psi)\ln K$$

at very large values of λ_0. The similarity with the fully plastic solution, equation (5.60), is of interest, but the mechanism of the process is governed by the changes of geometry in this case.

Other boundary conditions require recourse to the end load condition (equation 5.70). From equations (5.79) we may write:

$$p_z = (\phi+\psi\lambda_\theta^2)\left(\lambda_z^2-\frac{1}{\lambda_z^2\lambda_\theta^2}\right)+p_r$$

and by substitution from equation (5.81) and integrating we have:

$$F = \pi(p_a a^2-p_b b^2)+\left(\phi+\frac{\psi}{\lambda_z}\right)\left(\lambda_z-\frac{1}{\lambda_z^2}\right)\pi(b^2-a^2)$$

$$+\frac{A\pi}{2}\left(\psi-\frac{\phi}{\lambda_z^2}\right)\ln\left[\frac{a^2}{b^2}\left(\frac{\lambda_z b^2-A}{\lambda_z a^2-A}\right)\right] \tag{5.83}$$

where p_a and p_b are the values of p_r at $r = a$ and b respectively.

The end load for the plane strain case is simply determined by the substitution of $\lambda_z = 1$. For the closed end case

$$F = \pi P a^2$$

and $p_a = -P$, with $p_b = 0$ and λ_z is determined as the solution of the equation:

$$0 = \left(\phi+\frac{\psi}{\lambda_z}\right)\left(\lambda_z-\frac{1}{\lambda_z^2}\right)(b^2-a^2)+\frac{A}{2}\left(\psi-\frac{\phi}{\lambda_z^2}\right)\ln\left[\frac{a^2}{b^2}\left(\frac{\lambda_z b^2-A}{\lambda_z a^2-A}\right)\right]$$

This type of equation is best solved numerically and little is gained by pursuing it further analytically.

One set of boundary conditions which yields a simple form is the hose end coupling in which $a = a_0$ and we find the pressure at $r = a$, p_a, for a given value of b/b_0; i.e. λ_b. In this case the complete strain field is specified and we have:

$$\lambda_z = \frac{K^2-1}{\lambda_b^2K^2-1}$$

with

$$A = a_0^2(\lambda_z-1)$$

p_b may be eliminated between equations (5.81) and (5.83) for, say, $F = 0$ giving a closed form expression for p_a in terms of λ_b. The algebra becomes extremely laborious and again a computer program is the most efficient way of dealing with the problem. Indeed, once the basic equations have been established it is preferable to set these problems up in numerical form, including the integration, since analytical solutions become tedious and prone to errors.

5.3.6. *Compression of rubber cylinders*

An interesting example which has a practical application in the testing of rubber is the case of a cylinder bonded to two metal plates and compressed as shown in Figure 5.12. The method is of interest here since it involves the use of the equilibrium equations with axial variations and shear and also introduces

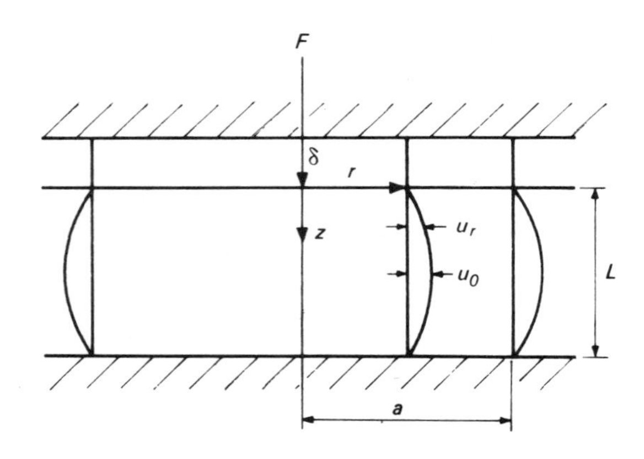

Figure 5.12 Compression of a bonded rubber cylinder.

the concept of the upper bound solution. The solution assumes small strains and linear elasticity and even with these a general solution is very complex. The method used is to assume the simplest possible compatible displacement system and then examine the predicted stresses. In Figure 5.12 a cylinder of radius a and length L is compressed a small amount δ and the radiar displacements at any given radius are assumed to be of parabolic form:

$$\text{i.e.} \qquad u_r = Az^2 + Bz + C$$

where A, B and C are functions of the radius only.

Clearly $u_r = 0$ at $z = 0$ and $z = L$ and hence:

$$u_r = A(z^2 - Lz)$$

If it is further assumed that the deformation takes place at constant volume ($\nu = 1/2$) then the volume of material bowed outwards at a radius r is equal to that displaced downwards. Thus by equating the two we have:

$$\pi r^2 \delta = \int_0^L 2\pi r u_r \, \mathrm{d}z$$

for small deflections. Hence by substituting for u_r we have:

$$r\delta = 2A \int_0^L (z^2 - Lz) \, \mathrm{d}z = -\frac{AL^3}{3}$$

i.e.
$$A = -\frac{3r\delta}{L^3}$$

giving:
$$u_r = \frac{3r\delta}{L}(z/L - z^2/L^2) \qquad (5.84)$$

Having assumed this distribution the direct strains are defined:

$$e_{\theta\theta} = \frac{u_r}{r} = \frac{3\delta}{L}(z/L - z^2/L^2)$$

and
$$e_{rr} = \frac{\partial u_r}{\partial r} = \frac{3\delta}{L}(z/L - z^2/L^2)$$

i.e. $e_{\theta\theta} = e_{rr}$ and since constant volume is assumed:

$$e_{zz} = -2e_{\theta\theta} = -\frac{6\delta}{L}(z/L - z^2/L^2) \qquad (5.85)$$

It should be noted that $e_{zz} = 0$ at $z = 0$ and $z = L$ because of bonding and the constant volume assumption. Also e_{zz} does not vary with r which is a reasonable condition for parallel plates. From equation (5.85) we may write:

$$e_{zz} = \frac{\partial u_z}{\partial z} = -\frac{6\delta}{L}(z/L - z^2/L^2)$$

and hence:

$$u_z = -\frac{6\delta}{L}(z^2/2L - z^3/3L^2) + R$$

where R is a function of r or a constant. The shear strain is given by:

$$e_{rz} = \frac{\partial u_r}{\partial z} + \frac{\partial u_z}{\partial r}$$

i.e.
$$e_{rz} = \frac{3r\delta}{L}(1/L - 2z/L^2) + \frac{\partial R}{\partial r}$$

Now
$$p_{rz} = Ge_{rz} = \frac{E}{3}e_{rz} \quad \text{for} \quad \nu = \tfrac{1}{2}$$

and hence:
$$p_{rz} = \frac{E\delta}{L}r(1/L - 2z/L^2) + \frac{E}{3}\frac{\partial R}{\partial r} \qquad (5.86)$$

The equality of hoop and radial strains when substituted in the elasticity equations gives:

$$p_{\theta\theta} - \tfrac{1}{2}(p_{rr} + p_{zz}) = p_{rr} - \tfrac{1}{2}(p_{\theta\theta} + p_{zz})$$

i.e.
$$p_{\theta\theta} = p_{rr}$$

The general equilibrium equations for this axisymmetric case in which all $\partial/\partial\theta$ terms are zero and the shear stress $p_{r\theta} = 0$ may be deduced from equations (1.28) and are:

$$\left.\begin{aligned} \frac{\partial p_{rr}}{\partial r} + \frac{\partial p_{rz}}{\partial z} + \frac{p_{rr} - p_{\theta\theta}}{r} &= 0 \\ \text{and} \qquad \frac{p_{rz}}{r} + \frac{\partial p_{rz}}{\partial r} + \frac{\partial p_{zz}}{\partial z} &= 0 \end{aligned}\right\} \tag{5.87}$$

The first of these gives:

$$\frac{\partial p_{rr}}{\partial r} = -\frac{\partial p_{rz}}{\partial z} = \frac{2Er\delta}{L^3}$$

and integrating we have:

$$p_{rr} = \frac{E\delta}{L^3} r^2 + S$$

where S is a function of z or a constant. Now $p_{rr} = 0$ at $r = a$ for all z and hence:

$$S = -\frac{E\delta}{L^3} a^3$$

i.e.

$$p_{rr} = -\frac{E\delta}{L^3} (a^3 - r^2) \tag{5.88}$$

Returning to the elasticity equations we have:

$$Ee_{\theta\theta} = p_{\theta\theta} - \tfrac{1}{2}(p_{rr} + p_{zz})$$

i.e.

$$p_{zz} = -2Ee_{\theta\theta} + p_{rr}$$

$\therefore$

$$p_{zz} = -\frac{6E\delta}{L}(z/L - z^2/L^2) - \frac{E\delta}{L^3}(a^2 - r^2) \tag{5.89}$$

Now the axial load is given by:

$$F = \int_0^a p_z 2\pi r \, \mathrm{d}r$$

and hence:

$$F = 2\pi\left[-\frac{6E\delta}{L}(z/L - z^2/L^2)\frac{a^2}{2} - \frac{E\delta}{L^3}\frac{a^4}{4}\right]$$

i.e.

$$F = -2\pi\frac{E\delta}{L}\left[3\left(\frac{z}{L} - \frac{z^2}{L^2}\right)a^2 + \frac{a^4}{4L^2}\right] \tag{5.90}$$

Clearly F cannot be a function of z since it must be the same at all sections and the solution does not satisfy equilbrium. Further, if the second equilibrium equation (5.87) is used with equation (5.86) we have:

$$\frac{\partial p_{zz}}{\partial z} = -\frac{2E\delta}{L}\left(\frac{1}{L}-\frac{2z}{L^2}\right)-\frac{E}{3}\left(\frac{\partial^2 R}{\partial r^2}+\frac{1}{r}\frac{\partial R}{\partial r}\right)$$

i.e.

$$p_{zz} = -\frac{2E\delta}{L}\left(\frac{z}{L}-\frac{z^2}{L^2}\right)-\frac{E}{3}\left(\frac{\partial^2 R}{\partial r^2}+\frac{1}{r}\frac{\partial R}{\partial r}\right)z+T$$

where T is a function of r or a constant. Comparison with equation (5.89) shows that the two expressions for p_{zz} are not compatible in functions of z^2.

The fact that equilibrium is not satisfied is not surprising considering the restraints put on the solution but the result is of practical value if the mean value of the z dependence is used:

i.e.

$$F = -\frac{6\pi E\delta a^2}{L}\left[\frac{1}{L}\int_0^L\left(\frac{z}{L}-\frac{z^2}{L^2}\right)\mathrm{d}z\right]-\frac{\pi}{2}\frac{E\delta a^4}{L^3}$$

$\therefore$

$$F = -\frac{\pi E\delta a^2}{L}-\frac{\pi}{2}\frac{E\delta a^4}{L^3} \tag{5.91}$$

Now for unbonded faces the load is given by:

$$F_0 = -\frac{\pi E\delta a^2}{L}$$

the first term, and hence the second provides a measure of the restraint effect. This is often expressed in terms of apparent modulus E_0 and the modulus determined on an unbonded cylinder, i.e.

$$-\frac{\pi E\delta a^2}{4} = -\frac{\pi E_0\, a^2}{4}-\frac{\pi}{2}\frac{E_0\delta a^4}{L^3}$$

$\therefore$

$$E = E_0\left(1+\frac{1}{2}\frac{a^2}{L^2}\right) \tag{5.92}$$

This result has been found to be an accurate representation of the behaviour of rubber cylinders. The reason is clearly that the assumed distributions are near to reality and hence although the answer is not rigorously correct it must be reasonably close. The difficulty is to determine how close and experimental verifications are invaluable in estimating this. The method can be applied to a whole range of geometries.

As a general point it should be noted that a solution, such as this, which satisfies compatibility but not equilibrium can be shown to be an upper bound to δ, i.e. the smallest value of F is bigger than the correct solution. The unbonded solution satisfies equilibrium with $p_r = p_\theta = p_{rz} = 0$ but not compatibility and is thus a lower bound. Thus taking the mean is likely to be near the solution although it is only known precisely that the solution for F is bounded by:

$$F_{\text{upper bound}} = -\pi \frac{E\delta a^2}{L}\left[\frac{3}{2}+\frac{1}{2}\frac{a^2}{L^2}\right]$$

and

$$F_{\text{lower bound}} = -\pi \frac{E\delta a^2}{L}$$

Detailed discussions of this method of analysis will be found in specialised texts.

5.4. Membranes

A membrane may be regarded as a thin sheet which will not resist shear forces and hence must always have zero bending moments at all points. Thus all loads are supported by direct forces in the plane of the sheet which is the reverse of the plate theory assumption discussed in chapter 4. In practice structures which approximate to membranes are common and include shells and thin sheet constructions and most structures using plastic film and thin sheet may be analysed in this way. Thermoforming operations are also closely modelled by membrane theory.

Real structures rarely correspond exactly to membranes but providing the edge restraints are not so severe so that bending is induced, most of those with one dimension very much smaller than the other two will approximate to a membrane. Thus we shall be concerned only with thin sections and a state of plane stress will be assumed throughout. It is possible to describe general membrane equations but lack of symmetry often induces bending or causes the section to buckle. For this reason axisymmetric systems are the most commonly encountered and will be the only ones discussed here.

5.4.1. *Equilibrium*

Consider an element cut at radius r from an axisymmetric membrane such that it subtends an angle of $\delta\theta$ at the axis and has a vertical height δv as shown in Figure 5.13(a). The dimensions of the element may be written in terms of the local slope β which changes over the length of the element as shown in Figure 5.13(b). Thus:

$$\delta l = \frac{\delta r}{\cos(\beta+\delta\beta/2)}$$

$$\text{Surface area} = \delta\theta \frac{\delta r}{\cos(\beta+\delta\beta/2)}\left(r+\frac{\delta r}{2}\right)$$

$$\text{Volume} = \delta\theta \frac{\delta r}{\cos(\beta+\delta\beta/2)}\left(r+\frac{\delta r}{2}\right)\left(h+\frac{\delta h}{2}\right)$$

The forces acting in the membrane wall are written as N, the force per unit length of membrane such that:

$$N = ph$$

where p is the local stress, and h is the local thickness.
As shown in Figure 5.13, N is taken in two directions; N_θ in the hoop direction

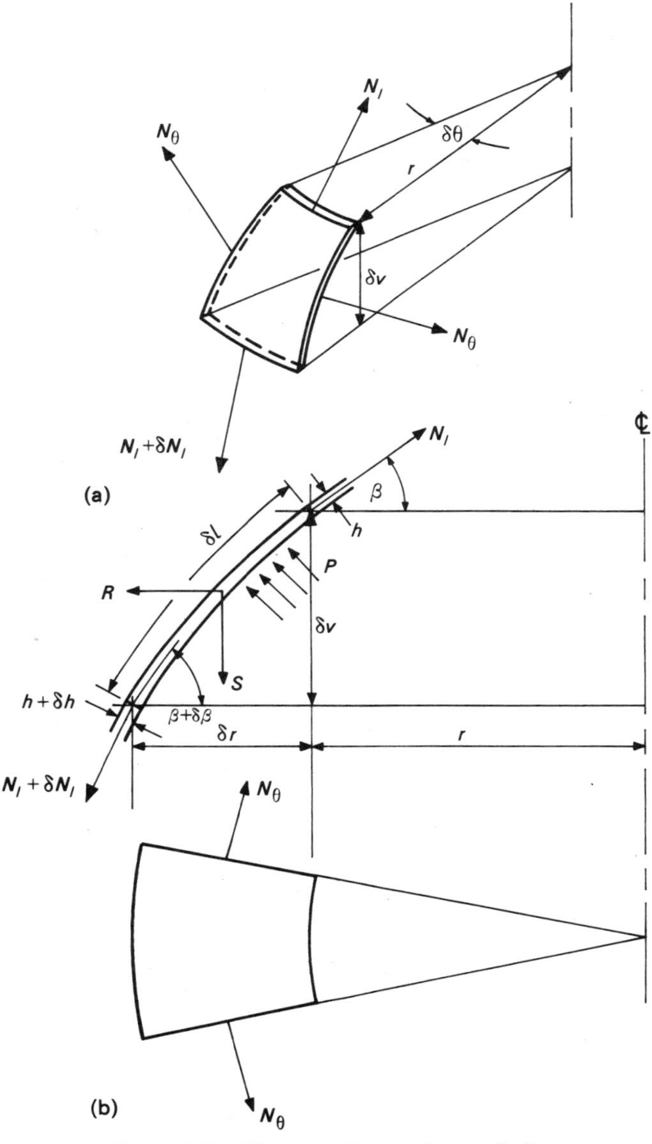

Figure 5.13 Element of a membrane shell.

and N_l in the meridional direction. The loads on the element are horizontal and vertical body forces per unit volume R and S respectively and a pressure P. The various parameters are allowed to vary in the usual way and equilibrium is considered. For examplè if we resolve the forces vertically we have:

$$N_l\, r\, \delta\theta \sin\beta - (N_l + \delta N_l)(r+\delta r)\,\delta\theta \sin(\beta+\delta\beta)$$

$$+P\left(r+\frac{\delta r}{2}\right)\delta r\,\delta\theta - S\,\frac{\delta\theta\left(r+\frac{\delta r}{2}\right)\delta r\left(h+\frac{Sh}{2}\right)}{\cos\left(\beta+\frac{\delta\beta}{2}\right)} = 0$$

Reducing this to the limit and ignoring all terms of higher than the second order we have:

$$\frac{\mathrm{d}}{\mathrm{d}r}(r\,N_l) + N_l\,\frac{r}{\tan\beta}\,\frac{\mathrm{d}\beta}{\mathrm{d}r} = \frac{Pr}{\sin\beta} - \frac{Srh}{\sin\beta\cos\beta}$$

or

$$\frac{\mathrm{d}}{\mathrm{d}r}(r\sin\beta . N_l) = Pr - \frac{Srh}{\cos\beta} \tag{5.93}$$

Similarly for horizontal equilibrium we have:

$$\frac{\mathrm{d}}{\mathrm{d}r}(r\,N_l) - N_l \tan\beta . r\,\frac{\mathrm{d}\beta}{\mathrm{d}r} = \frac{N_\theta}{\cos^2\beta} - Pr\,\frac{\sin\phi}{\cos^2\beta} - \frac{Rrh}{\cos^2\beta}$$

and

$$\frac{\mathrm{d}}{\mathrm{d}r}(r\cos\beta . N_l) = \frac{N_\theta}{\cos\beta} - Pr\tan\beta - \frac{Rrh}{\cos\beta} \tag{5.94}$$

We may eliminate $\mathrm{d}\beta/\mathrm{d}r$ from these two equations to give:

$$\frac{\mathrm{d}}{\mathrm{d}r}(r\,N_l) = N_\theta - r\,h(R + S\tan\beta) \tag{5.95}$$

and for the case of zero body forces we have:

$$\frac{\mathrm{d}N_l}{\mathrm{d}r} + \frac{N_l - N_\theta}{r} = 0 \tag{5.96}$$

which is the familiar form for axisymmetric equilibrium equations. An examination of the bending moment about any section of the element shows that if third order terms are ignored in comparison with second order then a zero bending moment condition of membrane theory is achieved.

5.4.2. *Small strain solutions*

For membranes undergoing small strains the shape does not change under load and all problems are statically determinate; i.e. the complete stress distribution may be obtained from equations (5.94) and (5.90). To obtain a solution it is simply necessary to define β in terms of r from the shape of the membrane and then deduce N_l and N_θ from the equilibrium equations, the loading and the boundary conditions. The stresses may be found by dividing the values of N by the thickness h and from these the strains and thus displacements may be found. Some examples are of value to illustrate the method.

(i) *Conical hopper*

Consider a conical hopper of half angle α and height H as shown in Figure 5.14 filled with a fluid of density ρ. If the vessel is supported at the upper rim the

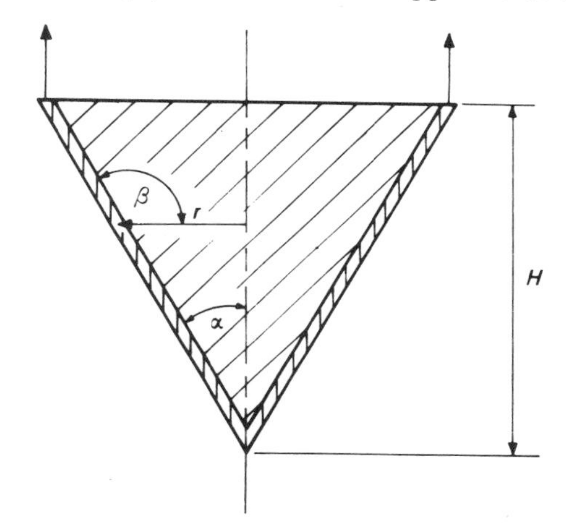

Figure 5.14 Conical hopper.

stress analysis problem is to find N_l and N_θ. Now the angle β used in the equilibrium equations is as shown in Figure 5.14 and:

$$\beta = \pi/2 + \alpha$$

i.e.

$$\frac{d\beta}{dr} = 0 \quad \text{for this case.}$$

The loading is due to pressure and at any radius r we have:

$$P = \rho g \left(H - \frac{r}{\tan \alpha} \right)$$

If we assume no body forces then from equations (5.93) and (5.95) we have:

$$\frac{d}{dr}(r\, N_l) = \frac{\rho g r}{\cos \alpha}\left(H - \frac{r}{\tan \alpha} \right) = N_\theta \tag{5.97}$$

since $\sin \beta = \sin(\pi/2 + \alpha) = \cos \alpha$

Integrating, this gives:

$$rN_l = \frac{\rho g}{\cos \alpha}\left(\frac{Hr^2}{2} - \frac{r^3}{3 \tan \alpha} \right) + A$$

where A is a constant and since r can be zero we have $A = 0$. The final result for N_l is thus:

$$N_l = \frac{\rho g r}{\cos \alpha}\left(H/2 - \frac{r}{3 \tan \alpha}\right) \tag{5.98}$$

The results are of some interest since the largest stresses they predict are not as one might first expect. For example N_l at the upper rim must support the whole weight of liquid and hence:

$$2\pi H \tan \alpha \, N_l \cos \alpha = \pi (H \tan \alpha)^2 \frac{H}{3} \rho g$$

i.e.

$$N_l = \frac{1}{6} \rho g \frac{H^2 \tan \alpha}{\cos \alpha} \qquad (r = H \tan \alpha)$$

This is the same result as from equation (5.98) with $r = H \tan \alpha$ as expected. However if equation (5.98) is differentiated and put to zero then the maximum value of N_l is given when $r = \frac{3}{4} H \tan \alpha$ and:

$$(N_l)_{\max} = \frac{3}{16} \rho g \, H^2 \frac{\tan \alpha}{\cos \alpha}$$

Further if we examine N_θ we find that this maximises at $r = \frac{1}{2} H \tan \alpha$ and:

$$(N_\theta)_{\max} = \frac{1}{4} \rho g \, H^2 \frac{\tan \alpha}{\cos \alpha}$$

which is the largest value of either force and is 1·5 times the rim value of N_l.

(ii) *Hemispherical cap under its own weight*
Figure 5.15 shows the cap which has a uniform thickness h and a density ρ so that the shape is defined by:

$$r = a \sin \beta$$

and the loading by:

$$S = \rho g, \qquad P = 0 \qquad \text{and} \qquad R = 0.$$

Thus from equation (5.93) we have:

$$\frac{\mathrm{d}}{\mathrm{d}r}(r \sin \beta N_l) = -\frac{\rho g r h}{\cos \beta}$$

which on rearranging gives:

$$\frac{\mathrm{d}}{\mathrm{d}r}(r^2 N_l) = -\frac{\rho g r h a}{(1 - r^2/a^2)^{1/2}}$$

Integrating we have:

$$N_l = \rho g a h \frac{(1 - r^2/a^2)}{r^2/a^2} + \frac{C}{r^2}$$

where C is a constant. Now at $r = 0$, N_l must be zero so that we have:

$$C = -\rho g a^3 h$$

and

$$N_l = \rho g a h \left[\frac{(1-r^2/a^2)^{1/2}-1}{r^2/a^2}\right] \tag{5.99}$$

N_θ can be calculated from equation (5.95) in the usual way. It is interesting to note that at $r = a$, N_l must balance the total weight of the hemisphere giving:

$$2\pi a N_l = -2\pi a^2 \rho g h$$

i.e.

$$N_l = -\rho g a h$$

which is given by equation (5.99) when $r = a$.

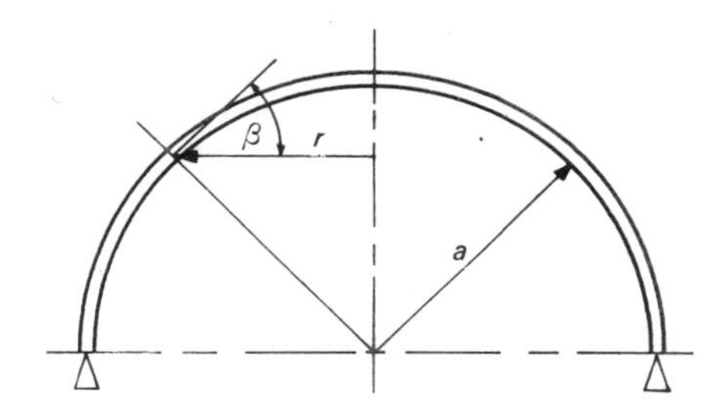

Figure 5.15 Hemispherical cap under its own weight.

5.4.3. *Large strain solutions*

We shall restrict attention in this section to the case where an initially flat sheet is forced into the shape of an axisymmetric shell of revolution. Thus in Figure 5.16 an annular ring originally at radius r_0, of width δr_0 and thickness h_0 moves to radius r, has a length δl and a thickness h. The extension ratios may be written as:

$$\left.\begin{aligned} \lambda_l &= \frac{\delta l}{\delta r_0} = \frac{\delta r}{\delta r_0}\frac{1}{\cos\beta} \\ \lambda_\theta &= \frac{r}{r_0} \\ \lambda_h &= \frac{h}{h_0} \end{aligned}\right\} \tag{5.100}$$

and using the constant volume assumption:

$$\lambda_l \lambda_\theta \lambda_h = 1$$

The general method of solution may be illustrated by the example, shown in Figure 5.17, of a uniform disc of radius a inflated by a lateral pressure P to form a dome. If we ignore body forces then we may write two equilibrium conditions from equations (5.93) and (5.95):

$$\frac{d}{dr}(r \sin \beta . N_l) = Pr \qquad \text{and} \qquad N_\theta = \frac{d}{dr}(rN_l)$$

The first equation may be integrated to give:

$$r \sin \beta \; N_l = \frac{Pr^2}{2} + C$$

where C is a constant. Since r can be zero, $C = 0$ to avoid singularities and we have:

$$\left.\begin{aligned} N_l &= \frac{Pr}{2 \sin \beta} \\ \text{and} \qquad N_\theta &= \frac{d}{dr}(r\, N_l) \end{aligned}\right\} \tag{5.101}$$

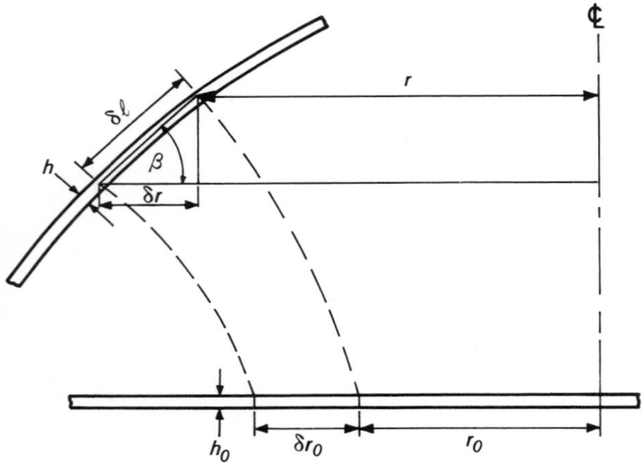

Figure 5.16 Large strains in a membrane.

If we take the sheet as being made of a rubber then the stress–strain relationships for the plane stress system (equation (2.35)):

$$p_l = (\phi + \psi\lambda_\theta^2)\left(\lambda_l^2 - \frac{1}{\lambda_l^2\lambda_\theta^2}\right)$$

and

$$p_\theta = (\phi + \psi\lambda_l^2)\left(\lambda_\theta^2 - \frac{1}{\lambda_l^2\lambda_\theta^2}\right)$$

Since N_l and N_θ are forces per unit length we may write:

$$N_l = p_l h = p_l \frac{h_0}{\lambda_l \lambda_\theta}$$

and similarly for N_θ and hence:

$$N_l = \frac{h_0}{\lambda_l \lambda_\theta}(\phi + \psi \lambda_\theta^2)\left(\lambda_l^2 - \frac{1}{\lambda_l^2 \lambda_\theta^2}\right)$$

and

$$N_\theta = \frac{h_0}{\lambda_l \lambda_\theta}(\phi + \psi \lambda_l^2)\left(\lambda_\theta^2 - \frac{1}{\lambda_l^2 \lambda_\theta^2}\right) \qquad (5.102)$$

From the definitions of the extension ratios, equation (5.100), r_0 may be eliminated to give:

$$\frac{d\lambda_\theta}{dr} = \frac{\lambda_\theta}{r}\left(1 - \frac{\lambda_\theta}{\lambda_l \cos \beta}\right) \qquad (5.103)$$

Thus equations (5.101), (5.102) and (5.103) give five independent equations and since there are five dependent variables, N_θ, N_l, λ_θ, λ_l and $\cos \beta$ then each can be expressed as a function of the independent variable r. It can be seen that the

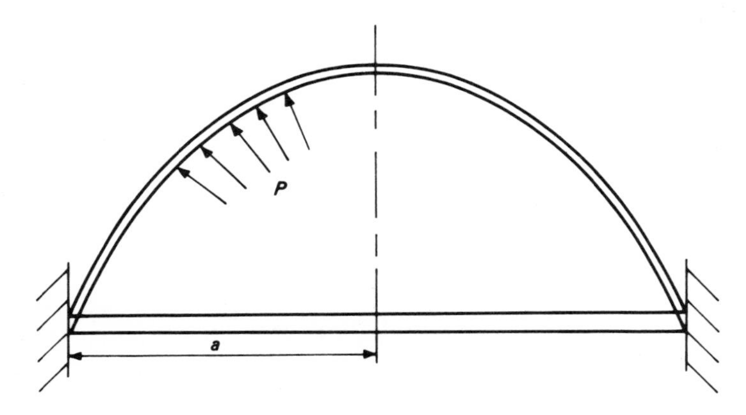

Figure 5.17 Laterial bulging of a sheet.

equations are very involved and their exact solution is an exercise in numerical analysis and need not be considered further here. It should be noted that the shape, β as a function of r, depends on the material properties and that only in special cases will simple geometries result. There are very few practical situations where the effort involved in a full solution is justified but approximate solutions do give useful results and they will be considered further.

(i) *Assumed shapes—bulging of a sheet*

If we consider again the bulging sheet problem then it is not an unreasonable assumption to consider the section as the cap of a sphere. Thus the shape may be taken as:

$$r = R \sin \beta \qquad (5.104)$$

where R is the radius of the sphere as shown in Figure 5.18. From equation (5.101) we have:

$$N_l = \frac{Pr}{2 \sin \beta} = \frac{PR}{2} \quad \text{a constant}$$

and hence $N_\theta = N_l$.
Thus the strains at all points must also be equal:

$$\lambda_\theta = \lambda_l$$

and hence:

$$\frac{r}{r_0} = \frac{\mathrm{d}r}{\mathrm{d}r_0} \frac{1}{\cos \beta}$$

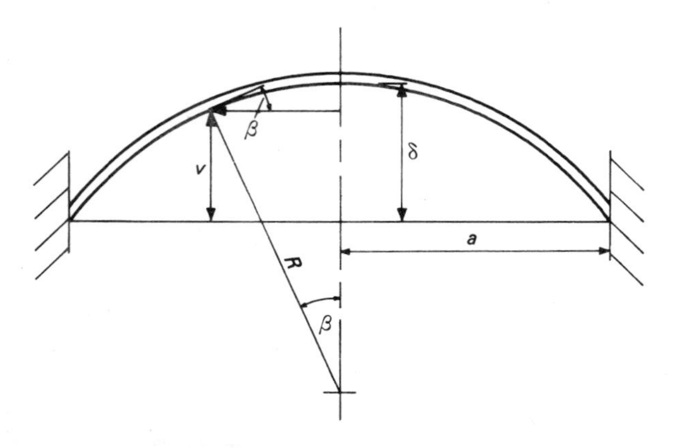

Figure 5.18 Assumed shape: cap of a sphere.

Substituting from equation (5.104) and integrating gives:

$$r_0 = \frac{Br}{R + \sqrt{(R^2 - r^2)}}$$

where B is a constant. Now at $r_0 = a$, $r = a$ and hence:

$$B = R + \sqrt{(R^2 - a^2)}$$

and we may write:

$$\lambda_\theta = \lambda_l = \frac{R + \sqrt{(R^2 - r^2)}}{R + \sqrt{(R^2 - a^2)}} \tag{5.105}$$

If we let the central deflection be δ then from the geometry we have:

$$R = \frac{a^2 + \delta^2}{2\delta} \tag{5.106}$$

and by rearranging equation (5.105) we have:

$$\lambda_\theta = \lambda_l = 1 + \frac{v\delta}{a^2} \tag{5.107}$$

It is interesting to note that:

$$\lambda_h = \frac{1}{\lambda_\theta \lambda_l} = \frac{1}{(1+v\delta/a^2)^2}$$

and at the top of the dome, $v = \delta$ gives the largest thickness reduction:

$$\hat{\lambda}_h = \frac{1}{(1+\delta^2/a^2)^2} \tag{5.108}$$

Thus the assumed form gives a strain distribution but if we examine the solution further we have:

$$N_l = p_l h = \frac{p_l h_0}{\lambda_l \lambda_\theta}$$

$$\therefore \qquad \frac{p_l h_0}{\lambda_l^2} = \frac{PR}{2}$$

i.e.
$$p_l = \left(\frac{PR}{2h_0}\right)\lambda_l^2 \tag{5.109}$$

The spherical shape can only be achieved by a material with this form of stress–strain curve. Since $\lambda_\theta = \lambda_l$ and $\lambda_\theta = 1$ at $r = a$ then there is no strain at the edge yet the stress–strain law must give a finite stress at $\lambda_l = 1$, as shown in equation (5.109), to maintain equilibrium. All real stress–strain relationships give a zero stress at $\lambda = 1$ and thus this solution is never correct near the edge. The shape is usually roughly spherical however and a reasonable answer may be obtained by taking a mean value of λ_h (i.e. assuming constant thickness overall):

$$\bar{\lambda}_h = \frac{1}{\delta}\int_0^\delta \frac{\mathrm{d}v}{(1+\delta v/a^2)^2}$$

i.e.
$$\bar{\lambda}_h = \frac{1}{1+\delta^2/a^2} \quad \text{and} \quad \bar{\lambda}_l^2 = 1+\delta^2/a^2 \tag{5.110}$$

If we now take an equivalent stress–strain relationship of the form:

$$\bar{p} = p_0 \bar{\varepsilon}^n$$

then for this problem we have, from equations (5.71):

$$\bar{p} = p_l \qquad \text{and} \quad \bar{\varepsilon} = 2\varepsilon_l = \ln \lambda_l^2$$

giving:

$$\bar{p} = \left(\frac{PR}{2h_0}\right)(1+\delta^2/a^2)$$

from equations (5.109) and (5.110) and hence:

$$\left(\frac{Pa}{2h_0 p_0}\right) = \frac{[\ln(1+\delta^2/a^2)]^n(2\delta/a)}{(1+\delta^2/a^2)^2} \tag{5.111}$$

This function has the form common to large strain solutions with a maximum pressure given by the condition:

$$n = \frac{3(\delta/a)^2 - 1}{2(\delta/a)^2} \ln\left[1+(\delta/a)^2\right] \tag{5.112}$$

It is of interest to note that for a perfectly plastic material $n = 0$:

$$(\delta/a)_{\max} = \frac{1}{\sqrt{3}} = 0{\cdot}577$$

and for $n = 1$,

$$(\delta/a)_{\max} = 1.182 \tag{5.113}$$

Thus for a practical range of n values the central deflection at instability $(\delta/a)_{\max}$ does not vary greatly and is usually in the region 0·6 to 0·8.

(ii) *Simplified stress–strain curves—thermoforming*
Many polymers are formed by heating them above their glass transition temperature so that they become rubbery. After forming in this condition they are then allowed to solidify to give the required shape in the glassy state material. It can be shown that a reasonable approximation to the stress–strain behaviour (section 2.2.2) may be obtained by using the rubber equations with $\psi = 0$ so that:

$$p_l = \phi\left(\lambda_l^2 - \frac{1}{\lambda_l^2\lambda_\theta^2}\right)$$

Now if we consider that forming operations are in general biaxial so that both λ_l and λ_θ are greater than unity then for large strains the expression may be approximated to:

$$p_l = \phi\lambda_l^2$$

and similarly,

$$p_\theta = \phi\lambda_\theta^2 \tag{5.114}$$

Solutions produced by these relationships are only precise as $\lambda_{l,\theta} \to \infty$ but in practice they provide a useful guide to deformation patterns. If we rewrite equations (5.114) in terms of the forces per unit length then from equation (5.100):

$$N_l = p_l h = \phi h_0 \frac{\lambda_l}{\lambda_\theta}$$

and

$$N_\theta = \phi h_0 \frac{\lambda_\theta}{\lambda_l}$$

Hence we have:

$$N_l N_\theta = (\phi h_0)^2 = N_0^2 \tag{5.115}$$

and considering only problems without body forces and using equation (5.96) gives:

$$\frac{dN_l}{dr} + \frac{1}{r}\left(N_l - \frac{N_0^2}{N_l}\right) = 0$$

Integrating we have a general expression for N_l of the form:

$$N_l^2 = N_0^2 - \frac{A}{r^2} \tag{5.116}$$

where A is a constant.

Clearly if $r \to 0$ as in the inflated dome case then $A = 0$ and $N_l = N_0$, a constant. This is, of course, the solution derived in the previous section since the stress–strain laws used, equations (5.114), are precisely those required to give a spherical cap. Thus, the previously derived analysis is the large strain rubber solution and from equation (5.109) we have the additional result:

$$\frac{PR}{2h_0} = \phi$$

and from equation (5.106):

$$\frac{Pa}{2h_0} = \phi \frac{2\delta/a}{1+(\delta/a)^2} \tag{5.117}$$

This function maximises when $\delta/a = 1$ which confirms the previous result of the insensitivity of the value to the shape of the stress–strain curve.

This pressure-deflection calculation can be greatly improved by using the mean value of extension ratio as before:

$$\bar{\lambda}_l^2 = 1 + \frac{\delta^2}{a^2}$$

and the more precise stress–strain curve derived from equation (5.29) to give:

$$\frac{Pa}{2h_0} = \frac{2(\bar{\lambda}_l^2 - 1)^{\frac{1}{2}}}{\bar{\lambda}_l^2}(\phi + \psi\bar{\lambda}_l^2)\left(1 - \frac{1}{\bar{\lambda}_l^6}\right) \tag{5.118}$$

For the special case of $\psi = 0$ we have a maximum for $\lambda_l = 1{\cdot}61$, i.e. $(\delta/a)_{max} = 1{\cdot}26$. The maximum pressure prediction of the large strain theory is much more accurate than the deflection prediction giving, from equation (5.117):

$$\left(\frac{Pa}{2h_0}\right)_{max} = \sqrt{3}\,\phi$$

while from equation (5.118),

$$\left(\frac{Pa}{2h_0}\right)_{max} = \sqrt{3}\,\phi(0{\cdot}94)$$

i.e. a difference of only 6%.

Some further examples illustrate the general method.

(a) Direct loading

Figure 5.19 shows a sheet of radius a which has a plunger of radius b forced upwards at the centre. The sheet will be assumed to be restrained such that $r = r_0 = a$ and b at the edges. If body forces and pressure are ignored then equation (5.93) gives:

$$r \sin \beta . N_l = B \tag{5.119}$$

where B is a constant.

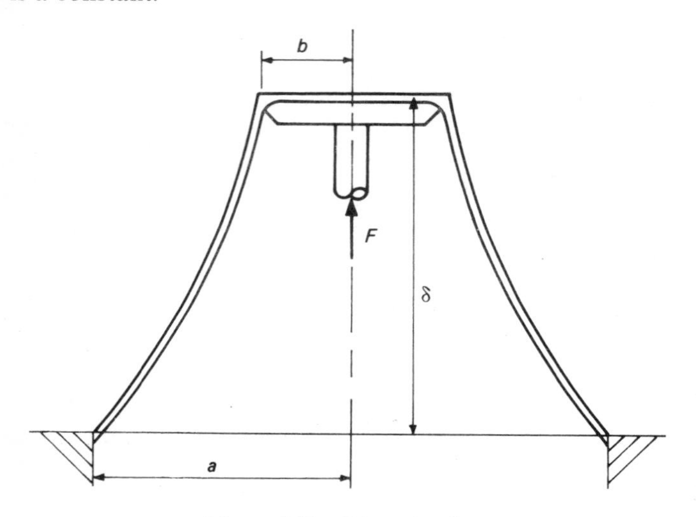

Figure 5.19 Direct loading.

Inspection of Figure 5.19 shows that equilibrium gives:

$$2\pi r \sin \beta . N_l = F$$

and hence:

$$B = F/2\pi$$

We may write N_l as:

$$N_l = N_0 \frac{\lambda_l}{\lambda_\theta}$$

and hence by substitution from equations (5.100) and (5.119) we have:

$$\frac{F}{2\pi} \frac{1}{r \sin \beta} = N_0 \frac{\mathrm{d}r}{\mathrm{d}r_0} \frac{1}{\cos \beta} \frac{r_0}{r}$$

and noting that $\tan \beta = -\mathrm{d}v/\mathrm{d}r$ this reduces to:

$$\frac{\mathrm{d}v}{\mathrm{d}r_0} = -\frac{F}{2\pi N_0} \frac{1}{r_0} \tag{5.120}$$

Integrating and using the boundary conditions $v = 0$ at $r_0 = a$ and $v = \delta$ at $r_0 = b$ we have:

$$\delta = \frac{F}{2\pi N_0} \ln \frac{a}{b} \tag{5.121}$$

Thus there is a linear relationship between F and δ for this simple stress–strain behaviour.

The strain distribution can be examined by formulating $\mathrm{d}v/\mathrm{d}r = -\tan\beta$ from equation (5.119) and then substituting for N_l from equation (5.116) to give:

$$\frac{\mathrm{d}v}{\mathrm{d}r} = \frac{F}{2\pi N_0} \frac{1}{\left[r^2 - \frac{1}{N_0{}^2}\left(A + \left(\frac{F}{2\pi}\right)^2\right)\right]^{1/2}}$$

Substituting from equation (5.120) and integrating gives:

$$r_0 = C\left[r + \left\{r^2 - \frac{1}{N_0{}^2}\left(A + \left(\frac{F}{2\pi}\right)^2\right)\right\}^{1/2}\right]$$

where C is a constant. Now since $r = r_0$ at both a and b the only solution is:

$$A = -\left(\frac{F}{2\pi}\right)^2 \quad \text{and} \quad C = \tfrac{1}{2}$$

giving $r_0 = r$ over the whole surface. Thus since the shape is logarithmic the thickness variation is given by:

$$\lambda_h = \frac{1}{\sqrt{\left\{1 + \left(\frac{\delta}{r \ln a/b}\right)^2\right\}}} \tag{5.122}$$

More accurate load predictions may be made by taking a mean value of λ_h and using the full stress–strain curves as described for the bulging case.

(b) Predetermined shapes

The method may be applied when sheets are formed by dies as illustrated in Figure 5.20 which shows a dome blown into a cone of wall angle α. Now the section not in contact is stretching in the usual way such that:

$$\lambda_l = \lambda_\theta$$

but when the material contacts the wall it is fixed at the slope of the wall and hence:

$$\beta = \alpha$$

giving, from equations (5.100):

$$\frac{\delta r}{\delta r_0} \frac{1}{\cos\alpha} = \frac{r}{r_0}$$

Since $r = r_0 = a$ we have:

$$r = a\left[\frac{r_0}{a}\right]^{\cos\alpha}$$

and the thickness variation becomes:

$$\lambda_h = \frac{1}{\lambda_\theta^2} = \left(\frac{r}{r_0}\right)^2 = \left[\frac{r}{a}\right]^{2\left(\frac{1}{\cos\alpha}-1\right)} \tag{5.123}$$

The pressure may be determined in the usual way to give:

$$P = 2\phi h_0 \frac{\sin\alpha}{r}\left[1 - \frac{\sin^6\alpha}{8(1-\cos\alpha)^3}\right] \tag{5.124}$$

using the more accurate stress–strain curve.

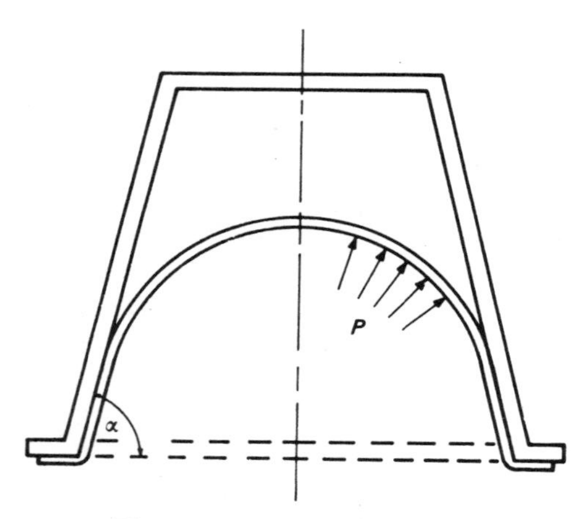

Figure 5.20 Forming in a die.

The special case of a vertical wall ($\alpha = \pi/2$) must be solved separately and gives:

$$\lambda_h = \exp\left(-\frac{2s}{a}\right) \tag{5.125}$$

where s is the distance up the wall.

Bibliography

1. Turner, C. E. *Introduction to Plate and Shell Theory*, Longmans, London (1965).
2. Faupel, J. H. *Engineering Design*, John Wiley and Sons Inc., New York (1964).
3. Green, A. E. and Adkins, J. E. *Large Elastic Deformations*, Oxford University Press (1960).
4. Williams, J. G. 'A Method of Calculation for Thermoforming Plastic Sheets', *Journ. Strain Anal.* (1970), 1, No. 5.

6 Problems of stress concentrations and cracks

6.1. Introduction

In this chapter we shall deal with the calculation of stresses due to the presence of holes, notches and cracks. In most practical situations the failure of a component stems from a crack which originates in the vicinity of some stress concentration and so it is of major importance in design studies to estimate the local stress state. The analysis is essentially that of a three-dimensional stress and strain field with variations of both in all three directions. Complete analytical solutions to such complex situations are very few in number and the discussion here will be limited to two-dimensional solutions for the case of plane stress. One exception to this will be the case of plane strain which is derived from the plane stress case.

The analsysis is difficult even in the case of small strains and linear elastic materials, and all the solutions discussed here will be subject to these limitations. Comments on the relevance to visco-elastic materials will be made but the results will be confined to the linear elastic case and some simple plasticity. It is felt that the essential nature of the behaviour of stress concentrations is described by linear elasticity and that its extension to time dependent materials is adequately achieved by the use of the time dependent modulus concept.

For the purposes of polymer stress analysis it is not considered essential to have a detailed understanding of the methods of solution used in these problems. They involve powerful general mathematical techniques and could not be described adequately here. For this reason an outline only of the method used is given and it is recommended that specialised texts be followed if further details are required. It is hoped, however, that the reader will gain a sufficient appreciation of the analysis and that the missing steps will not be too difficult to accept.

As a further point of interest it should be noted that many of the results given here are used in the analysis of fracture behaviour. However, the discussion will be largely limited to the stress analysis and the detailed relevance to fracture criteria is left to specialised texts. Once the nature of the solutions given has been fully grasped the significance of much fracture theory will be more easily appreciated.

6.2. Stress functions

The basic method of analysis uses the stress function concept. Let us consider plane stress so that using a cartesian coordinate system we have:

$$p_{zz} = p_{xz} = p_{yz} = 0$$

The general solution of all plane stress, linear elasticity problems may be deduced from the usual basic steps of equilibrium, compatibility and the stress–strain relationships so that we must solve the following equations:

(i) Equilibrium, ignoring body forces (equations (1.27)):

$$\left.\begin{aligned} \frac{\partial p_{xx}}{\partial x} + \frac{\partial p_{xy}}{\partial y} = 0 \\ \frac{\partial p_{yy}}{\partial y} + \frac{\partial p_{xy}}{\partial x} = 0 \end{aligned}\right\} \tag{6.1}$$

(ii) Compatibility, from equations (1.52) and (1.53):

$$e_{xx} = \frac{\partial u}{\partial x}, \qquad e_{yy} = \frac{\partial v}{\partial y}, \qquad e_{xy} = \frac{\partial u}{\partial y} + \frac{\partial v}{\partial x}$$

and hence, by eliminating u and v

$$\frac{\partial^2 e_{xx}}{\partial y^2} + \frac{\partial^2 e_{yy}}{\partial x^2} = \frac{\partial^2 e_{xy}}{\partial x \, \partial y} \tag{6.2}$$

from equation (1.56).

(iii) Stress–strain relationships, equations (2.4):

$$\left.\begin{aligned} e_{xx} &= \frac{1}{E}(p_{xx} - \nu p_{yy}) \\ e_{yy} &= \frac{1}{E}(p_{yy} - \nu p_{xx}) \\ e_{xy} &= \frac{2(1+\nu)}{E} p_{xy} \end{aligned}\right\} \tag{6.3}$$

The use of stress functions involves the choice of a new variable ϕ which satisfies these equations and solutions are then considered in terms of this variable. Let us define ϕ such that:

$$p_{xx} = \frac{\partial^2 \phi}{\partial y^2}, \qquad p_{yy} = \frac{\partial^2 \phi}{\partial x^2} \qquad \text{and} \quad p_{xy} = -\frac{\partial^2 \phi}{\partial x \, \partial y} \tag{6.4}$$

Clearly these definitions have been chosen such that the equilibrium conditions, equations (6.1), are satisfied. The further conditions of compatibility and the stress–strain relationship may be considered by substituting for the strains in equation (6.2) from equation (6.3) which gives:

$$\frac{\partial^2}{\partial y^2}(p_{xx}-\nu p_{yy})+\frac{\partial^2}{\partial x^2}(p_{yy}-\nu p_{xx}) = 2(1+\nu)\frac{\partial^2 p_{xy}}{\partial x\,\partial y}$$

If the stresses are now replaced by ϕ we have:

$$\frac{\partial^4\phi}{\partial y^4}+2\frac{\partial^4\phi}{\partial x^2\,\partial y^2}+\frac{\partial^4\phi}{\partial x^4} = 0$$

It is usual to use the operator ∇^2, defined as:

$$\nabla^2 = \frac{\partial^2}{\partial x^2}+\frac{\partial^2}{\partial y^2}$$

so that,

$$\nabla^2\phi = \frac{\partial^2\phi}{\partial x^2}+\frac{\partial^2\phi}{\partial y^2}$$

and operating a second time we have:

$$\nabla^4\phi = \left(\frac{\partial^2}{\partial x^2}+\frac{\partial^2}{\partial y^2}\right)\nabla^2\phi$$

$$= \frac{\partial^4\phi}{\partial x^4}+\frac{2\partial^4\phi}{\partial x^2\partial y^2}+\frac{\partial^4\phi}{\partial y^4}$$

Thus we may write the condition for which ϕ satisfies all three basic conditions as:

$$\nabla^4\phi = 0 \tag{6.5}$$

The solution of this equation solves all linear elastic small strain problems in plane stress without body forces. The particular form of ϕ is determined by the boundary conditions of the problem considered. Problems such as the stress distributions in beams, for example, have been solved in this way. The discussion here, however, will be conducted, not in cartesian coordinates, but in terms of cylindrical polar coordinates, since this system is more convenient for the examples given later.

6.2.1. *Cylindrical polar coordinates*

Figure 6.1(a) shows the stress components in cartesian coordinates as used in the basic derivation and Figure 6.1(b) shows the coordinate system to which the result is to be transposed. The coordinates are related by:

$$x = r\cos\theta \text{ and } y = r\sin\theta$$

so that,

$$x^2+y^2 = r^2 \text{ and } \tan\theta = y/x$$

Hence we may deduce the derivative relationships:

e.g. $$2x = 2r\frac{r}{\partial x}, \therefore \frac{\partial r}{\partial x} = \frac{x}{r} = \cos\theta \text{ and similarly}$$

$$\frac{\partial \theta}{\partial x} = -\frac{\sin\theta}{r}, \frac{\partial \theta}{\partial y} = \frac{\cos\theta}{r}, \frac{\partial r}{\partial x} = \cos\theta \text{ and } \frac{\partial r}{\partial y} = \sin\theta \quad (6.6)$$

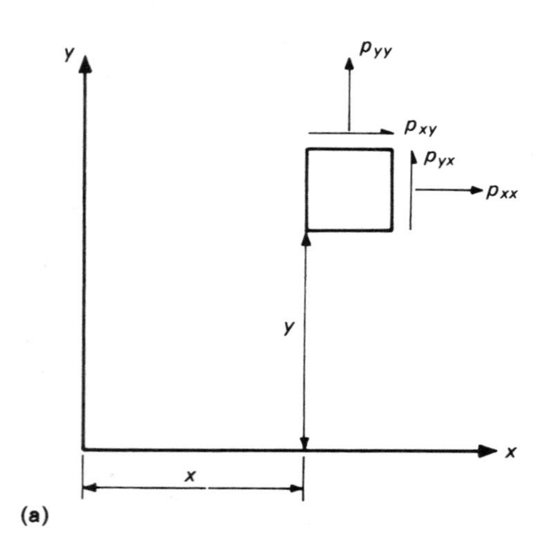

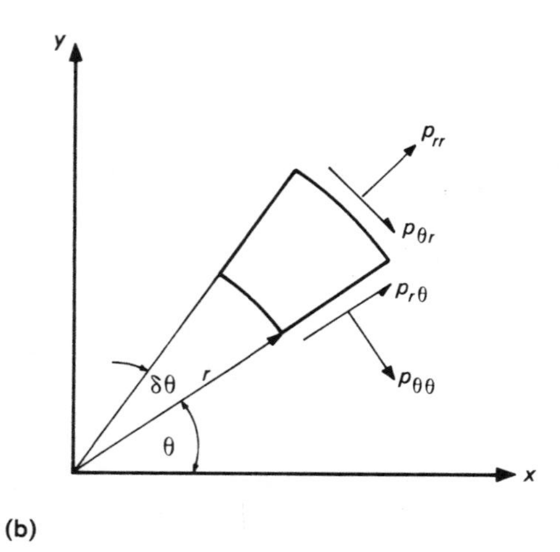

Figure 6.1 Cartesian and cylindrical polar coordinates.

Now the derivative of ϕ with respect to y may be written as:

$$\frac{\partial \phi}{\partial y} = \frac{\partial \phi}{\partial \theta}\frac{\partial \theta}{\partial y} + \frac{\partial \phi}{\partial r}\frac{\partial r}{\partial y}$$

i.e.

$$\frac{\partial \phi}{\partial y} = \frac{\cos\theta}{r}\frac{\partial \phi}{\partial \theta} + \sin\theta \frac{\partial \phi}{\partial r}$$

Differentiating a second time and substituting for the derivatives with respect to x and y we have:

$$\frac{\partial^2 \phi}{\partial y^2} = \frac{\cos^2\theta}{r}\frac{\partial^2 \phi}{\partial \theta^2} - \frac{2\sin\theta\cos\theta}{r^2}\frac{\partial \phi}{\partial \theta} + \sin^2\theta \frac{\partial^2 \phi}{\partial r^2} + \frac{\cos^2\theta}{r}\frac{\partial \phi}{\partial r} + \frac{2\sin\theta\cos\theta}{r}\frac{\partial^2 \phi}{\partial \theta \partial r} \tag{6.7}$$

Similarly we may derive:

$$\frac{\partial^2 \phi}{\partial x^2} = \cos^2\theta \frac{\partial^2 \phi}{\partial r^2} - \frac{2\sin\theta\cos\theta}{r}\frac{\partial^2 \phi}{\partial \theta \partial r} + \frac{\sin^2\theta}{r}\frac{\partial \phi}{\partial r} + \frac{\sin^2\theta}{r^2}\frac{\partial^2 \phi}{\partial \theta^2} + \frac{2\sin\theta\cos\theta}{r^2}\frac{\partial \phi}{\partial \theta} \tag{6.8}$$

and by adding equations (6.7) and (6.8) we have:

$$\nabla^2 \phi = \frac{\partial^2 \phi}{\partial x^2} + \frac{\partial^2 \phi}{\partial y^2} = \frac{\partial^2 \phi}{\partial r^2} + \frac{1}{r}\frac{\partial \phi}{\partial r} + \frac{1}{r^2}\frac{\partial^2 \phi}{\partial \theta^2}$$

Thus the basic equation may be transposed to:

$$\nabla^4 \phi = \left(\frac{\partial^2}{\partial r^2} + \frac{1}{r}\frac{\partial}{\partial r} + \frac{1}{r^2}\frac{\partial^2}{\partial \theta^2}\right)\left(\frac{\partial^2 \phi}{\partial r^2} + \frac{1}{r}\frac{\partial \phi}{\partial r} + \frac{1}{r^2}\frac{\partial^2 \phi}{\partial \theta^2}\right) = 0 \tag{6.9}$$

The various stress components may be derived from the usual relationships (equation (1.21))

$$p_{rr} = \left(\frac{p_{xx} + p_{yy}}{2}\right) + \left(\frac{p_{xx} - p_{yy}}{2}\right)\cos 2\theta + p_{xy}\sin 2\theta \tag{6.10}$$

and substituting from equation (6.4):

$$p_{rr} = \frac{1}{2}\left(\frac{\partial^2 \phi}{\partial y^2} + \frac{\partial^2 \phi}{\partial x^2}\right) + \frac{\cos 2\theta}{2}\left(\frac{\partial^2 \phi}{\partial y^2} - \frac{\partial^2 \phi}{\partial x^2}\right) - \sin 2\theta \frac{\partial^2 \phi}{\partial x \partial y}$$

By the same process as used in deriving equations (6.7) and (6.8) we have:

$$\frac{\partial^2 \phi}{\partial x \partial y} = \frac{\sin 2\theta}{2}\left(\frac{\partial^2 \phi}{\partial r^2} - \frac{1}{r}\frac{\partial \phi}{\partial r} - \frac{1}{r^2}\frac{\partial^2 \phi}{\partial \theta^2}\right) + \cos 2\theta \left(\frac{1}{r}\frac{\partial^2 \phi}{\partial r \partial \theta} - \frac{1}{r^2}\frac{\partial \phi}{\partial \theta}\right)$$

and by substitution into equation (6.10) together with equations (6.7) and (6.8) we have:

$$\left.\begin{aligned} p_{rr} &= \frac{1}{r}\frac{\partial \phi}{\partial r}+\frac{1}{r^2}\frac{\partial^2 \phi}{\partial \theta^2} \\ \text{and similarly:}\quad & \\ p_{\theta\theta} &= \frac{\partial^2 \phi}{\partial r^2} \\ \text{and}\quad p_{r\theta} &= \frac{1}{r^2}\frac{\partial \phi}{\partial \theta}-\frac{1}{r}\frac{\partial^2 \phi}{\partial r \partial \theta} = -\frac{\partial}{\partial r}\left(\frac{1}{r}\frac{\partial \phi}{\partial \theta}\right) \end{aligned}\right\} \quad (6.11)$$

The use of these equations is best illustrated by some special cases.

6.2.2. *The line load*

Consider the case of a line load of F per unit length normal to a flat surface as shown in Figure 6.2(a). If we take θ from the vertical direction as shown in Figure 6.2(b) then it can be seen immediately that there is a boundary condition: $p_{r\theta} = p_{\theta\theta} = 0$ at $\theta = \pm\pi/2$ for all values of r. Now it is usual to assume that solutions will be of the form:

$$\phi = R\,\Theta$$

where R is a function of r and Θ a function of θ. This problem has a particularly simply solution since from equations (6.11) we may write:

$$p_{\theta\theta} = \frac{\partial^2 \phi}{\partial r^2} = \Theta\frac{\mathrm{d}^2 R}{\mathrm{d}r^2} = 0 \quad \text{at } \theta = \pm\pi/2$$

and if we assume that $\Theta \neq 0$ at $\pm\pi/2$ then we have:

$$\frac{\mathrm{d}^2 R}{\mathrm{d}r^2} = 0 \quad \text{for all values of } r$$

and

$$R = A_1 r + A_2$$

where A_1 and A_2 are constants. If we consider $p_{r\theta}$ we have:

$$p_{r\theta} = \frac{\mathrm{d}\Theta}{\mathrm{d}\theta}\left(\frac{R}{r^2}-\frac{1}{r}\frac{\mathrm{d}R}{\mathrm{d}r}\right)$$

and substituting for R gives the result:

$$p_{r\theta} = \frac{\mathrm{d}\Theta}{\mathrm{d}\theta}\left(\frac{A_1}{r}+\frac{A_2}{r^2}-\frac{A_1}{r}\right) = \frac{\mathrm{d}\Theta}{\mathrm{d}\theta}\frac{A_2}{r^2}$$

Now if we choose $A_2 = 0$ then we have the condition that $p_{r\theta} = 0$ which satisfies the boundary conditions. However $p_{\theta\theta}$ and $p_{r\theta}$ are then not only zero at $\theta = \pm\pi/2$ but at all values of θ and hence only p_{rr} exists. This stress distribution is called pure radial loading and is the fundamental stress distribution for a line load.

To determine the function Θ we must return to equation (6.9) and using:

$$\phi = A_1 r \,\Theta$$

we have:

$$\frac{d^4\Theta}{d\theta^4} + 2\frac{d^2\Theta}{d\theta^2} + \Theta = 0$$

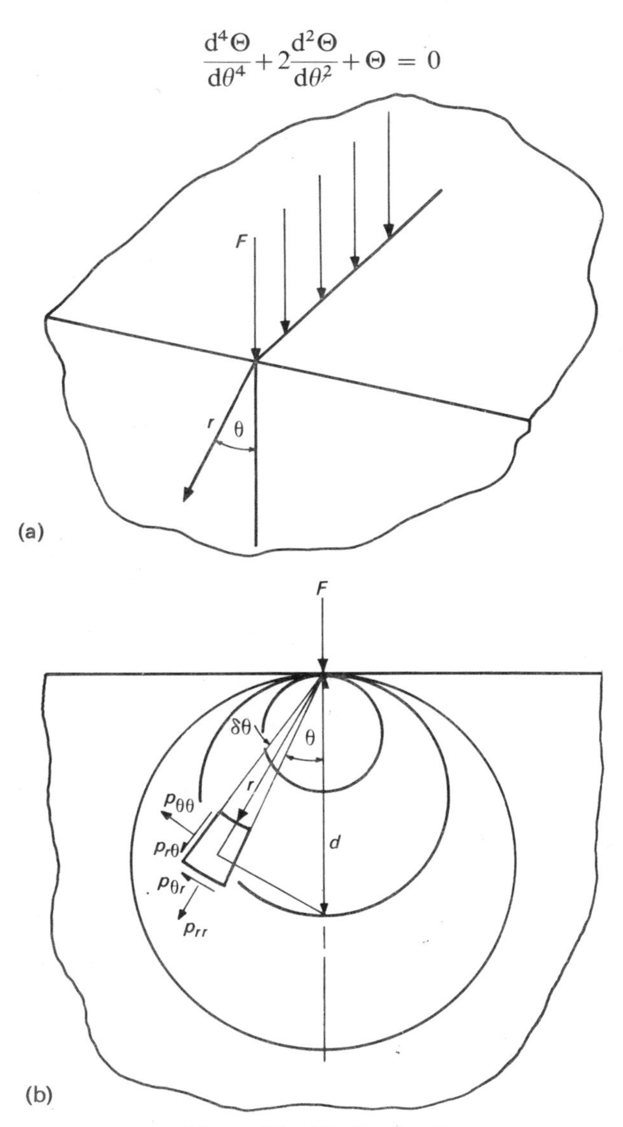

Figure 6.2 The line load.

This equation has a general solution of the form:

$$\Theta = A_3 \sin\theta + A_4 \cos\theta + A_5\theta \sin\theta + A_6\theta \cos\theta$$

and the coefficients may be deduced by considering the remaining boundary condition that:

$$-F = \int_{-\pi/2}^{+\pi/2} p_{rr}\, r \cos\theta \, \mathrm{d}\theta \tag{6.12}$$

i.e. the vertical resultant of p_{rr} since $p_{\theta\theta}$ and $p_{r\theta}$ are zero, must equal F. Now:

$$p_{rr} = \frac{A_1}{r}\left(\Theta + \frac{\mathrm{d}^2\Theta}{\mathrm{d}\theta^2}\right)$$

and hence

$$p_{rr} = \frac{2A_1}{r}(A_5 \cos\theta - A_6 \sin\theta)$$

Substituting in equation (6.12) and integrating we have:

$$-F = A_1 A_5 \pi$$

giving a final result:

$$p_{rr} = -\frac{2F}{\pi}\frac{\cos\theta}{r} \tag{6.13}$$

with A_3, A_4 and A_6 put to zero since they are indeterminate for this boundary condition.

This solution is correct even though many assumptions were made in its derivation since the stress function used:

$$\phi = \text{const. } r\theta \sin\theta$$

satisfies all the boundary conditions. It is not, in general, possible to deduce the form directly in this way and a trial and error procedure is necessary to determine the correct terms. A discussion of the general solution will be given in a later section but at this stage it is sufficient to regard the method as one of guessing a form and then determining its suitability with the ∇^4 relationship and the boundary conditions.

The line load is of particular interest because the stress system is simple. Clearly, at any θ, as $r \to 0$, the stress tends to infinity under the point load. In practice the material will yield and there will be a plastic zone under the point of application. If we consider the shape of the stress distribution then a family of circles of diameter d all tangential at the load point, as shown in Figure 6.2(b), may be drawn such that:

$$d = \frac{r}{\cos\theta}$$

and hence:

$$p_{rr} = -\frac{2F}{\pi d} \tag{6.14}$$

Thus p_{rr} is a constant around a circle as shown which can be confirmed in a simple photo-elastic test since these circles appear as isoclinics. If it is assumed that the area of yielding is small so that the presence of the yielded material does not affect the elastic solution appreciably then the diameter of the circular plastic zone is given by:

$$d_P = \frac{2F}{\pi p_Y} \tag{6.15}$$

By a similar method of analysis solutions may be obtained for loads applied at different angles and for wedges.

6.2.3. *Stresses around a circular hole*

Consider now a round hole of radius a in an infinite plate subjected to a uniform stress p at infinity, as shown in Figure 6.3. The usual method of deducing the stress function for this case is to consider a radius $b \gg a$ such that the effect of the hole is negligible and then to resolve the applied stress p on the boundary. Thus by the usual analysis, equations (1.21), we have:

$$\left.\begin{aligned} p_{rr} &= p/2+(p/2)\cos 2\theta \\ \text{and} \qquad p_{r\theta} &= -(p/2)\sin 2\theta \end{aligned}\right\} \tag{6.16}$$

The analysis is then conducted in two stages since the first part of the expression for p_{rr} is a uniform radial tension $p/2$ and from the Lamé equations (5.44) we may write down the expression for the stresses due to this term:

$$p'_{rr} = \frac{p}{2}\left(\frac{b^2}{b^2-a^2}\right)\left(1-\frac{a^2}{r^2}\right), \quad p'_{\theta\theta} = \frac{p}{2}\left(\frac{b^2}{b^2-a^2}\right)\left(1+\frac{a^2}{r^2}\right) \tag{6.17}$$

and $p'_{r\theta} = 0$ since we have axial symmetry.

The θ dependence of the remaining terms in equation (6.16) suggests a form of stress function:

$$\phi = R\cos 2\theta$$

since this satisfies the θ dependence of both p_{rr} and $p_{r\theta}$. Substitution of this in equation (6.9) gives:

$$\left(\frac{\partial^2}{\partial r^2}+\frac{1}{r}\frac{\partial}{\partial r}+\frac{1}{r^2}\frac{\partial^2}{\partial\theta^2}\right)\left(\frac{\mathrm{d}^2R}{\mathrm{d}r^2}+\frac{1}{r}\frac{\mathrm{d}R}{\mathrm{d}r}-\frac{4R}{r^2}\right)\cos 2\theta = 0$$

and hence:

$$\left(\frac{\mathrm{d}^2}{\mathrm{d}r^2}+\frac{1}{r}\frac{\mathrm{d}}{\mathrm{d}r}-\frac{4}{r^2}\right)\left(\frac{\mathrm{d}^2R}{\mathrm{d}r^2}+\frac{1}{r}\frac{\mathrm{d}R}{\mathrm{d}r}-\frac{4R}{r^2}\right) = 0$$

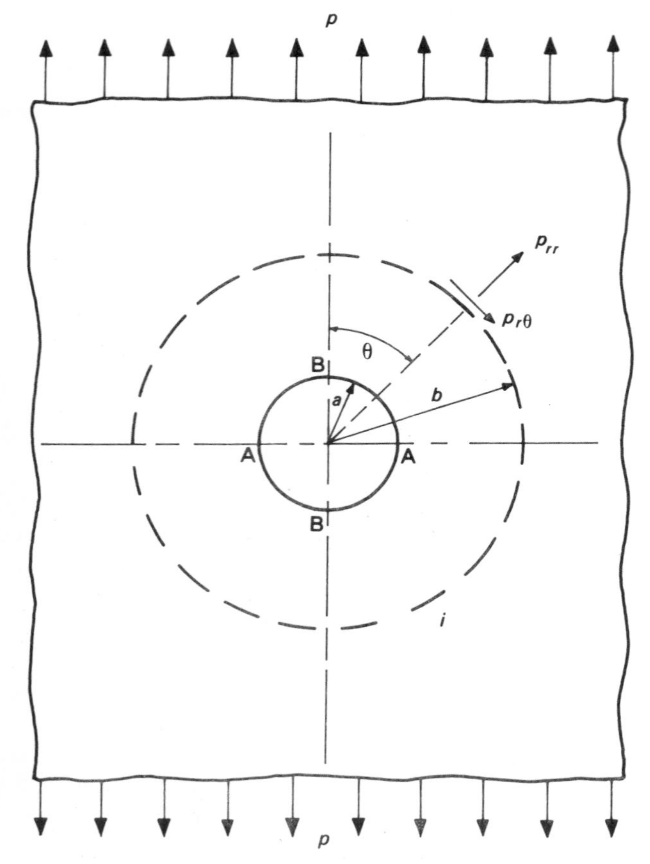

Figure 6.3 Stresses around a circular hole.

The general solution to this equation has the form:

$$R = A_1 r^2 + A_2 r^4 + A_3 \frac{1}{r^2} + A_4 \tag{6.18}$$

where A_1, A_2, A_3 and A_4 are constants. Substitution of R and Θ in the expressions for the stresses gives the results:

$$\left.\begin{aligned} p''_{rr} &= -\left(2A_1 + \frac{6A_3}{r^4} + \frac{4A_4}{r^2}\right)\cos 2\theta \\ p''_{\theta\theta} &= \left(2A_1 + 12A_2 r^2 + \frac{6A_3}{r^4}\right)\cos 2\theta \\ p''_{r\theta} &= \left(2A_1 + 6A_2 r^2 - \frac{6A_3}{r^4} - \frac{2A_4}{r^2}\right)\sin 2\theta \end{aligned}\right\} \tag{6.19}$$

The final stress distribution is given by:

$$p_{rr} = p'_{rr} + p''_{rr}, \quad p_{\theta\theta} = p'_{\theta\theta} + p''_{\theta\theta}, \quad p_{r\theta} = p'_{r\theta} + p''_{r\theta}$$

and we have the boundary conditions:

$$\left.\begin{aligned} p_{rr} &= 0 \text{ at } r = a \\ p_{r\theta} &= 0 \quad \text{at } r = a \\ p_{rr} &= (p/2)\cos 2\theta \quad \text{at } r = b \\ \text{and} \qquad p_{r\theta} &= (-p/2)\sin 2\theta \quad \text{at } r = b \end{aligned}\right\} \tag{6.20}$$

If we assume that $b \gg a$, then the expressions for the constants reduce to

$$A_1 = -\frac{p}{4}, \; A_2 = 0, \; A_3 = \frac{pa^4}{4} \text{ and } A_4 = \frac{pa^2}{2}$$

giving the final result:

$$\left.\begin{aligned} p_{rr} &= \frac{p}{2}\left(1 - \frac{a^2}{r^2}\right) + \frac{p}{2}\left(1 + \frac{3a^4}{r^4} - \frac{4a^2}{r^2}\right)\cos 2\theta \\ p_{\theta\theta} &= \frac{p}{2}\left(1 + \frac{a^2}{r^2}\right) - \frac{p}{2}\left(1 + \frac{3a^4}{r^4}\right)\cos 2\theta \\ \text{and} \qquad p_{r\theta} &= -\frac{p}{2}\left(1 - \frac{3a^4}{r^4} + \frac{2a^2}{r^2}\right)\sin 2\theta \end{aligned}\right\} \tag{6.21}$$

It should be noted that the stresses die out very quickly away from the edge of the hole. The points of interest are on the surface of the hole ($r = a$) where:

$$p_{r\theta} = p_{rr} = 0 \quad \text{and} \quad p_{\theta\theta} = p - 2p\cos 2\theta$$

and the limiting values for $p_{\theta\theta}$ are:

$$\begin{aligned} \theta &= \pi/2 \quad \text{and} \quad 3\pi/2, \qquad p_\theta = 3p \\ \text{and} \qquad \theta &= 0 \quad \text{and} \quad \pi, \qquad p_\theta = -p \end{aligned} \tag{6.22}$$

Thus at the sides of the hole at points A, Figure 6.3, we have a stress of $3p$ and at points B one of $-p$. It is the factor of 3 which is the main cause of weakening in bodies containing holes. When $r = 2a$, i.e. one radius from the side of the hole, at $\theta = \pi/2$, the factor has fallen to 1·22 so the effect is very localised. It is also of interest to note that if a crack forms at some value of $p_{\theta\theta}$ then in tension it will form at the side of the hole and be normal to the applied stress. Under compression (applied stress $-p$) the largest tension is at B and is p giving cracks parallel to the applied stress of three times the tensile value.

6.2.4. *Stresses around an elliptical hole*

The solution of this problem is of particular interest since results derived from it form the basis of much of fracture theory. The general solution is laborious and best carried out using a complex function form of analysis. The method will be omitted here but the final result for stresses along the major axis of the ellipse will be given. Figure 6.4 shows the system described with an elliptical

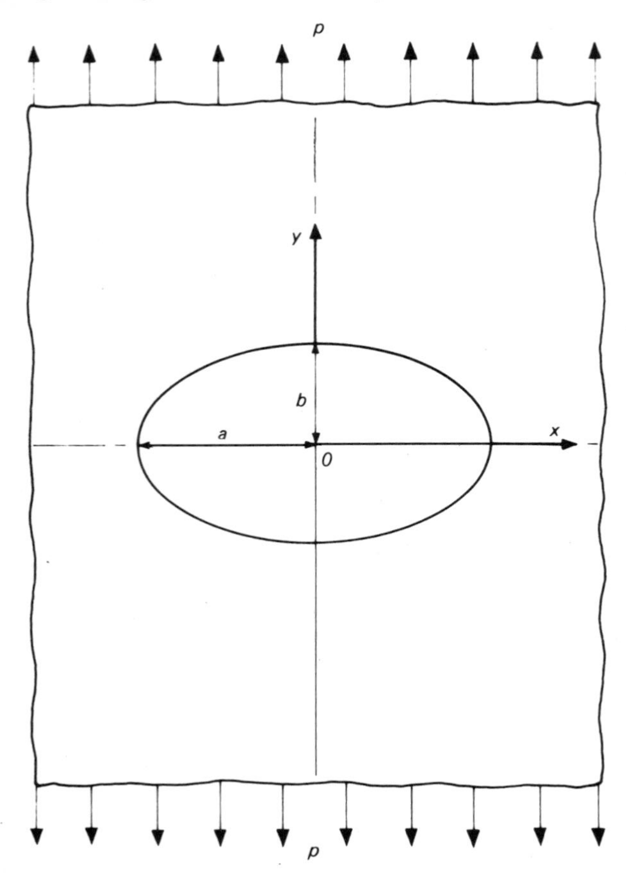

Figures 6.4 Stresses around an elliptical hole.

hole of major axis $2a$ and minor axis $2b$ subjected to a uniform tension at infinity normal to the major axis. The result is expressed in terms of the variables:

$$2B = a+b \quad \text{and} \quad m = \frac{a-b}{a+b} \tag{6.23}$$

and a distance parameter:

$$\alpha = \frac{x}{2B}+\sqrt{\left(\frac{x}{2B}\right)^2-m} \tag{6.24}$$

where x is the distance from the centre of the ellipse. The stresses along the major axis of the ellipse ($y = 0$) are given by:

$$\left.\begin{aligned} p_{yy} &= \frac{Z_2}{2}+Z_1 \\ \text{and} \qquad p_{xx} &= \frac{Z_2}{2}-Z_1 \end{aligned}\right\} \tag{6.25}$$

$$\left.\begin{aligned} \text{where} \qquad Z_1 &= \frac{p}{2}\left\{1+\left(\frac{m^2-1}{\alpha^2-m}\right)\left[1+\left(\frac{m-1}{\alpha^2-m}\right)\left(\frac{3\alpha^2-m}{\alpha^2-m}\right)\right]\right\} \\ \text{and} \qquad Z_2 &= p\left\{1+\frac{2(1+m)}{\alpha^2-m}\right\} \end{aligned}\right\} \tag{6.26}$$

A simple special case of this is the circle in which:

$$b = a \quad \text{giving} \quad 2B = 2a \quad \text{and} \quad m = 0.$$

The variable α now becomes x/B and:

$$Z_1 = \frac{p}{2}\left[1-\frac{1}{\alpha^2}\left(1-\frac{3}{\alpha^2}\right)\right]$$

$$Z_2 = p\left[1+\frac{2}{\alpha^2}\right]$$

Thus the final stress distributions are:

$$\left.\begin{aligned} p_{yy} &= \frac{p}{2}\left[2+\left(\frac{a}{x}\right)^2+3\left(\frac{a}{x}\right)^4\right] \\ \text{and} \qquad p_{xx} &= \frac{3p}{2}\left[\left(\frac{a}{x}\right)^2-\left(\frac{a}{x}\right)^4\right] \end{aligned}\right\} \tag{6.27}$$

Comparison with equations (6.21) for $\theta = \pi/2$ and noting that $p_{\theta\theta} \equiv p_{yy}$, $p_{rr} \equiv p_{xx}$ and $x \equiv r$ shows that they are in agreement.

A result of major importance from this solution is the stress at the ends of the hole on the major axis, i.e. $x = \pm a$. Substituting for $2B$, m and $x = a$ in equation (6.24) gives:

$$\alpha = \frac{a}{a+b}+\sqrt{\left\{\frac{a^2}{(a+b)^2}-\frac{a-b}{(a+b)}\right\}} = 1$$

Hence from equations (6.26) we have:

$$Z_1 = \frac{p}{2}\left\{1+\frac{m^2-1}{1-m}\left[1+\left(\frac{m-1}{1-m}\right)\left(\frac{3-m}{1-m}\right)\right]\right\}$$

i.e.

$$Z_1 = \frac{p}{2}\left(\frac{3+m}{1-m}\right)$$

and similarly:

$$Z_2 = p\left(\frac{3+m}{1-m}\right)$$

Substitution in the expressions for the stresses gives the results that $p_{xx} = 0$ as expected and:

$$p_{yy} = p\left(1 + \frac{2a}{b}\right) \tag{6.28}$$

It is usual to express this result in terms of the radius of curvature of the ellipse at the end of the major axis given by:

$$\rho = b^2/a$$

and hence:

$$p_{yy} = p\left(1 + 2\sqrt{\frac{a}{\rho}}\right) \tag{6.29}$$

This is a valuable result since many notches can be quite accurately approximated by the elliptical shape and this expression provides a measure of the stress concentration factor:

$$N = \frac{p_{yy}}{p}$$

Specialised texts provide tabulated solutions for a wide range of special geometries, and they can be consulted when required. Some special cases are of interest, e.g. $\rho = a$ is the solution for the round hole and also as $\rho \to 0$, $N \to \infty$. In fact all of the tabulated solutions present this fundamental problem that as $\rho \to 0$ (the sharp crack) the stresses tend to infinity. It is the study of this tendency which constitutes most of the analysis of fracture mechanics which we will examine in the following sections.

6.3. Stresses around cracks

The simplest first approach to this problem is to examine the solution for a sharp crack of length 2a in an infinite plate with a uniform stress p applied at infinity. This may be obtained directly from the elliptical hole problem since the crack may be regarded as an ellipse with the minor axis as zero:

i.e. $$b = 0$$

in which case:

$$2B = a \quad \text{and} \quad m = 1$$

Substitution in equations (6.25) and (6.26) gives the results:

$$\left.\begin{aligned} p_{xx} &= \frac{2p}{\alpha^2 - 1} \\ \text{and} \qquad p_{yy} &= p\,\frac{\alpha^2 + 1}{\alpha^2 - 1} \end{aligned}\right\} \tag{6.30}$$

From equation (6.24) we have:

$$\alpha = \frac{x}{a} + \sqrt{\left(\frac{x}{a}\right)^2 - 1}$$

and the expressions for the stresses become:

$$\left.\begin{aligned} p_{xx} &= p\left(\frac{x}{\sqrt{x^2 - a^2}} - 1\right) \\ \text{and} \qquad p_{yy} &= \frac{px}{\sqrt{x^2 - a^2}} \end{aligned}\right\} \qquad (6.31)$$

As expected both expressions tend to infinity as $x \to a$ and p_{yy} tends to p for large values of x as shown in Figure 6.5. The form can be further investigated

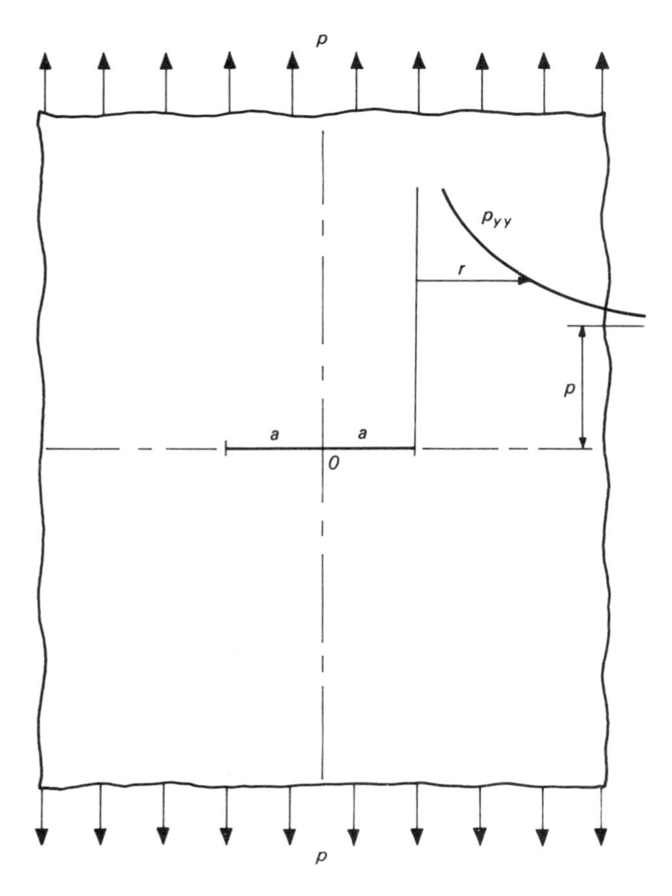

Figures 6.5 Stresses at the tip of a crack.

by rewriting the expressions in terms of r the distance from the crack tip such that:

$$x = a+r$$

and equations (6.30) may be expanded as a series to give:

$$p_{yy} = p\left[\left(\frac{a}{2r}\right)^{1/2} + \frac{3}{8}\left(\frac{2r}{a}\right)^{1/2} - \frac{5}{128}\left(\frac{2r}{a}\right)^{3/2} + \cdots\right] \tag{6.32}$$

Clearly, as $r \to 0$ the stress system is dominated by the first term and we may write:

$$\left.\begin{aligned} p_{yy} &\simeq \frac{p\sqrt{a}}{\sqrt{2r}} \\ \text{and similarly} \qquad p_{xx} &\simeq \frac{p\sqrt{a}}{\sqrt{2r}} \end{aligned}\right\} \tag{6.33}$$

6.3.1. *Fracture criteria*

No attempt will be made here to examine criteria of fracture in any detail but it is useful to define some terms in current use. In terms of establishing a criterion equations (6.33) present obvious problems since p_{yy} and p_{xx} tend to infinity as r tends to zero. It is possible to describe the stress state at the crack tip by considering the parameter:

$$p_{yy}\sqrt{2r}$$

since this remains finite as $r \to 0$. To conform to normal convention we shall introduce a factor of π and write:

$$K_{\mathrm{I}} = p_{yy}\sqrt{2\pi r} \tag{6.34}$$

and equations (6.33) may be written as:

$$K_{\mathrm{I}} = p\sqrt{\pi a} \tag{6.35}$$

Thus in this case for any values of p and a there is a value of K_{I}, the stress intensity factor. It can be postulated that fracture will take place when this parameter, which typifies the state of stress, reaches a critical value K_{IC} so that the fracture criterion becomes:

$$K_{\mathrm{IC}} = p\sqrt{\pi a} \tag{6.36}$$

This is the same form as the well known Griffith fracture criterion based on an energy release rate analysis and the two are interchangeable. The conventional Griffith theory considers that if an infinite plate with a uniform stress p has a crack of length $2a$ introduced normal to the stress, then the change of strain energy per unit thickness is given by:

$$U = \frac{\pi p^2 a^2}{2E} \tag{6.37}$$

This is equivalent to saying that the introduction of the crack results in a fall in the energy, i.e. some strain energy will be released. As the crack grows the strain energy release rate is given by:

$$G = \frac{\partial U}{\partial a} = \frac{\pi p^2 a}{E} \tag{6.38}$$

If it is postulated that the formation of a new crack of length δa results in the absorption of energy given by:

$$G_c \, \delta a$$

then the crack will be unstable if the released energy $G \, \delta a$ is greater than that absorbed:

i.e. $$G \, \delta a \geqslant G_c \, \delta a$$

Thus the critical condition is given by:

$$G_c = \frac{\pi p^2 a}{E} \tag{6.39}$$

In some descriptions the energy per unit area of new crack face γ is used so that:

$$G_c = 2\gamma$$

giving the Griffith equation:

$$p = \sqrt{\frac{2E\gamma}{\pi a}}$$

at fracture.

Comparison of equations (6.38) and (6.35) gives the result:

$$K_I{}^2 = EG \tag{6.40}$$

which can be shown to be true for all plane stress systems. The result can be converted to plane strain by dividing the right-hand side by $(1-\nu^2)$. Equation (6.37) is sometimes useful for derivations of K_I expressions in particular systems (see section (6.3.8)).

The value of the parameter K_{IC} (or G_c) lies very much in it being a simple fracture criterion. It can be determined experimentally by measuring p at fracture for a range of values of a. If these crack lengths are very much smaller than the size of plate tested then the infinite plate solution given here may be employed and say, πp^2 plotted versus $1/a$ giving K_{IC}^2 as the slope of the resulting line.

It is sometimes useful to regard K_{IC} as made up of two parameters, i.e. it may be taken as a critical stress $p_{yy} = p_c$ at a critical distance $r = c$ so that:

$$K_{IC} = p_c\sqrt{2\pi c} \tag{6.41}$$

This is equivalent to saying that the fracture originates ahead of the crack tip, a widely observed phenomenon, and then breaks back to the crack tip. The stress state at this point is of interest since from equations (6.33) $p_{yy} = p_{xx}$ and if we postulate plane strain then:

$$p_{zz} = \nu(p_{yy} + p_{xx})$$

and for $\nu = \frac{1}{2}$ we have hydrostatic tension.

6.3.2. *Blunt cracks*

It is of interest at this point to consider the case of a crack with a finite radius of curvature at the tip. If we take the case when the radius is very small, i.e. a crack but with a finite tip radius, then we may consider the elliptical hole solution for the special case of $\rho \ll a$. If, in addition, we confine our attention to a small distance r away from the crack tip ($r \ll a$) then we may write:

$$2B = a\left(1 + \sqrt{\frac{\rho}{a}}\right) \quad \text{and} \quad m = \frac{1 - \sqrt{\dfrac{\rho}{a}}}{1 + \sqrt{\dfrac{\rho}{a}}}$$

and

$$\alpha = 1 - \sqrt{\frac{\rho}{a}} + \sqrt{\frac{2r + \rho}{a}}$$

Substituting these results in equations (6.25) and (6.26) we have the expression:

$$p_{yy} = \frac{p\sqrt{a}}{\sqrt{2r}} \frac{1 + \dfrac{\rho}{r}}{\left(1 + \dfrac{\rho}{2r}\right)^{3/2}} \tag{6.42}$$

When $\rho = 0$ we have the sharp crack solution and when $r = 0$

$$p_{yy} = p\,2\sqrt{\frac{a}{\rho}}$$

which is the same as equation (6.29) for $\rho \ll a$. If we postulate the fracture criterion of $p_{yy} = p_c$ and $r = c$ then equation (6.42) may be rewritten as:

$$\frac{p\sqrt{\pi a}}{p_c\sqrt{2\pi c}} = \frac{K_{IB}}{K_{IC}} = \frac{\left(1 + \dfrac{\rho}{2c}\right)^{3/2}}{\left(1 + \dfrac{\rho}{c}\right)} \tag{6.43}$$

For $\rho/c \gg 1$ this expression tends to:

$$\frac{K_{IB}}{K_{IC}} = \frac{1}{2\sqrt{2c}}\sqrt{\rho}$$

so that a graph K_{IB}/K_{IC} versus $\sqrt{\rho}$ tends to a straight line of slope $1/(2\sqrt{2c})$ and hence c may be determined. Figure 6.6 shows the expression plotted and

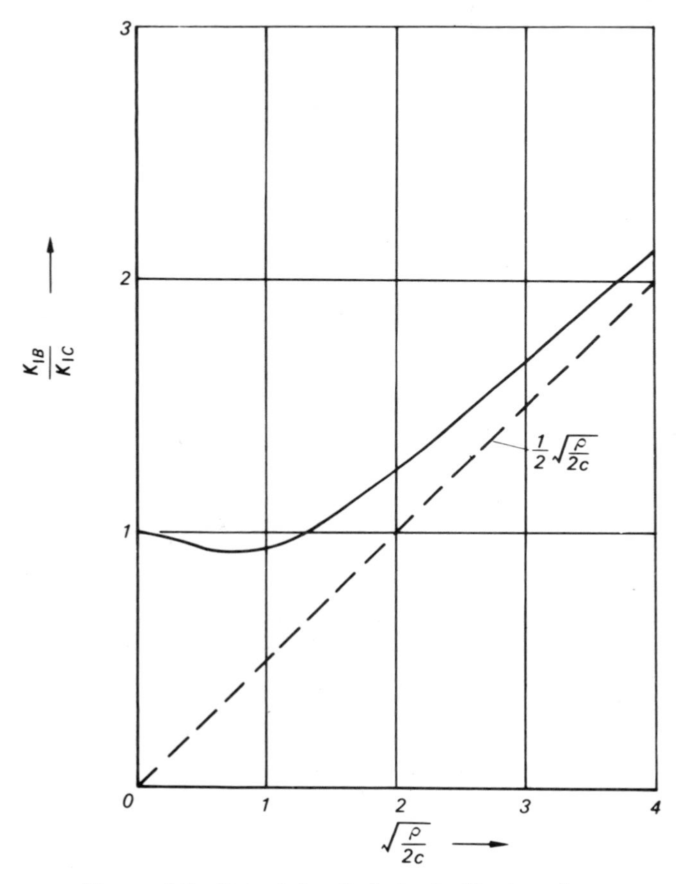

Figures 6.6 Stress intensity factor for blunt cracks.

it is interesting to note that K_{IB} falls below K_{IC} for small radii of curvature and then rises again. The minimum value is given when:

$$\frac{\rho}{2c} \doteq \frac{1}{2}$$

and

$$\left(\frac{K_{IB}}{K_{IC}}\right)_{\min} = \frac{3\sqrt{3}}{4\sqrt{2}} = 0{\cdot}916$$

This use of blunt cracks is well suited to the determination of p_c and c if they are required.

6.3.3. *General expressions for stresses around cracks*

We may now consider the general case of stress distributions around the tip of a crack and for this purpose analyse a wedge as shown in Figure 6.7. Consider

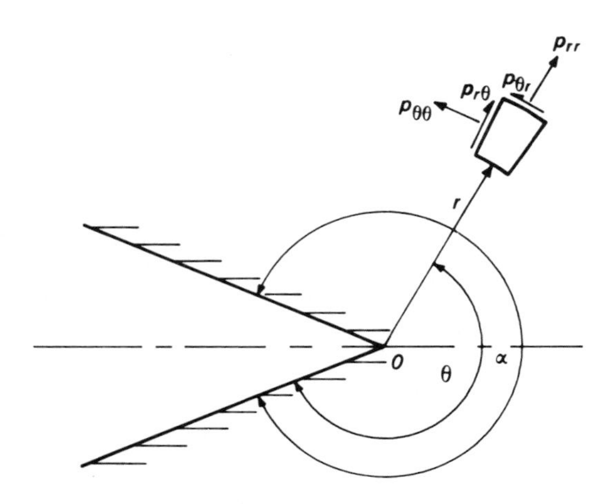

Figure 6.7 The wedge crack.

the angle of the wedge to be α and the origin of the polar coordinate system to be at the tip of the wedge as shown. If we consider the faces of the wedge to be stress free then this general problem has the boundary conditions:

$$p_{\theta\theta} = p_{r\theta} = 0 \quad \text{at} \quad \theta = 0 \quad \text{and} \quad \alpha$$

If we now turn to the stress function in polar coordinates we may use the variable separable form:

$$\phi = R\Theta$$

where R is a function of r and Θ is a function of θ.
The expressions for the stresses from equations (6.11) then become:

$$\left.\begin{aligned} p_{rr} &= \frac{\Theta}{r}\frac{\mathrm{d}R}{\mathrm{d}r} + \frac{R}{r^2}\frac{\mathrm{d}^2\Theta}{\mathrm{d}\theta^2} \\ p_{\theta\theta} &= \Theta\frac{\mathrm{d}^2R}{\mathrm{d}r^2} \\ p_{r\theta} &= \frac{\mathrm{d}\Theta}{\mathrm{d}\theta}\left(\frac{R}{r^2} - \frac{1}{r}\frac{\mathrm{d}R}{\mathrm{d}r}\right) \end{aligned}\right\} \qquad (6.44)$$

and from equation (6.9) the functions must satisfy:

$$\left(\frac{\partial^2}{\partial r^2}+\frac{1}{r}\frac{\partial}{\partial r}+\frac{1}{r^2}\frac{\partial^2}{\partial \theta^2}\right)\left(\Theta\frac{\mathrm{d}^2 R}{\mathrm{d}r^2}+\frac{\Theta}{r}\frac{\mathrm{d}R}{\mathrm{d}r}+\frac{R}{r^2}\frac{\mathrm{d}^2\Theta}{\mathrm{d}\theta^2}\right)=0$$

Possible solutions to this equation can be written as:

$$\begin{aligned}\phi &= C_0 r^2+(C_1+C_2 r^2)\ln r+(C_3+C_4 r^2)\theta \\ &\quad +(C_5 \sin\theta+C_6\cos\theta\times\theta r+C_7 r\ln r) \\ &\quad +\sum r^n[a_n \sin n\theta+a'_n\sin(n-2)\theta+b_n\cos(n-2)\theta+b'_n\cos n\theta] \quad (6.45)\end{aligned}$$

where n may take any value. The first two terms are the solution to axial symmetry problems and will give, for example, the Lamé solutions. Since this problem is not symmetrical they may be ignored. The third term will not meet the boundary conditions and may be ignored. The fourth term is that which gives the line load solution and, while it satisfies the boundary conditions, is only applicable to point loads at the wedge tip. This sort of loading is of little practical significance and will be excluded.

Each of the terms under the summation sign is capable of satisfying the boundary conditions and if we take the general term:

$$R = r^n\ \Theta = a_n\sin n\theta+a'_n\sin(n-2)\theta+b_n\cos(n-2)\theta+b'_n\cos n\theta$$

we may consider first the conditions at $\theta = 0$. Since both $p_{\theta\theta}$ and $p_{r\theta}$ are zero then:

$$\Theta = 0 \qquad \text{and} \qquad \frac{\mathrm{d}\Theta}{\mathrm{d}\theta}=0$$

and hence:

$$b'_n = -b_n \qquad \text{and} \quad a'_n = -\frac{n}{n-2}a_n$$

Using these relationships with $p_{\theta\theta} = p_{r\theta} = 0$ at $\theta = \alpha$ we have:

$$\frac{\cos n\alpha-\cos(n-2)\alpha}{\sin n\alpha-\dfrac{n}{n-2}\sin(n-2)\alpha}=\frac{(n-2)\sin(n-2)\alpha-n\sin n\alpha}{n\cos n\alpha-n\cos(n-2)\alpha}$$

This may be simplified to give the condition:

$$\sin(n-1)\alpha = (n-1)\sin\alpha \qquad (6.46)$$

We shall limit our attention here to the line crack for which $\alpha = 2\pi$ and hence:

$$\sin(n-1)2\pi = 0$$

giving valid solutions of:

$$n = 0,\quad \pm\tfrac{1}{2},\quad \pm 1,\quad \pm\tfrac{3}{2},\ \ldots$$

The stress function is:

$$\phi_n = r^n\left[a'_n\left(\sin(n-2)\theta - \frac{n-2}{n}\sin n\theta\right) + b'_n(\cos n\theta - \cos(n-2)\theta)\right]$$

and the expressions for the stresses are:

$$p_{rr} = (1-n)r^{n-2}[a'_n((n-4)\sin(n-2)\theta - (n-2)\sin n\theta) + b'_n(n\cos n\theta - (n-4)\cos(n-2)\theta)]$$

$$p_{\theta\theta} = n(1-n)r^{n-2}\left[a'_n\left(\sin(n-2)\theta - \frac{n-2}{n}\sin n\theta\right) + b'_n(\cos n\theta - \cos(n-2)\theta)\right]$$

$$p_{r\theta} = (1-n)r^{n-2}[a'_n((n-2)\cos(n-2)\theta - (n-2)\cos n\theta) - b'_n(n\sin n\theta - (n-2)\sin(n-2)\theta)]$$

The fact that all of these terms have a dependence on r of the form r^{n-2} puts restrictions on values of n if the displacements are to remain finite at the crack tip. Thus if we consider e_{rr} we have:

$$e_{rr} = \frac{1}{E}(p_r - \nu p_\theta)$$

and hence $e_{rr} \propto r^{n-2}$ and since e_{rr} is related to the radial displacement by:

$$e_{rr} = \frac{\partial u_r}{\partial r}$$

we have:

$$u_r \propto \frac{r^{n-1}}{n-1}$$

Thus for u_r to remain finite as $r \to 0$ we have:

$$n > 1$$

($n = 1$ is the trivial solution with all stresses zero) and hence the values of n are limited to:

$$n = \tfrac{3}{2},\quad 2,\quad \tfrac{5}{2},\ \ldots$$

It is also convenient to replace θ at this stage by the angle from the line of the crack ψ as shown in Figure 6.8 so that:

$$\psi = \theta - \pi$$

and hence we may write for $p_{\psi\psi}$

$$p_{\psi\psi} = \frac{3}{4r^{1/2}}\left[\frac{a'_1}{3}\left(-\cos\frac{3\psi}{2} - 3\frac{\cos\psi}{2}\right) + b'_1\left(\sin\frac{3\psi}{2} + \sin\frac{\psi}{2}\right)\right]$$

$$-2b'_2(1-\cos 2\psi)+\frac{15}{4}r^{1/2}\left[\frac{a'_3}{5}\left(5\cos\frac{\psi}{2}-\cos\frac{5\psi}{2}\right)+b'_3\left(\sin\frac{\psi}{2}-\sin\frac{5\psi}{2}\right)\right]+\cdots \quad (6.47)$$

Similar expressions may be written for p_{rr} and $p_{r\psi}$ and clearly the full stress system is defined as a series with unknown coefficients a'_n and b'_n. These coefficients are determined by the boundary conditions of the particular problem and the

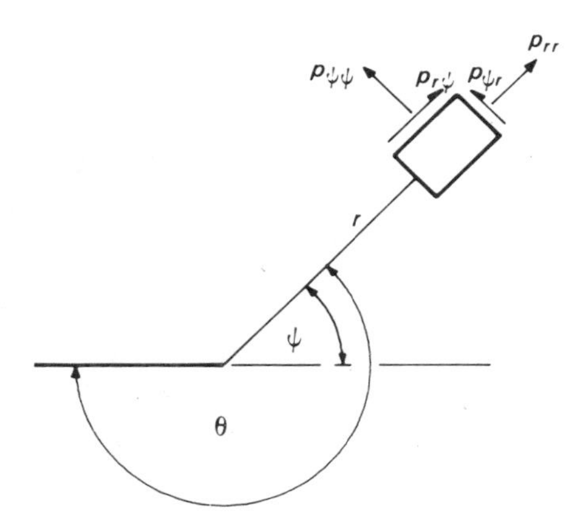

Figure 6.8 Stresses at a line crack.

series can be used in numerical methods. The general form is instructive in that the first term contains the only singularity, an $r^{-1/2}$ as derived from the elliptical hole result, and that this dominates the solution near the crack tip. The second term has no r dependence ($n = 2$) and it is interesting to deduce the meaning of the constant b'_2. The stress components from the second term are:

$$\begin{aligned} p_{\psi\psi} &= -2b'_2(1-\cos 2\psi) \\ p_{rr} &= -2b'_2(1+\cos 2\psi) \\ p_{r\psi} &= +2b'_2 \sin 2\psi \end{aligned} \quad (6.48)$$

If we now consider the stresses at the point r and ψ as produced by applied stresses at infinity of p_{yy}^{∞}, p_{xx}^{∞} and p_{xy}^{∞} as shown in Figure 6.9 then by the usual resolving methods we have:

$$p_{\psi\psi} = -p_{xy}^{\infty}\sin 2\psi+\frac{p_{xx}^{\infty}}{2}(1-\cos 2\psi)+\frac{p_{yy}^{\infty}}{2}(1+\cos 2\psi)$$

$$p_{rr} = p_{xy}^{\infty}\sin 2\psi+\frac{p_{xx}^{\infty}}{2}(1+\cos 2\psi)+\frac{p_{yy}^{\infty}}{2}(1-\cos 2\psi)$$

and

$$p_{r\psi} = p_{xy}^{\infty}\cos 2\psi-\frac{p_{xx}^{\infty}}{2}\sin 2\psi+\frac{p_{yy}^{\infty}}{2}\sin 2\psi$$

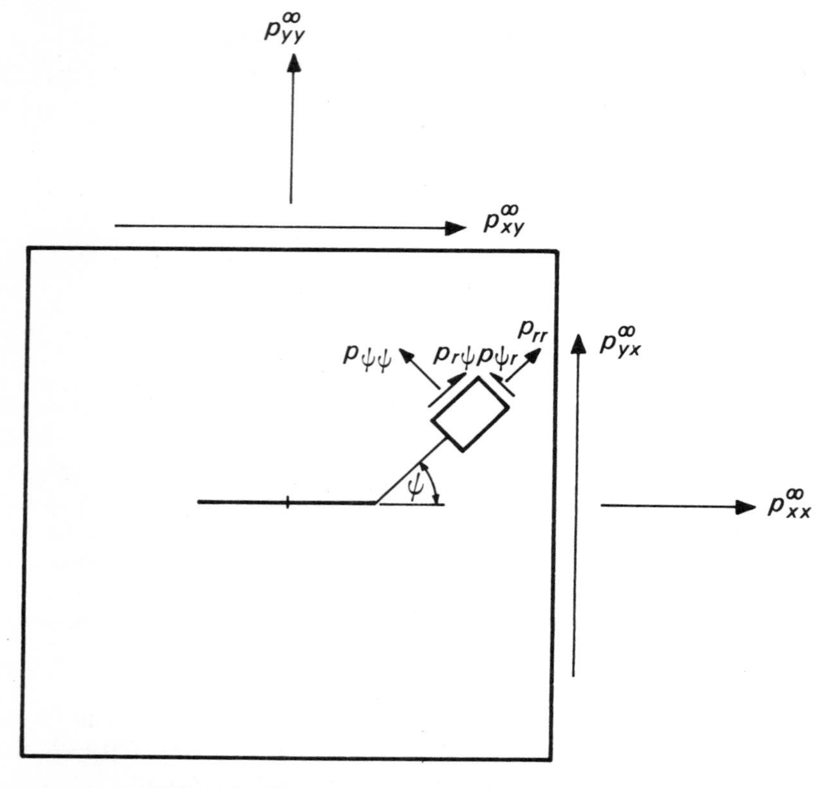

Figure 6.9 Stress distributions from applied stresses.

Comparison of these results with equations (6.48) shows that $b_2' = -p_{xx}^{\infty}/4$. This is to be expected since p_{xx}^{∞} is parallel to the crack and undergoes no change due to its presence. Thus we may conclude that the term independent of r is that stress which is unaffected by the stress concentration and may be added as required. If we restrict our attention to small values of r then we have the singular terms:

$$\left.\begin{aligned}
p_{rr} &= -\frac{1}{2r^{1/2}}\left[\frac{a_1'}{2}\left(-\cos\frac{3\psi}{2}+5\cos\frac{\psi}{2}\right)+b_1'\left(\frac{3}{2}\sin\frac{3\psi}{2}-\frac{5}{2}\sin\frac{\psi}{2}\right)\right]\\
p_{\psi\psi} &= -\frac{3}{4r^{1/2}}\left[\frac{a_1'}{3}\left(+\cos\frac{3\psi}{2}+3\cos\frac{\psi}{2}\right)-b_1'\left(\sin\frac{3\psi}{2}+\sin\frac{\psi}{2}\right)\right]\\
p_{r\psi} &= -\frac{1}{2r^{1/2}}\left[\frac{a_1'}{2}\left(\sin\frac{3\psi}{2}+\sin\frac{\psi}{2}\right)+b_1'\left(\frac{3}{2}\cos\frac{3\psi}{2}+\frac{1}{2}\cos\frac{\psi}{2}\right)\right]
\end{aligned}\right\} \quad (6.49)$$

The fundamental value of this result is that it illustrates that the stress distribution around a crack tip has the same form whatever the loading on the body and that it is only the constants of the equations which vary. This form of solution is not well suited to the determination of a'_1 and b'_1 and they are best found by considering solutions for the whole body such as that for the infinite plate given earlier. Comparison with this solution may be made with $p_{\psi\psi}$ and p_{rr} which is equivalent to p_{yy} and p_{xx} when $\psi = 0$:

$$p_{\psi\psi} = -\frac{a'_1}{\sqrt{r}} \quad \text{and} \quad p_{rr} = -\frac{a'_1}{\sqrt{r}}$$

and comparison with equations (6.33) gives the result:

$$a'_1 = -\frac{p^{\infty}_{yy}\sqrt{a}}{\sqrt{2}} = -\frac{K_{\mathrm{I}}}{\sqrt{2\pi}} \tag{6.50}$$

for this set of boundary conditions. Examination of equations (6.49) shows that the a'_1 terms for p_{rr} and $p_{\psi\psi}$ are symmetrical about $\psi = 0$; i.e. there are displacements of the same sign on both sides and the crack opening mode is as shown in Figure 6.10(a) and is termed mode I and hence denoted by K_{I}. The terms in b'_1, on the other hand, give stresses and hence displacements which change sign (skew symmetric) resulting in shearing as shown in Figure 6.10(b) which is termed mode II. A similar ellipitcal hole analysis for a shear stress applied at infinity and comparison of the shear stress $p_{r\psi}$ and p_{xy} shows that:

$$b'_1 = -\frac{p^{\infty}_{xy}\sqrt{a}}{\sqrt{2}} = -\frac{K_{\mathrm{II}}}{\sqrt{2\pi}} \tag{6.51}$$

A further mode, mode III, as shown in Figure 6.10(c), does exist but is not considered here since it is produced by stresses normal to the plane of the sheet. The relevant stress intensity factor is K_{III}.

We may finally rewrite equations (6.49) in terms of K_{I} and K_{II} and rearranging terms in ψ gives:

$$\left.\begin{aligned}
p_{rr} &= \frac{K_{\mathrm{I}}}{\sqrt{2\pi r}}\frac{1}{2}\cos(\psi/2)(3-\cos\psi) + \frac{K_{\mathrm{II}}}{\sqrt{2\pi r}}\frac{1}{2}(3\cos\psi - 1)\sin(\psi/2) \\
p_{\psi\psi} &= \frac{K_{\mathrm{I}}}{\sqrt{2\pi r}}\frac{1}{2}\cos(\psi/2)(1+\cos\psi) - \frac{K_{\mathrm{II}}}{\sqrt{2\pi r}}\frac{1}{2}(1+\cos\psi)\sin(\psi/2) \\
p_{r\psi} &= \frac{K_{\mathrm{I}}}{\sqrt{2\pi r}}\frac{1}{2}\cos(\psi/2)\sin\psi + \frac{K_{\mathrm{II}}}{\sqrt{2\pi r}}\frac{1}{2}(3\cos\psi - 1)\cos(\psi/2)
\end{aligned}\right\} \tag{6.52}$$

6.3.4. *Crack propagation*

The results of the general stress analysis of the crack tip can be used to predict the direction of crack propagation under an applied stress system. Let us consider first a pure mode I system with an applied stress p^{∞}_{yy}. From equations

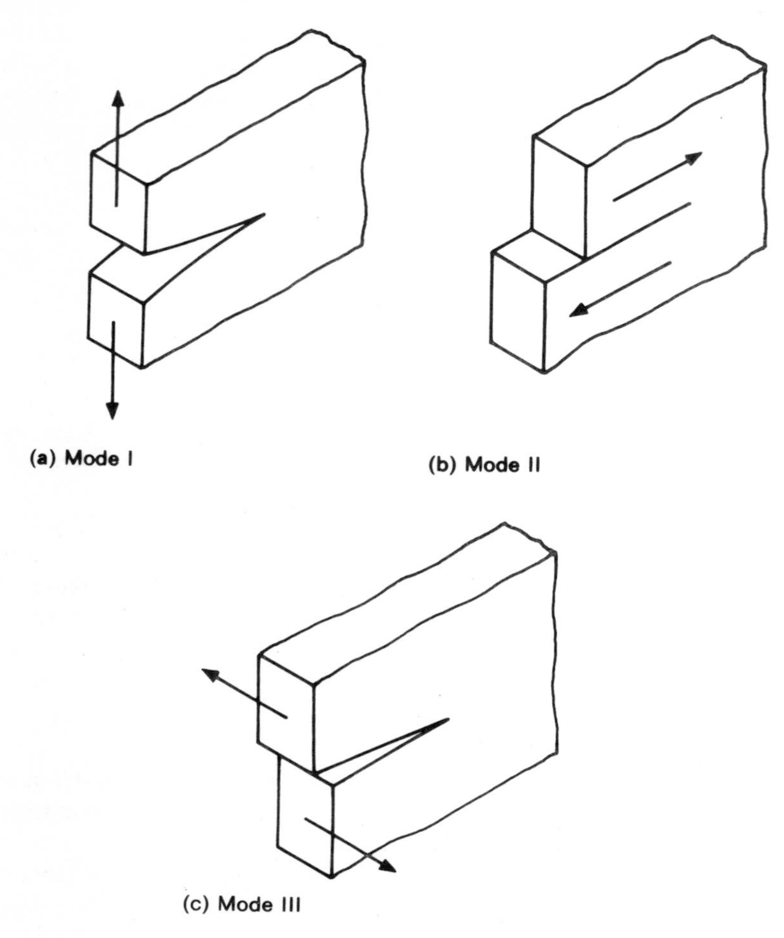

Figure 6.10 Fracture modes.

(6.52) we may calculate the expressions for the principal stresses from:

$$p_{1,2} = \frac{p_{\psi\psi} + p_{rr}}{2} \pm \sqrt{\left(\frac{p_{rr} - p_{\psi\psi}}{2}\right)^2 + p_{r\psi}^{\,2}}$$

which gives the result:

$$p_{1,2} = \frac{K_{\mathrm{I}}}{\sqrt{2\pi r}} \cos(\psi/2)[1 \pm \sin(\psi/2)] \tag{6.53}$$

and the angle which the greatest principal stress makes with the r direction is deduced from:

$$\tan 2\theta = \frac{2p_{r\psi}}{p_{rr} - p_{\psi\psi}}$$

i.e.
$$\theta = \frac{\pi}{4} - \frac{\psi}{4} \tag{6.54}$$

The largest principal stress may be derived from equation (6.53) by differentiation which gives the result $\psi = \pi/3$ at this condition and:

$$(p_1)_{\max} = \frac{3\sqrt{3}}{2}\frac{K_{\mathrm{I}}}{2\sqrt{2\pi r}} = \frac{1{\cdot}3K_{\mathrm{I}}}{\sqrt{2\pi r}} \tag{6.55}$$

In the plane of the crack, $\psi = 0$ we have:

$$(p_1)_{\psi=0} = \frac{K_{\mathrm{I}}}{\sqrt{2\pi r}}$$

It would not be unreasonable to suppose that the crack direction would be governed by $(p_1)_{\max}$ but clearly this is not so since the crack runs along $\psi = 0$ and not $\psi = \pi/3$. The reason may be found in equation (6.54) where the direction of the $(p_1)_{\max}$ is given by:

$$\theta_{\max} = \pi/4 - \pi/12 = \pi/6$$

and the angle to the horizontal is:

$$\theta_{\max} + \psi = \pi/6 + \pi/3 = \pi/2$$

i.e. the maximum stress is in the vertical direction. Figure 6.11 shows the locus of a fixed value of p_1 together with the directions at various points. Clearly if a crack is propagated normal to p_1 at any point other than $\psi = 0$ it will not be a continuation of the existing crack tip at O. The stress which does fulfil this condition is $p_{\psi\psi}$ since, by definition, lines normal to its direction will always pass through the crack tip and in this case this occurs at $\psi = 0$.

This concept seems to have a wide applicability in polymers since mode II fractures are very rare. It appears that all stress systems will propagate a crack along the maximum $p_{\psi\psi}$ direction in mode I as opposed to mode II. Thus if we consider a crack under a shear stress p_{xy}^{∞} only we have:

$$p_{\psi\psi} = \frac{p_{xy}^{\infty}\sqrt{\pi a}}{\sqrt{2\pi r}}\frac{3}{2}(1+\cos\psi)\sin(\psi/2) \tag{6.56}$$

and
$$p_{r\psi} = \frac{p_{xy}^{\infty}\sqrt{\pi a}}{\sqrt{2\pi r}}\frac{1}{2}(3\cos\psi - 1)\cos(\psi/2) \tag{6.57}$$

The maximum value of $p_{\psi\psi}$ may be determined by differentiating equation (6.56) and equating to zero which is equivalent to the condition $p_{r\psi} = 0$ from equation (6.57):

$$(3\cos\psi - 1)\cos(\psi/2) = 0$$

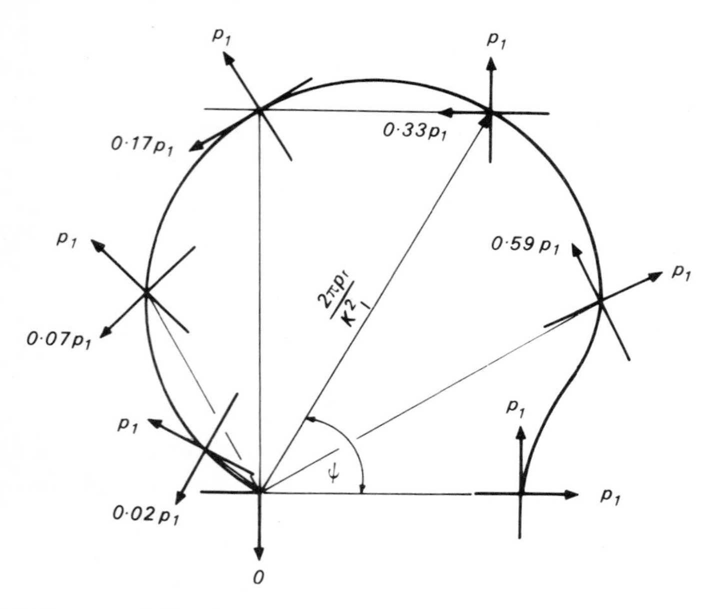

Figure 6.11 Locus and directions of a constant principal stress p_1 and the corresponding values of p_2.

i.e. $\cos\psi = \frac{1}{3}$ or -1 and the first is the relevant condition since $p_{\psi\psi} = 0$ for the second. Substitution in equation (6.56) gives:

$$(p_{\psi\psi})_{\max} = \frac{+p^{\infty}_{xy}\sqrt{\pi a}}{\sqrt{2\pi r}}\;\frac{2}{\sqrt{3}}$$

and since this is in mode I we have-

$$(p_{\psi\psi})_{\max}\sqrt{2\pi r} = K_{IC}$$

at fracture. Hence we have a fracture condition:

$$p^{\infty}_{xy}\sqrt{\pi a} = \frac{\sqrt{3}}{2}\,K_{IC} \tag{6.58}$$

and an effective K_{IC} in shear of $\sqrt{3}/2$ times the tensile value. The cracks run at angles of $-70{\cdot}5^\circ$ (i.e. $\cos^{-1}\frac{1}{3}$) to the original direction. The effect may be easily confirmed with torsion tests on cracked tubes or any attempt to produce mode II failures.

The results may be generalised for any applied K_I and K_{II} values since the condition $p_{r\psi} = 0$ gives the condition:

$$K_I \sin\psi + K_{II}(3\cos\psi - 1) = 0 \tag{6.59}$$

An example of this is the angled crack problem shown in Figure 6.12 in which a crack is set at an angle β to the applied stress p. To determine the angle at which the crack will run, ψ_0, we can resolve p into p_{yy}^{∞}, p_{xx}^{∞}, p_{xy}^{∞} applied to the crack to give:

$$p_{yy}^{\infty} = p \sin^2 \beta, \quad p_{xx}^{\infty} = p \cos^2 \beta, \quad p_{xy}^{\infty} = p \sin \beta \cos \beta$$

and hence:

$$K_{\mathrm{I}} = p \sin^2 \beta . \sqrt{\pi a} \qquad \text{and} \quad K_{\mathrm{II}} = p \sin \beta \cos \beta . \sqrt{\pi a}$$

Substituting into equation (6.59) gives the result:

$$\cot \beta = \frac{\sin \psi_0}{3 \cos \psi_0 - 1}$$

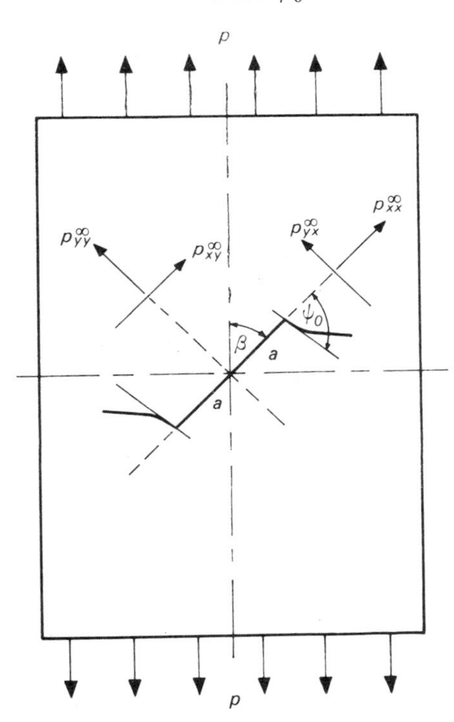

Figure 6.12 The angled crack.

This is shown in Figure 6.13 compared with the direction normal to the applied stress (i.e. horizontal). ψ_0 is negative because the crack moves downwards. For $\beta > 30°$ the crack is above the horizontal while for $\beta < 30°$ it is below. The solution presents problems as $\beta \rightarrow 0$ since it gives the simple shear case which is clearly not true. The difficulty is that the p_{xx}^{∞} component becomes important as $\beta \rightarrow 0$ and this is not included in the analysis. In the limit, when $\beta \rightarrow 0$, it must be predominant and the fracture will be with $\psi_0 = -90°$ so that the solution is a vertical line along the $-\psi_0$ axis from $-70{\cdot}5°$ to $-90°$ as shown in Figure 6.13.

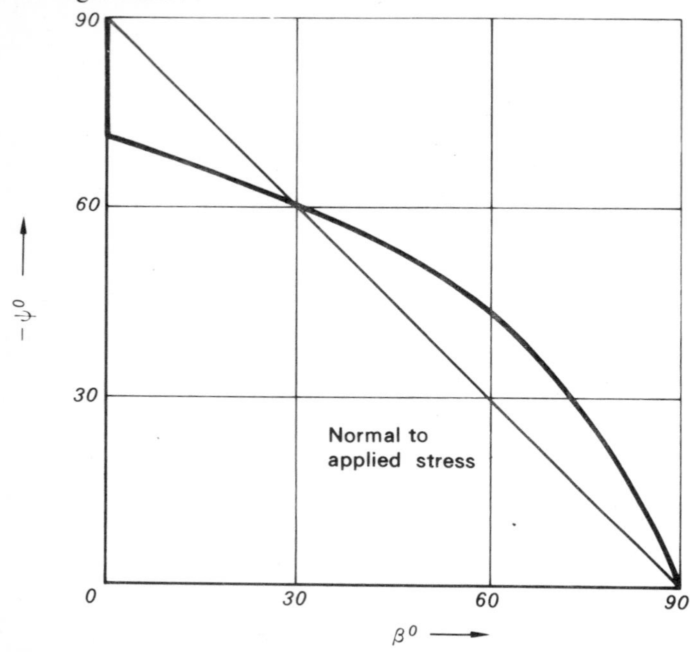

Figure 6.13 Crack propagation angles in the angled crack problem.

6.3.5. *Small scale yielding*

It is clear that, as the stresses increase near a crack tip, plastic yielding will occur so that the actual stress achieved will remain finite. The general elastic-plastic solution is difficult and such solutions which do exist are beyond the scope of this discussion. However, providing that the size of the plastic zone at the tip of the crack is very small then its presence will not influence the elastic stress system significantly. This being so, the size and shape of the plastic zone may be determined from say, the Von Mises yield criterion, and the stress distribution. In this case we will consider only mode I but we will use general plane strain ($e_{zz} = 0$) so that the third principal stress is given by:

$$p_3 = \nu(p_1 + p_2)$$

Substituting in the yield criterion gives:

$$p_Y{}^2 = (p_1{}^2 + p_2{}^2)[1 - \nu(1-\nu)] - p_1 p_2[1 + 2\nu(1-\nu)] \tag{6.60}$$

where p_Y is the tensile yield stress. Substituting for the principal stresses from equations (6.53) we have:

$$p_Y{}^2 = \frac{K_I{}^2}{2\pi r_p} \cos^2\frac{\psi}{2}\left[4[1 - \nu(1-\nu)] - 3\cos^2\frac{\psi}{2}\right] \tag{6.61}$$

Figure 6.14 shows the locus r_p for $\nu = \frac{1}{2}, \frac{1}{3}$ and 0. For the constant volume case ($\nu = \frac{1}{2}$) there are two symmetrical lobes with $r_p = 0$ at $\psi = 0$ where $p_1 = p_2 = p_3$, i.e. hydrostatic tension. A typical minimum value, $\nu = \frac{1}{3}$, gives a small value of r_p at $\psi = 0$ but in general the zone is similar to the constant volume case. $\nu = 0$

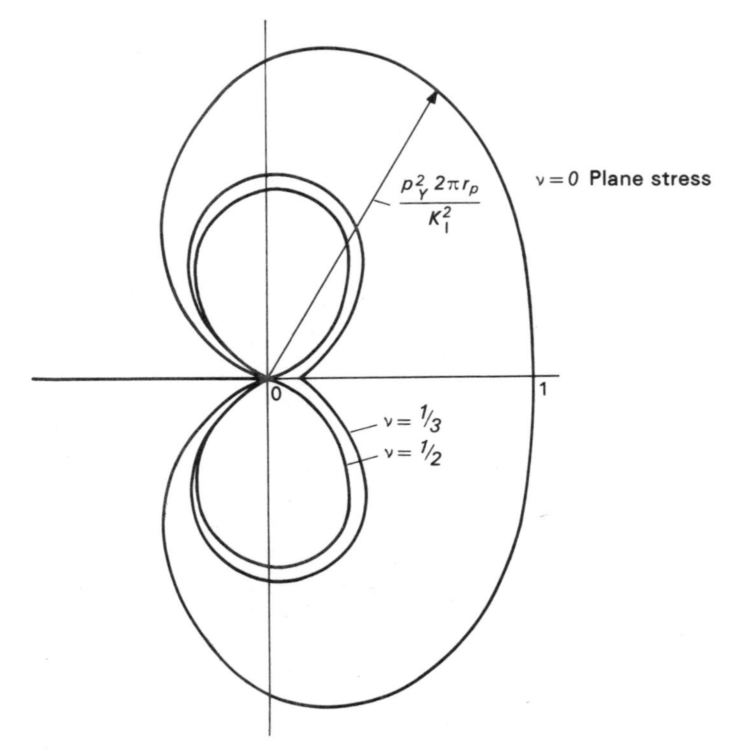

Figure 6.14 Plastic zone shapes.

is of special interest since it is the plane stress situation ($p_3 = 0$) and it can be seen that there is a substantially larger zone. The plastic zone size in a given situation is usually given as the plane stress value of r_p at $\psi = 0$, i.e. the length along the crack direction and:

$$(r_p)_0 = \frac{1}{2\pi}\left(\frac{K_I}{p_Y}\right)^2 \tag{6.62}$$

The significance of the parameter may be illustrated by considering the shape of the zone through the thickness of a specimen as shown in Figure 6.15(a). At the middle section plane strain operates but at the edges there is plane stress. If the thickness $h \gg r_p$ then the plane strain system dominates because the zone is restrained but if $h \ll r_p$ then the zone will be in plane stress. The transition from one to the other is therefore governed by the ratio of r_p to h and, as shown in Figure 6.15(b), this occurs when there is a clear path for 45° plastic slip through the thickness. The maximum depth of the zone can be shown to be at $\psi = 80°$ and to have a value:

$$\hat{x} = 2{\cdot}04(r_p)_0$$

and thus the condition is:

$$\hat{x} = h_{\text{crit}}$$

i.e.

$$h_{\text{crit}} = \frac{2{\cdot}04}{2\pi}\left(\frac{K_{\text{I}}}{p_{\text{Y}}}\right)^2 = 0{\cdot}42\left(\frac{K_{\text{I}}}{p_{\text{Y}}}\right)^2 \tag{6.63}$$

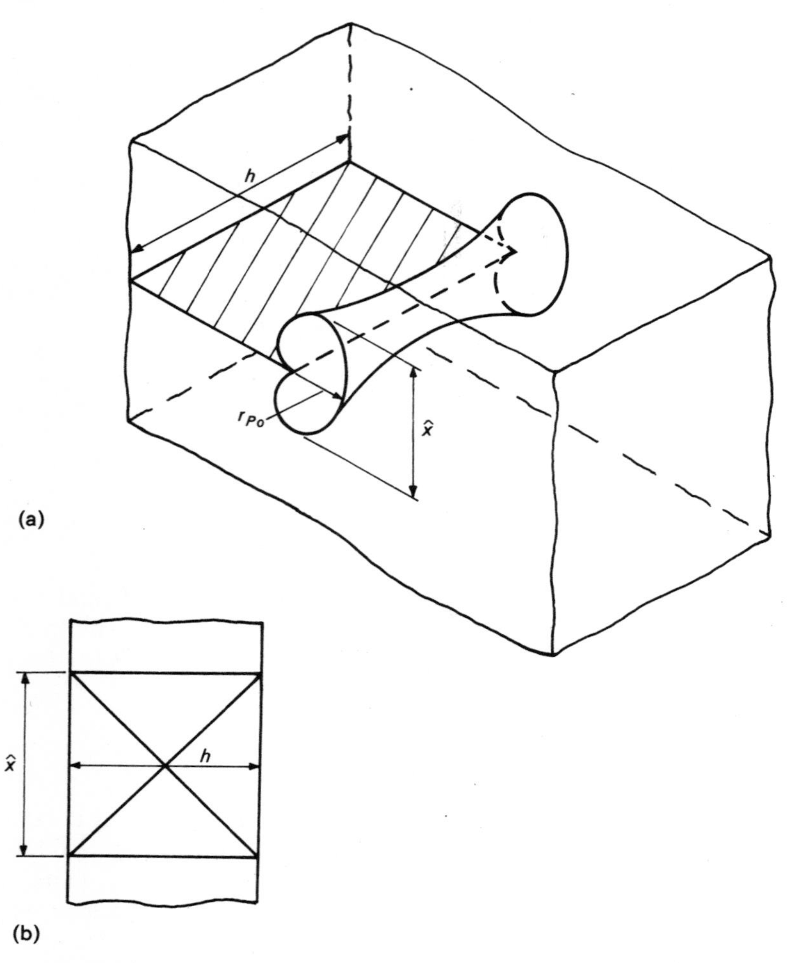

Figure 6.15 Plastic zone shapes and the plane strain-plane stress transition.

6.3.6. *Loaded crack faces*

The expressions for K so far discussed have all been for infinite plates loaded at infinity but we shall now consider the case where the loads are applied at the crack faces. The results have a wide range of practical applications some of which will be described. The basic solution from which others may be derived is for a crack of length $2a$, as shown in Figure 6.16, loaded with a pair of splitting forces F at $\pm b$. The stress p_{yy} along the crack line, $y = 0$, is given by the function:

$$p_{yy} = \frac{2F}{\pi} \frac{x}{\sqrt{(x^2-a^2)}} \frac{\sqrt{(a^2-b^2)}}{x^2-b^2} \tag{6.64}$$

The result is obtained from a complex stress function analysis.

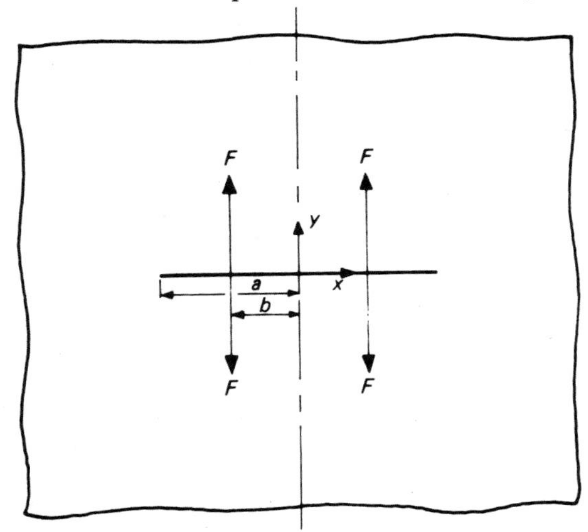

Figure 6.16 Concentrated loads on a crack face.

If we consider only the vicinity of the crack tip and introduce r, the distance from the crack tip, then:

$$p_{yy} = \frac{2F}{\pi} \frac{1}{\sqrt{2r}} \frac{\sqrt{a}}{\sqrt{(a^2-b^2)}}$$

If we now substitute for K_I from equation (6.34) we have:

$$K_I = \frac{2F}{\sqrt{\pi}} \frac{\sqrt{a}}{\sqrt{(a^2-b^2)}} \tag{6.65}$$

A practical test specimen sometimes used has the load applied at the centre, $b = 0$, and if we take the central load $F' = 2F$, since the two are coincident, then:

$$K_I = \frac{F'}{\sqrt{\pi a}} \tag{6.66}$$

The value of this arrangement is that as the crack grows K_I decreases and the load must be increased to maintain a fracture K_{IC}. This is an inherently stable system as opposed to the usual uniform stress at infinity in which

$$K_I = p\sqrt{\pi a}$$

and so K_I increases with increasing a. Hence the uniform stress case will fail in an unstable manner under a constant load while the centrally loaded crack will not.

The basic solution may be used to solve problems involving stress applied to part of a crack face as shown in Figure 6.17. Here a stress p_b is applied between

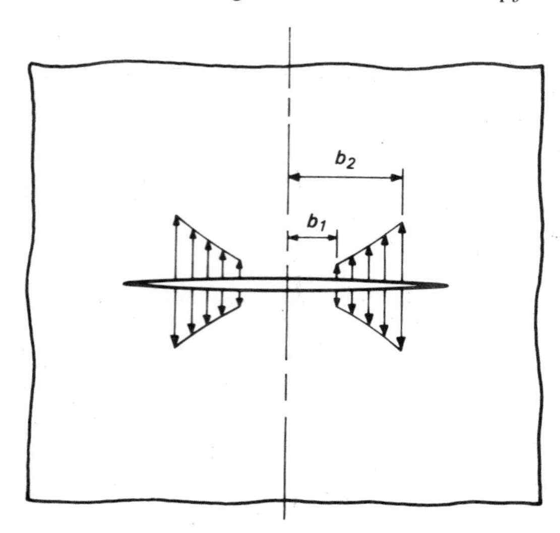

Figure 6.17 Stress on part of a crack face.

b_1 and b_2 and if we consider that at some point on the loaded section the stress can be represented by a point load p_b db then K_I for the distribution may be obtained from equation (6.65):

$$K_I = \frac{2\sqrt{a}}{\sqrt{\pi}} \int_{b_1}^{b_2} \frac{p_b \, db}{\sqrt{a^2 - b^2}} \tag{6.67}$$

The special case of uniform pressure p over the whole face then gives the result:

$$K_I = \frac{2\sqrt{a}}{\sqrt{\pi}} \int_{b}^{a} \frac{p \, db}{\sqrt{a^2 - b^2}}$$

i.e.

$$K_I = \frac{2\sqrt{a}}{\sqrt{\pi}} p \left[-\cos^{-1} \frac{b}{a} \right]_0^a = p\sqrt{\pi a} \tag{6.68}$$

This result shows that a crack loaded with a stress over its faces is equivalent to one loaded at infinity with the same stress and the faces left stress-free.

6.3.7. *Crazing*

Crazing is a phenomenon which occurs in polymers when crack-like discontinuities are formed, in which fibrils connect the two faces of the crack. The restraining of the faces may be described by a uniform stress $-p_c$ over the crack faces and from equation (6.68) we have:

$$K'_{\mathrm{I}} = -p_c\sqrt{\pi a}$$

where $2a$ is the length of the craze and K'_{I} is at the craze tip. If a uniform stress p is applied at infinity then this also gives a stress intensity factor:

$$K''_{\mathrm{I}} = p\sqrt{\pi a}$$

so that the net stress intensity factor is:

$$K_{\mathrm{I}} = K''_{\mathrm{I}} + K'_{\mathrm{I}} = (p - p_c)\sqrt{\pi a} \tag{6.69}$$

Thus there is no effective value for $p < p_c$ and propagation is unlikely to occur. If an initial crack of length $2a$ is present which has a craze extension as shown in Figure 6.18 then the stress intensity factor due to the craze may be derived

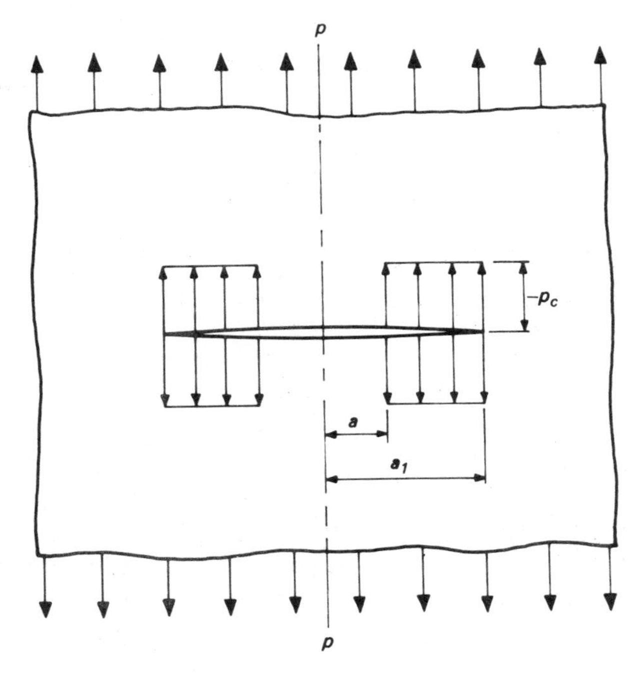

Figure 6.18 Stresses in a craze.

from equation (6.67):

$$K'_{\mathrm{I}} = \frac{2\sqrt{a_1}}{\sqrt{\pi}} \int_a^{a_1} \frac{-p_c \, db}{\sqrt{(a_1^2 - b^2)}}$$

i.e.

$$K'_{\mathrm{I}} = -\frac{2\sqrt{a_1}}{\sqrt{\pi}} p_c \cos^{-1}\left(\frac{a}{a_1}\right) \tag{6.70}$$

The applied stress p is the same as before and hence the net value of the stress intensity factor is:

$$K_{\mathrm{I}} = p\sqrt{\pi a_1}\left[1 - \frac{2}{\pi}\frac{p_c}{p}\cos^{-1}\frac{a}{a_1}\right] \tag{6.71}$$

If it is supposed that crazes will cease to grow when there is no stress intensity factor at the crack tip, i.e. the stress remains finite at p_c, then this condition is given when:

$$\frac{p}{p_c} = \frac{2}{\pi}\cos^{-1}\frac{a}{a_1} \tag{6.72}$$

This is the well-known Dugdale model of a line plastic zone and if we consider small values of stress, i.e. $p \ll p_c$ and $r = a_1 - a \ll a$ then we have:

$$r_c = \frac{p^2 \pi^2 a}{8 p_c^2}$$

If we let $(K_{\mathrm{I}})_0 = p\sqrt{\pi a}$ then:

$$r_c = \frac{\pi}{8}\left(\frac{(K_{\mathrm{I}})_0}{p_c}\right)^2 \tag{6.73}$$

which compares with the usual plane stress result, equation (6.62):

$$(r_p)_0 = \frac{1}{2\pi}\left(\frac{K_{\mathrm{I}}}{p_{\mathrm{Y}}}\right)^2$$

i.e. for a given value of p_c or p_{Y} the line zone is larger by a factor of $\pi^2/4$, i.e. 2·47. For the case when r_c is small, equation (6.71) can be expressed as:

$$K_{\mathrm{I}} = p\sqrt{\pi a} - \frac{2}{\pi} p_c \sqrt{2\pi r_c} \tag{6.74}$$

and the measured value of K_{I} at fracture, K_{IC}, can be corrected by the term shown.

The value of K_{I} at a craze tip as derived from equation (6.71) is shown plotted in Figure 6.19. For $p < p_c$, K_{I} decreases to zero and reaches the equilibrium condition as given in equation (6.72). For $p > p_c$, K_{I} falls initially but then rises again and for $a_1 > a$ tends to the completed craze solution, equation (6.69). The relationship between applied stress and craze length at the equilibrium

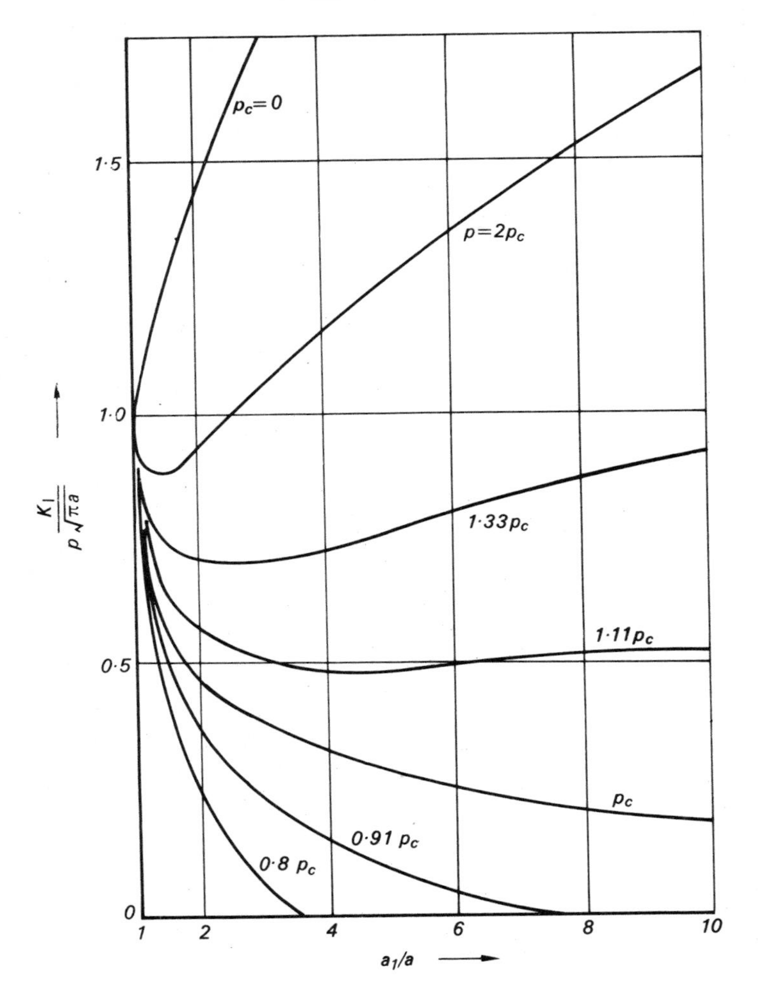

Figure 6.19 Stress intensity factors at a craze tip.

condition is shown in Figure 6.20. There is clearly a very rapid change in a_1 for p greater than about $0{\cdot}8p_c$ which results in the length of craze being very sensitive to changes in p in this range.

6.3.8. *Compliance and cleavage tests*

A method of determining K_I for a cracked body may be described in terms of the compliance of the body. If we assume that the material is linearly elastic then for any given crack length there will be a linear relationship between the

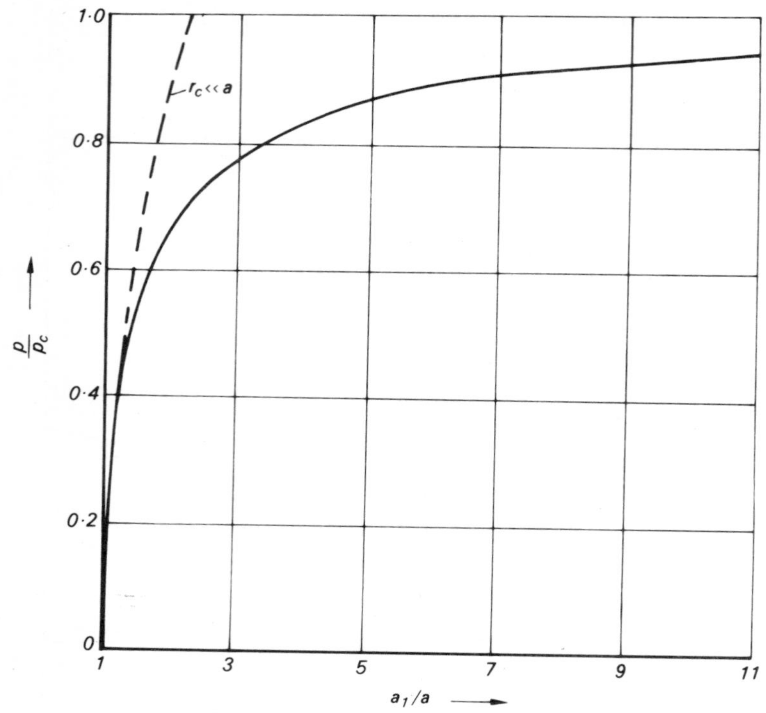

Figure 6.20 Equilibrium craze lengths.

applied load F and the resulting overall deflection δ. The slope of this line may be taken as C, the compliance:

i.e.
$$C = \delta/F$$

Now at any given load the stored elastic energy is given by:

$$U = \tfrac{1}{2}F\delta = \frac{F^2}{2}C$$

and G, the strain energy release rate per unit thickness of crack is where h_c

$$G = \frac{1}{h_c}\frac{\partial U}{\partial a} = \frac{F^2}{2h_c}\frac{\partial C}{\partial a} \tag{6.75}$$

is the thickness of the cracked section. The inclusion of h_c enables specimens to be considered where h_c is different from the overall specimen thickness h. Equation (6.75) may be related to K_I by using equation (6.40) giving a final result:

$$K_I^2 = \frac{EF^2}{2h_c}\frac{\partial C}{\partial a} \tag{6.76}$$

This expression may be used to determine K_I experimentally since C can be determined for various crack lengths and an experiment C versus a curve constructed. This can then be differentiated graphically to give $\partial C/\partial a$ and hence K_I. The method is particularly useful when using specimens of finite width for which exact theoretical solutions are not known.

As an example of its use as an analytical tool we may consider the parallel cleavage specimen shown in Figure 6.21. This specimen is quite convenient for

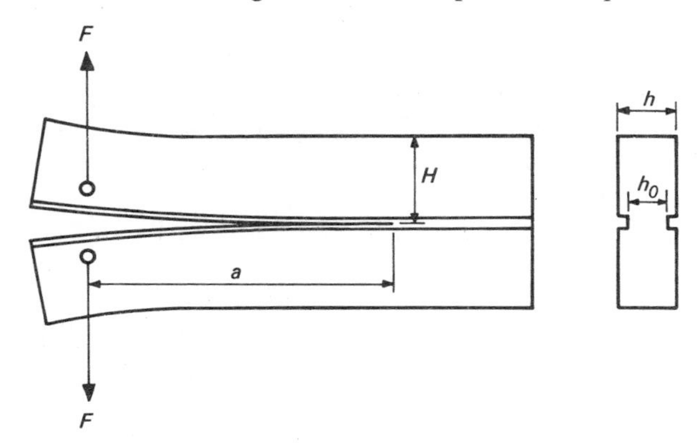

Figure 6.21 Parallel cleavage specimen.

testing and has found wide use in crack speed work. The two arms may be treated, to a first approximation, as two slender cantilevers which may be analysed using conventional beam theory (section 4.2). Thus the compliance for both arms is given by:

$$C = \frac{2\delta}{F} = \frac{2}{3}\frac{a^3}{EI}$$

since

$$\delta = \frac{1}{3}\frac{Fa^3}{EI}$$

Substituting for I in the usual way and differentiating we have:

$$\frac{\partial C}{\partial a} = \frac{24a^2}{EhH^3}$$

and hence:

$$K_I{}^2 = \frac{4F^2}{hh_c}\left(\frac{3a^2}{H^3}\right) \tag{6.77}$$

Side grooves are frequently used to guide the crack so that $h_c < h$. Because of the assumption that the specimen has slender arms and the usual beam theory limitations this result often requires refinement. For example we may include shear deflection which gives:

$$K_I{}^2 = \frac{4F^2}{hh_c}\left[\frac{3a^2}{H^3} + \frac{3}{4}\frac{(1+\nu)}{H}\right] \tag{6.78}$$

If we take the special case of $\nu = \frac{1}{3}$ and $h = h_c$ we have:

$$\frac{K_I h H^{3/2}}{Fa} = 2\sqrt{3}\left(1+\frac{1}{3}\frac{H^2}{a^2}\right)^{1/2} \tag{6.79}$$

Corrections for the fact that the cantilevers are not truly built-in can also be made numerically and a proven form of correction is:

$$\frac{K_I h H^{3/2}}{Fa} = 3{\cdot}46+2{\cdot}38(H/a) \tag{6.80}$$

For fracture testing it would be of great value if a specimen could be designed in which K_I did not vary with crack length. For the cleavage test this can be achieved if, in equation (6.77), we have:

$$\frac{3a^2}{H^3} = k \qquad a \text{ constant.}$$

Such a specimen could be manufactured but it is difficult and a simplification is given if we consider a tapered cleavage specimen as shown in Figure 6.22.

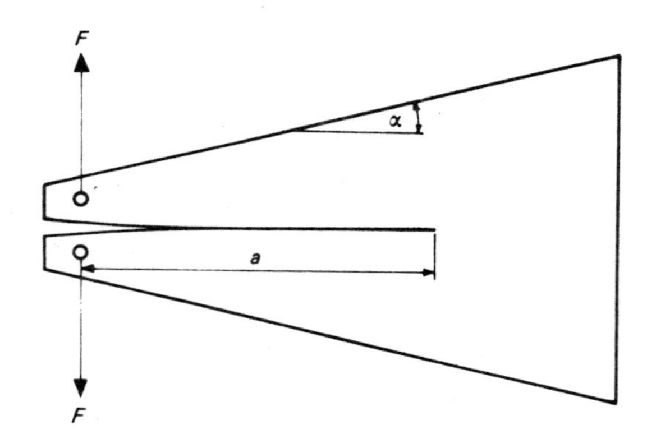

Figure 6.22 Tapered cleavage specimen.

If α is half the included tapered angle, and H_0 the depth of one arm at the loading pins, then:

$$H = H_0 + a \tan \alpha$$

and hence:

$$H_0 k = 3\left(\frac{a}{H_0}\right)^2 \frac{1}{\left(1+\tan\alpha\cdot\dfrac{a}{H_0}\right)^3} \tag{6.81}$$

Figure 6.23 shows $H_0 k$ plotted versus a/H_0 for various values of $\tan \alpha$ and it can be seen that for $\tan \alpha = 0{\cdot}2$, $\alpha \simeq 11°$, $H_0 k$ remains constant at around 10·5 for the range $5 < a/H_0 < 20$. Providing the crack growth is limited in range, a constant K_I can be achieved,

i.e.
$$K_I^2 \simeq \frac{42F^2}{hh_cH_0} \tag{6.82}$$

Experimental calibration or numerical analysis is advisable in practical cases to determine the constant precisely.

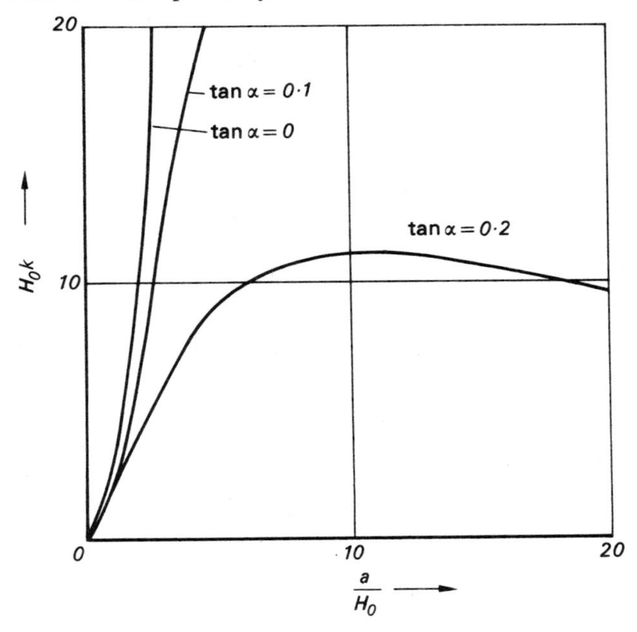

Figure 6.23 Compliance parameter for tapered cleavage.

6.3.9. *Finite width correction factors*

All the solutions for K_I so far discussed have been for a crack of length $2a$ in an infinite body. In many practical cases, and particularly in testing, the crack is not usually small in comparison with the specimen width and there are also distortions of the infinite plate stress field because of the free edges. An example of the free edge effect is shown in Figure 6.24 in which a specimen has a crack of length a in one edge. If it is assumed that $a \ll W$ the usual infinite plate solution is altered by a constant factor giving a result:

$$K_I = 1{\cdot}12\, p \sqrt{\pi a} \tag{6.83}$$

For $a/W >$ about 0·1 the solution will be influenced by the increase of the nominal stress on the cracked section. This nominal stress augments p to give an effectively higher value of K_I. It is generally not possible to obtain analytical solutions for these corrections but in the absence of these, or precise numerical solutions, approximations can be made. One such method is to consider that we may write:

$$K_{\mathrm{I}} = p' \sqrt{\pi a} \tag{6.84}$$

where p' is the augmented applied stress. p' is made up of p_m, the mean stress over the cracked section assuming no crack is present and p_n, the nominal

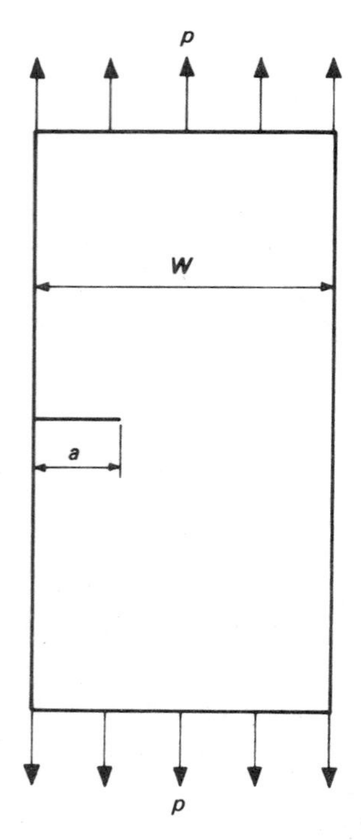

Figure 6.24 A single edge notch specimen without bending.

stress at the crack tip ignoring stress concentration effects. The new mean stress at the crack tip can then be taken as:

$$\frac{p_m + p_n}{2}$$

but at the centre is still p_m. Thus the mean over the whole crack is:

$$\frac{1}{2}\left(p_m + \frac{p_m + p_n}{2}\right)$$

which may be used as the augmented stress:

$$p' = \frac{3p_m + p_n}{4} \tag{6.85}$$

The concept of the use of a mean stress over the crack face can be justified using equation (6.67):

$$K_\mathrm{I} = \sqrt{\pi a}\left(\frac{2}{\pi}\int_0^a \frac{p_b \mathrm{d}b}{\sqrt{a^2 - b^2}}\right)$$

where the term in brackets may be approximated to:

$$\frac{1}{a}\int_0^a p_b \,\mathrm{d}b$$

the mean stress p_m. This average value provides a reasonable approximation for most distributions of p_b.

As an example of the approximate method consider a centre-cracked plate as shown in Figure 6.25. The relevant stresses are:

$$p_m = p \quad \text{and} \quad p_n = \frac{p}{1 - a/W}$$

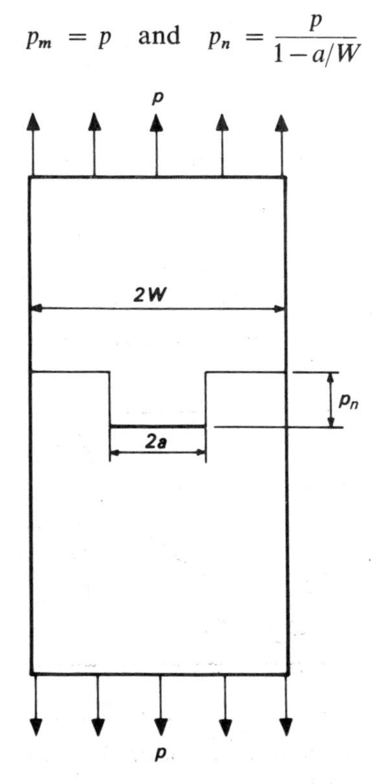

Figure 6.25 Centre-cracked plate.

and hence:

$$p' = \frac{p}{2}\,\frac{2-\dfrac{3}{2}\dfrac{a}{W}}{1-\dfrac{a}{W}}$$

i.e.

$$\frac{K_I}{p\sqrt{\pi a}} = \frac{1}{2}\,\frac{2-\dfrac{3}{2}\dfrac{a}{W}}{1-\dfrac{a}{W}} \tag{6.86}$$

This result is shown in Figure 6.26 and indicates a rapidly increasing factor for $a/W > 0{\cdot}1$. The method may be applied to a wide range of geometries and loadings to give simple, if not very precise, solutions.

For testing purposes it is essential to have precise results and a number of numerical solutions are given here. The same factor:

i.e.

$$\frac{K_I}{p\sqrt{\pi a}}$$

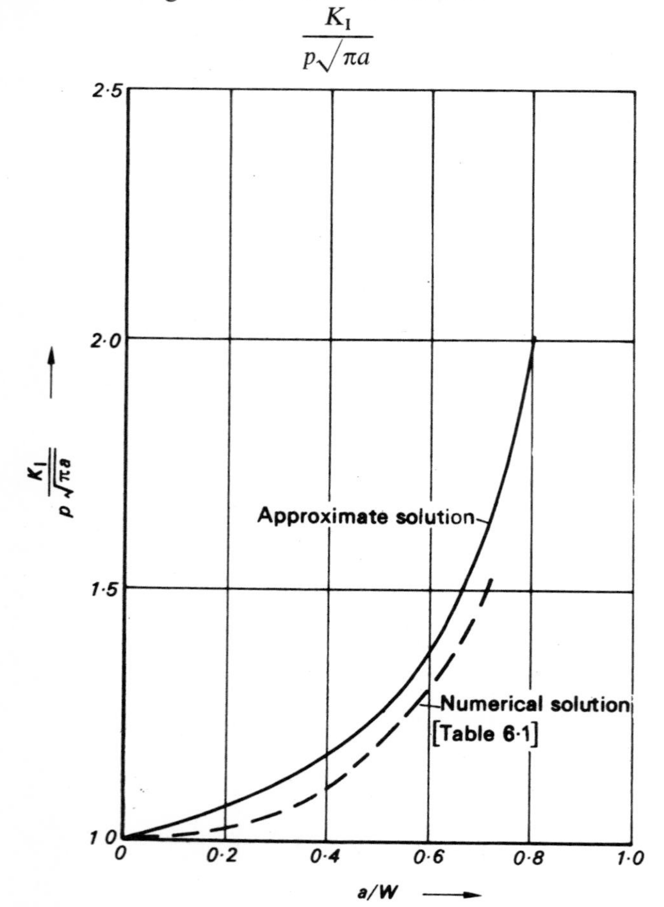

Figure 6.26 Finite width correction factors for a centre-cracked plate.

is used and is given for the geometries shown in Figure 6.27 in table 6.1. The stress p is the gross value for the tension specimens:

i.e.
$$p = \frac{F}{Wh}$$

and in bending is the maximum bending stress:

$$p = \frac{6M}{W^2h}$$

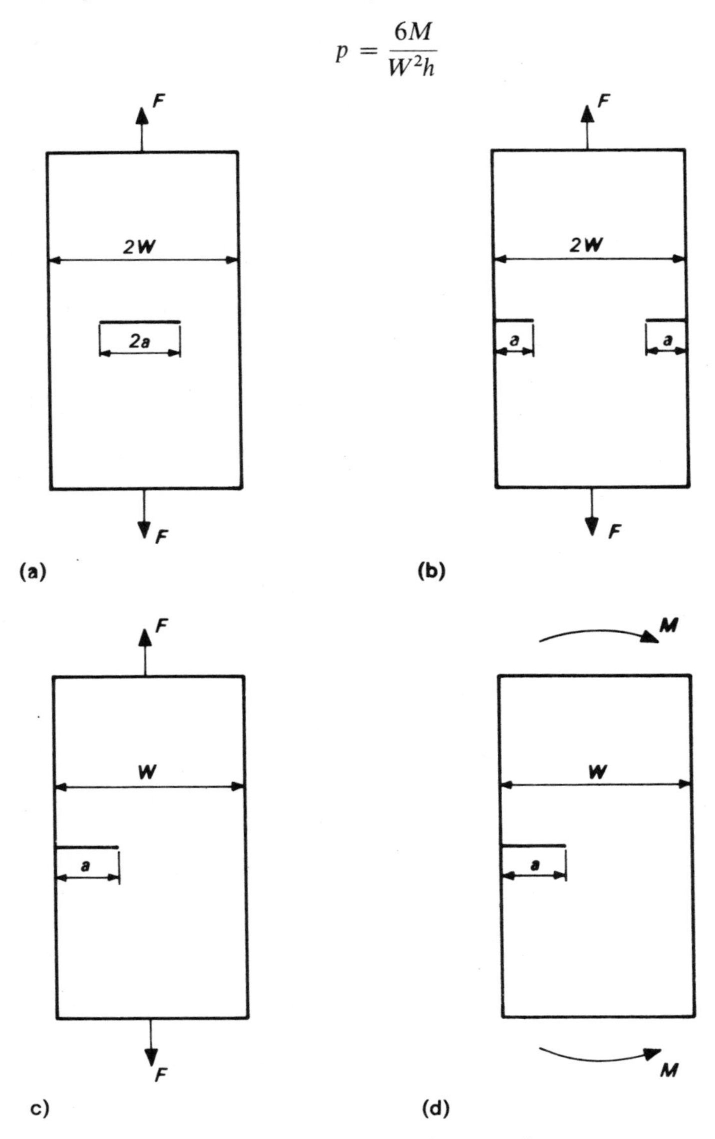

Figure 6.27 Geometries of test specimens.

The centre-cracked specimen results are shown plotted with the approximate solution in Figure 6.26. It can be seen to be below the approximate values and to tend to unity at $a/W = 0$. All the other three tend to 1·12 at $a/W = 0$ with the bending result passing through a minimum. It is of interest to note that the single edge notch specimen in tension has the largest correction factor because of the induced bending which increases with crack length.

Table 6.1 *Finite Width Correction Factors*

$$\frac{K_I}{p\sqrt{\pi a}}, \quad p = \frac{6M}{hW^2} \quad \text{for bending}$$

$\frac{a}{W}$	(a) Centre cracked plates	(b) Double edge notched	(c) Single edge notch tension	(d) Single edge notch bending
0	1·000	1·12	1·120	1·120
0·1	1·010	1·12	1·186	1·045
0·2	1·025	1·12	1·373	1·054
0·3	1·052	1·13	1·663	1·124
0·4	1·101	1·14	2·107	1·258
0·5	1·181	1·15	2·830	1·496
0·6	1·301	1·22	—	—
0·7	1·470	1·34	—	—
0·8	—	1·57	—	—

Bibliography

1. Timoshenko, S. P. and Goodier, J. W. *Theory of Elasticity*, Third Edition, McGraw Hill (1970)
2. Peterson, R. E. *Stress Concentration Design Factors*, John Wiley and Sons Inc., New York (1953)
3. Liebowitz, H. (ed.) *Fracture: an Advanced Treatise*, Vol. 2., *Mathematical Fundamentals*, Academic Press (1968)
4. Williams, M. L. 'On the Stress Distributions at the Base of a Stationary Crack', *J. App. Mechs.*, (Mar. 1957) 24, No. 1.

Index

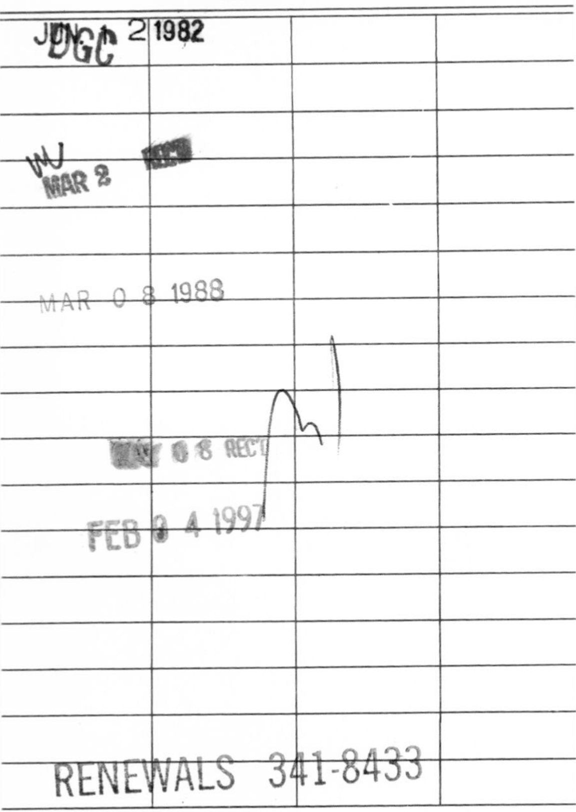
Phone Renewal: 321-7621
DATE DUE
JUN 2 1982
DGC
MAR 2
MAR 0 8 1988
FEB 0 4 1997
RENEWALS 341-8433
DEMCO 38-297